70 Topics in Current Chemistry

Fortschritte der Chemischen Forschung

W0246017

Structural Theory of Organic Chemistry

N. D. Epiotis
W. R. Cherry
S. Shaik
R. Yates
F. Bernardi

Springer-Verlag
Berlin Heidelberg GmbH 1977

Nicolaos D. Epiotis
William R. Cherry
Sason Shaik
Ronald L. Yates

Department of Chemistry,
University of Washington,
Seattle Washington, 98195 U.S.A.

Fernando Bernardi

Istituto di Chimica Organica,
Universita di Bologna, ltaly

This series presents critical reviews of the present position and future trends in modern chemical research. It is addressed to all research and industrial chemists who wish to keep abreast of advances in their subject.

As a rule, contributions are specially commissioned. The editors and publishers will, however, always be pleased to receive suggestions and supplementary information. Papers are accepted for "Topics in Current Chemistry" in English.

ISBN 978-3-662-15837-1   ISBN 978-3-540-37472-5 (eBook)
DOI 10.1007/978-3-540-37472-5

Library of Congress Cataloging in Publication Data. Main entry under title: Structural theory of organic chemistry. (Topics in current chemistry ; 70). Includes indexes. 1. Molecular orbitals. 2. Chemistry, Physical organic. I. Epiotis, N. D., 1944- II. Series. QD1.F58 vol. 70 [QD461] 540'.8s [547'.1'28] 76-57966

© by Springer-Verlag Berlin Heidelberg 1977
Originally published by Springer-Verlag Berlin Heidelberg New York in 1977
Softcover reprint of the hardcover 1st edition 1977

Typesetting and printing: Schwetzinger Verlagsdruckerei GmbH, 6830 Schwetzingen. Bookbinding: Konrad Triltsch, Graphischer Betrieb, 8700 Würzburg

Preface

While MO theory has had a profound impact on the way in which chemists think about reactivity problems, a corresponding influence in the area of structural chemistry has been absent. Typically, "steric effects" provide the basis for rationalizing trends conforming to ordinary intuition and "attractive Van der Waals forces" are invoked to rationalize trends which are opposite to those expected on the basis of the concept of "steric repulsion".

However, molecules which exist preferentially in a "crowded" geometry are not mere aberrations of an order dominated by "steric effects". Indeed, in some classes of molecules, preference for "crowdedness" seems to be the rule rather than the exception. This observation stimulated an initial research of geometric isomerism which later blossomed into a theoretical investigation of structural chemistry. This book constitutes an abbreviated account of our experiences in dealing with such problems during the time period June 1972 – June 1976.

The aim of our work has been to arrive at a *qualitative* understanding of the key factors which determine the preferred geometry of a molecule. The specific procedure involves three principal stages:

a) The *analysis* stage, at which a one electron Hückel-type Molecular Orbital (MO) model is applied to the problem at hand. MO interaction diagrams are the conveyors of the key theoretical deliberations.

b) The *test* stage, at which explicit quantum mechanical computations are carried out to test the validity of the model. It is important to emphasize that the target of attention is not only the final numerical answer but, more importantly, the printout of the MO's and the density matrix. A key electronic effect leaves its mark on the MO's and the density matrix resulting from a calculation while the balance of *all* electronic effects is related to the computed total energy. A *dominant* electronic effect (*e.g.*, a symmetry imposed barrier) is almost always identifiable, regardless of the quality of computation. By contrast, the balance of *all* effects is a more sensitive problem and the answer depends on the quality of computation in a manner which is anything but predictable. This realization constitutes the basis for our preoccupation with *electronic effects* and a *qualitative* understanding of structural chemistry.

c) The *application* stage, at which the predictions of the theoretical model as well as the additional insights provided by the explicit calculations are compared with the available experimental data or form the basis for the design of a new experiment. In short, the triptych espoused in this work is ANALYSIS-TEST-APPLICATION, or, ANALYSIS-COMPUTATION-PREDICTION. In the opinion of the authors, *this method constitutes the only realistic way of approaching complex problems*

V

of chemical structure at this point in time. Related philosophical dispositions are evident in the publications of Gimarc, Hehre, Hoffmann, Lowe, and Salem.

The present work does not answer all questions regarding "why" a given molecule exhibits a geometrical preference. Additional problems, such as the effect of "correlation energy" on structural trends and the related challenge of developing a qualitative understanding of configuration interaction within the context of a model, are currently under investigation. Nonetheless, we believe that the overview of structural chemistry developed herein is sufficiently satisfactory to arouse the interest of theoreticians and experimentalists alike, especially because the vast majority of chemists has been exposed to ways of thinking which are substantially different from those espoused in this work.

June 1976   Nicolaos Demetrios Epiotis
                                             William R. Cherry
                                             Sason Shaik
                                             Ronald L. Yates
                                             Fernando Bernardi

Contents

Introduction 1

Part I. Theory 3
1.   General Theory 3
1.1.   Orbital Energies and Interaction Matrix Elements 7
1.2.   The Concept of Matrix Element and Energy Gap Controlled Orbital
        Interactions 17
1.3.   Examples of Matrix Element Control of Orbital Interactions 19
1.4.   Computational Tests of the OEMO Approach 20

Part II. Nonbonded Interactions 23
2.   Theory of Nonbonded Interactions 23
2.1.   Pi Nonbonded Interactions and Pi Aromatic, Nonaromatic, and
        Antiaromatic Geometries 23
2.2.   Sigma Nonbonded Interactions 36
2.3.   Indices of Nonbonded and Steric Interactions 40

3.   The Effect of Nonbonded Interactions on Molecular Structure . . . 48
3.1.   Nonbonded Interaction Control of the XCX Angle in X_2CH_2 and
        $X_2C=Y$ Molecules 49
3.2.   Conformational Isomerism of CH_3CH_3 and CH_2XCH_2X Molecules . 54
3.3.   Conformational Isomerism of X_4Y_2 and $X_2H_2Y_2$ Molecules . . . 62
3.4.   Conformational Isomerism of X–Y–Y–X and W–X–Y–Z Molecules . 64
3.5.   Conformational Isomerism of $CH_3CH=X$ Molecules 66
3.6.   Torsional Isomerism of CHX=CHX and CHX=CHY Molecules . . . 69
3.7.   Torsional Isomerism of $CH_3CH=CHX$ Molecules 74
3.8.   Torsional Isomerism of CXY=CXY Molecules 77
3.9.   Conformational Isomerism of CH_3COX Molecules 81
3.10.   Conformational Isomerism of XCH_2COY Molecules 85
3.11.   Conformational Isomerism of R–X–R Molecules 85
3.12.   Conformational Isomerism of R–X–R' Molecules 91
3.13.   Torsional Isomerism of Cations and Anions 95
3.14.   Torsional Isomerism of Substituted Benzenes 99
3.15.   Conformational Isomerism of Diene Systems 102
3.16.   Nonbonded Interactions in Peptides and Polypeptides 108
3.17.   Torsional Isomerism in Ring Systems 110
3.18.   The Concept of the Isoconjugate Series 114
3.19.   Assorted Systems 114

Contents

4.   Tests of Nonbonded Interactions 115
4.1.   Physical Manifestations of Nonbonded Interactions 115
4.2.   Spectroscopic Probes of Nonbonded Interactions 123
4.3.   Reactivity Probes of Nonbonded Interactions 126

Part III. Geminal Interactions 131
5.   Definitions 131
5.1.   Theory of Lone Pair-Sigma Bond Geminal Interactions 131
5.2.   The Pyramidality of AX_3 Molecules 140
5.3.   Miscellaneous Problems 143

Part IV. Conjugative Interactions 147
6.   Donor and Acceptor Molecular Fragments and the Question of *Syn*
        vs. *Anti* Overlap 147

7.   Structural Effects of n-π, σ-π, and π-π Interactions 156
7.1. n-π Interactions. 156
7.2. σ-π Interactions 157
7.3. σ-p^+ Interactions 158
7.4. π-π Interactions 160
7.5.   Competitive n-π, σ-π, and π-π Effects 160

8.   Tests of n-π, σ-π, and π-π Interactions 161
8.1.   Physical Manifestations of n-π, σ-π, and π-π Interactions 161
8.2.   Spectroscopic Probes of n-π, σ-π and π-π Interactions 162
8.3.   Reactivity 163

9.   Structural Effects of n-σ Interactions 163
9.1.   Possible Examples of Matrix Element Control of n-σ Interactions . . 182

10.   Tests of n-σ Interactions 183
10.1.   Spectroscopic Probes of n-σ Interactions 183
10.2.   Reactivity Probes of n-σ Interactions 185

11.   Structural Effects of σ-σ Interactions 189

Part V. Bond Ionicity Effects 199
12.   Theory 199
12.1.   The LCFC Approach to Geminal Interactions 201
12.2.   Structural Isomerism 207
12.3.   Bond Strengths 215
12.4.   Valence Isomerism 217

13.   Other Approaches 219

14.   Conclusion 229

References 223

Author Index 243

VIII

Introduction

In the past four years, we published various papers with the purpose of drawing the attention of chemists to the following possibilities:

a) Nonbonded interactions and their influence on torsional isomerism can be understood within the context of simple one electron MO theory[1-8]. These ideas are discussed in Parts I and II.

b) Sigma interactions, designated geminal interactions, may affect the shape of molecules and associated shape-related properties in a manner which is also understandable within the context of one electron MO theory[9]. These concepts are presented in Part III.

c) Chemical reactivity can be understood in terms of donor-acceptor interactions with definite trends being expected as one reactant becomes an increasingly better donor and the other an increasingly better acceptor[10-14]. These ideas have now been applied in an explicit manner to problems of molecular structure and are discussed in Part IV under the heading of conjugative interactions.

d) Bond ionicity effects can be best understood by means of an effective one electron configuration interaction approach in a way which is suitable for the formulation of general predictive rules[15]. These ideas are discussed in Part V.

These sense in which terms like conjugative interactions, nonbonded interactions, etc., are meant will become clear when we discuss each individual type of interaction or effect. Suffice to say that, in many instances, conjugative interactions as well as geminal interactions or bond ionicity effects contain implicitly the idea of nonbonded interactions. Thus, it should be emphasized that the labels of the basic types of interactions proposed here reflect the way in which the problem is formulated rather than different electronic principles.

In Parts I and II, a molecule is viewed as a composite of submolecular fragments each described by delocalized MO's. Nonbonded and geminal interactions can be simply formulated in this manner. In Part IV, a molecule is viewed as a composite of submolecular fragments each described by hybrid bond MO's. Conjugative interactions are best understood in this fashion. Finally, in Part V a one electron configuration interaction approach involving a Linear Combination of Fragment Configurations (LCFC) is used to reveal bond ionicity effects. Indeed, this book can be titled "The Chemist's Handbook of MO Interactions and Their Implications for Molecular Architecture".

Throughout the entire work, we have tried to present experimental as well as computational results pertinent to cases under scrutiny. The calculation results are especially significant since they provide tests for the proposed theoretical models.

It is our hope that the reader will be stimulated to delve deeper into the world of orbital interactions and, hopefully, find a chance to apply the key concepts to problems of direct interest.

Part I. Theory

1. General Theory

The theoretical analysis to be employed throughout Part I of this work is based upon One-Electron Molecular Orbital (OEMO) theory. In our approach, a given molecule in a specified geometry is constructed by a sequential union of molecular fragments. A typical construction is illustrated below:

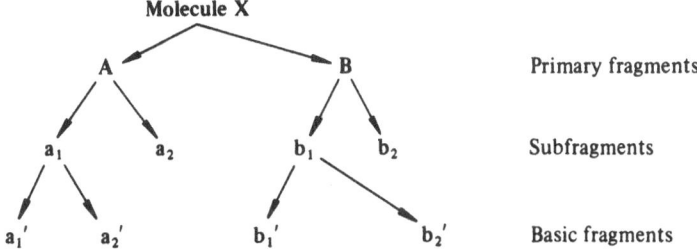

In certain problems of torsional isomerism we shall employ the dissection of a molecule A–B into two closed shell fragments A and B. On the other hand, most problems of torsional isomerism which we shall be dealing with in the first part of this work can be treated by employing the dissection shown below. The appropriate definitions are specified in parenthesis.

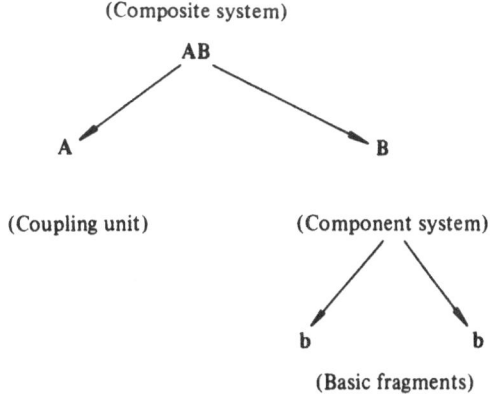

Finally, an alternative approach employs a dissection into two radical fragments as shown below:

3

(Composite system)

AB

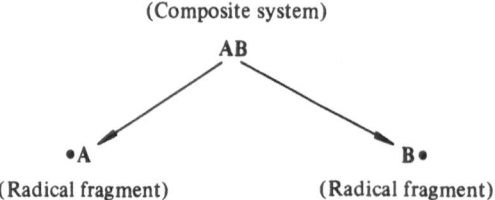

•A   B•

(Radical fragment)   (Radical fragment)

The delocalized group MO's of each fragment can be constructed by means of perturbation theory[16-19] or explicit calculations. Once one knows the MO's of basic fragments, he can construct the MO's of any molecule by means of relatively simple operations. Here, we note that a recent publication of Salem and Jørgensen[20] is a welcome addition to the library of any organic chemist since it includes an extensive compilation of basic fragments and their MO's and provides illustrative examples of the theoretical manipulations involved in the construction of a total system from subunits.

The union of any two molecular fragments is accompanied by an energy change which depends upon the interaction between the MO manifolds of the two fragments. This energy change is evaluated with respect to an effective one-electron Hamiltonian operator, the choice of the operator being such as to confer maximal simplicity to the analysis. The energy change of a MO belonging to one manifold due to its interaction with a number of MO's belonging to a second manifold is a simple sum of the energy changes resulting from each interaction. In this work, we need to distinguish between two types of MO interactions:

(a) The interaction of a doubly occupied MO, ϕ_i, with a vacant nondegenerate MO, ϕ_j, leads to two electron stabilization, ΔE_i^2, which is inversely proportional to the energy separation of the two MO's, $\epsilon_i - \epsilon_j$, and directly proportional to the square of their interaction matrix element, H_{ij}. This is a well known result of perturbation theory and the assumptions involved in its derivation are valid for most systems studied in this work. The algebraic expression for the two electron stabilization is given below:

$$\Delta E_i^2 = \frac{2(H_{ij} - \epsilon_i S_{ij})^2}{\epsilon_i - \epsilon_j} \tag{1}$$

If overlap is neglected, i.e. $S_{ij} = 0$, the above expression takes the form shown below.

$$\Delta E_i^2 = \frac{2H_{ij}^2}{\epsilon_i - \epsilon_j} \tag{2}$$

The interaction of a singly occupied MO, ϕ_i, with a singly occupied degenerate MO, ϕ_j, leads to two electron stabilization given by the following equation:

$$\Delta E_i^2 = \frac{2(H_{ij} - \epsilon_i S_{ij})}{1 + S_{ij}} \tag{3}$$

The corresponding expression where overlap is neglected is shown below.

$$\Delta E_i^2 = 2H_{ij} \tag{4}$$

(b) The interaction of two doubly occupied nondegenerate MO's, ϕ_i and ϕ_j, leads to net four electron destabilization, ΔE_{ij}^4, which increases as the overlap integral of the two MO's, S_{ij}, and the mean of their energies, ϵ_0, increase. This result is obtained by application of the variational method to the case of a two orbital-four electron interaction and involves no special assumptions[6, 21, 22]. The four electron destabilization energy is given by the equation:

$$\Delta E_{ij}^4 = \frac{4(\epsilon_i S_{ij}^2 - H_{ij} S_{ij})}{1 - S_{ij}^2} \tag{5}$$

If the doubly occupied MO's, ϕ_i and ϕ_j, are degenerate the four electron destabilization energy is given by the equation:

$$\Delta E_{ij}^4 = \frac{4(\epsilon_i S_{ij}^2 - H_{ij} S_{ij})}{1 - S_{ij}^2} \tag{6}$$

It is clear that when overlap is neglected the four electron destabilizing interaction becomes zero, i.e. $\Delta E_{ij}^4 = 0$.

At this point a discussion of the approximation of the interaction matrix element H_{ij}, where i and j are MO's, is appropriate. In general, we distinguish two situations:

a) In problems which involve comparisons of two systems having identical atomic constitution, the interaction matrix element can be simply approximated as indicated below[23].

$$H_{ij} = k\,S_{ij}, \; k = -39.7 \text{ eV} \tag{7}$$

Here, S_{ij} is the overlap integral of the two interacting MO's, i and j.

b) In problems which involve comparisons of two systems which differ in atomic constitution, the interaction matrix element, H_{ij}, is expanded in terms of interaction matrix elements between AO's, h_{mn}, which are approximated as indicated below:

$$h_{mn} = \frac{1}{2}(\beta_A + \beta_B)\,S_{mn} = \beta_{AB}^0 \, S_{mn} \tag{8}$$

The above equation is the one employed in the CNDO parametrization developed by Pople and co-workers[24]. Here, β_A and β_B are specific to the atoms A and B and S_{mn} is the overlap integral between two AO's m and n of A and B. The interaction matrix element between AO's, h_{mn}, is frequently called the resonance integral of the m and n AO's[a].

---

[a] The terms resonance integral, h_{mn}, and coulomb integral, h_{mm}, conform to the nomenclature pertinent to Hückel theory. It should be pointed out that the parametrization suggested in Table 1 is not neccessarily the best one but is adequate for illustrating qualitative principles. Obviously, when a given trend arises from conflicting variations of energy gaps and interaction matrix elements, a prudent choice of the β_{AB}^0 parameters is essential for obtaining reliable results.

Table 1. Two center interaction matrix elements

A_m, B_n	r_{mn}	S_{mn}[a]	β^0_{AB} (eV)[b]	h_{mn}(eV)
F2s-C2s	1.381	.2473	−30.00	−7.419
O2s-C2s	1.430	.2713	−26.00	−7.054
C13s-C2s	1.767	.2670	−21.62	−5.783
F2p-C2p, σ	1.381	.2571	−30.00	−7.713
O2p-C2p, σ	1.430	.2855	−26.00	−7.423
C13p-C2p, σ	1.767	.3255	−21.66	−7.050
F2p-C2s, σ	1.381	.2053	−30.00	−6.159
O2p-C2s, σ	1.430	.2455	−26.00	−6.383
C13p-C2s, σ	1.767	.3124	−21.66	−6.766
F2s-C2p, σ	1.381	.3443	−30.00	−10.329
O2s-C2p, σ	1.430	.3568	−26.00	−9.277
C13s-C2p, σ	1.767	.3064	−21.66	−6.637
F2p-C2p, π	1.381	.1206	−30.00	−3.620
O2p-C2p, π	1.430	.1392	−26.00	−3.620
C13p-C2p, π	1.767	.1392	−21.66	−3.015
Csp^3-Osp^3, σ	1.430	.5427	−26.00	−14.110
Csp^3-Ssp^3, σ	1.820	.6017	−19.57	−11.775
Csp^3-Fsp, σ	1.381	.5431	−30.00	−16.293
Csp^3-$Clsp$, σ	1.767	.5841	−21.66	−12.652
Csp^2-Fp, σ	1.381	.5670	−30.00	−17.010
Csp^2-Clp, σ	1.767	.6510	−21.66	−14.101

[a] Overlap integrals were calculated with a CNDO/2 program.
[b] β^0_{AB} values were taken from Ref.[24]

Representative values of β^0_{AB}, S_{mn} and h_{mn} are provided in Table 1. An examination of these data leads to the formulation of the following qualitative rules:

a) With only one exception, the resonance integral, h_{mn}, decreases in absolute magnitude as X varies *down* a column of the Periodic Table, *i.e.* as X becomes less electronegative along a column.

b) With only one exception, the resonance integral h_{mn} decreases in absolute magnitude as X varies *to the left* of a row of the Periodic Table, *i.e.* as X becomes less electronegative along a row. It should be noted, however, that in the cases of overlap of pure AO's the differences are very small.

At various points, we shall be interested in the effect of substitution on the strength of a given orbital interaction. In general, a substituent will alter the energy gap and the interaction matrix element of two levels and produce a change in the strength of the interaction. Hence, we would like to know how fast the strength of an interaction changes as a result of a change in $\epsilon_i - \epsilon_j$ and H_{ij} introduced by the substituent. The following differential forms of the equations for two electron stabilization, neglecting overlap, will be useful in our future discussions.

$$\frac{d(\Delta E_i^2)}{d(H_{ij})} = \frac{4H_{ij}}{\epsilon_i - \epsilon_j} \tag{9}$$

$$\frac{d(\Delta E_i^2)}{d(H_{ij})} = 2 \qquad \text{(degenerate MO's)} \tag{10}$$

$$\frac{d(\Delta E_i^2)}{d(\epsilon_i - \epsilon_j)} = \frac{-2H_{ij}^2}{(\epsilon_i - \epsilon_j)^2} \tag{11}$$

$$\frac{d(\Delta E_i^2)}{d(\epsilon_i - \epsilon_j)} = 0 \qquad \text{(degenerate MO's)} \tag{12}$$

1.1. Orbital Energies and Interaction Matrix Elements

At this point, we have completed the presentation of the key equations which will be crucial to the development of a predictive theory of molecular structure. These equations will form the basis for determining the *relative* stability of isomers, the *relative* stabilization of a cationic, radical or anionic center by substituents, etc. On the other hand, the differential expressions (9) to (12) will form the basis for determining how substitution affects the *relative* stability of isomers, the *relative* stabilization of cationic, radical and anionic centers, etc. It is then obvious that a working knowledge of Eqs. (1) to (6) presupposes a great familiarity with the key quantities involved in these equations, namely, orbital energies and interaction matrix elements.

We shall first consider the effect of atomic replacement on sigma or pi orbital energies.

$$CH_3-F \xrightarrow[\text{Replacement}]{\text{Atomic}} CH_3-Cl$$

$$CH_2=CH_2 \xrightarrow[\text{Replacement}]{\text{Atomic}} CH_2=SiH_2$$

Here, we shall inquire as to how the energy of a given MO is related to features like atom electronegativity and/or bond distance. In quantum mechanical terms, the problem amounts to determining how the energy of a ith MO is altered if we change one (or more) atomic coulomb integral, h_{mm}, and/or one (or more) resonance integral, h_{mn}, by a small amount. The pertinent expression, derived on the basis of simple perturbation theory, is the following:

$$\delta \epsilon_i = \sum_m a_{im}^2 \, \delta h_{mm} + 2 \sum_{m<n} \sum a_{im} \cdot a_{in} \, \delta h_{mn} \tag{13}$$

On the basis of Eq. (13), we distinguish the following cases:

a) Replacement of an atom is accompanied by a greater change in the first term of Eq. (13). In such a case, the energy of the ith MO will be reduced if $\delta h_{mm} < 0$

and increased if $\delta h_{mm} > 0$. The former situation obtains when an atom is replaced by a more electronegative one and the latter when an atom is replaced by a more electropositive one. In general, experience shows that the coulomb integral term variation dominates the resonance integral term variation whenever an atom is replaced by another atom of the same row of the Periodic Table. Thus, for example, the energy of a given MO will progressively decrease as carbon is replaced by a more electronegative atom of the same row, *i.e.* the MO energy depression will increase in the order $F > O > N$.

b) Replacement of an atom is accompanied by a greater change in the second term of Eq. (13). In such a case, the energy of the ith MO will be raised or lowered depending upon the sign of the coefficient product $a_{im} a_{in}$ as well as the sign of δh_{mn}. The various possibilities are summarized below:

δh_{mn}	$a_{im} a_{in}$	δe_i
+	+	+
+	−	−
−	+	−
−	−	+

Experience shows that, in most cases, the resonance integral term variation dominates the coulomb integral term variation whenever an atom is replaced by another atom of the same column of the Periodic Table. Thus, for example, the energy of the σ^* antibonding MO of HX, where X is a halogen, will progressively decrease as fluorine is substituted by a less electronegative atom of the same column, *i.e.* the MO energy will increase in the order $\sigma_{HF}^* > \sigma_{HCl}^* > \sigma_{HBr}^* > \sigma_{HI}^*$.

An interesting and important corollary of the above analysis is that in systems where the Highest Occupied Molecular Orbital (HOMO) and Lowest Unoccupied Molecular Orbital (LUMO) have identical or similar electron densities, replacement of an atom so that h_{mn} becomes less negative results in shrinkage of the HOMO-LUMO energy gap while replacement of an atom so that h_{mn} becomes more negative has the opposite effect. These results are valid regardless of the direction in which the coulomb integral of the variable atom changes and may have wide applicability to ultraviolet spectroscopy[25].

	Calculated HOMO-LUMO Gap (eV)
$CH_2{=}CH_2$	15.22
$CH_2{=}O$	17.90

The HOMO-LUMO gap calculated by an *ab initio* method using an STO-3G basis set[26] seems to support these ideas. A more definitive test will be possible after the nature of the lowest excited state in carbon unsaturated systems is understood[27].

We shall next consider the effect of substitution on sigma and pi orbital energies.

$$CH_3{-}F \xrightarrow{\text{Substitution}} NC{-}CH_2{-}F$$

$$CH_2{=}CH_2 \xrightarrow{\text{Substitution}} NC{-}CH{=}CH_2$$

We distinguish two different effects of the substituent:

a) *The Inductive Effect.* Here, the substituent X is assumed to modify the coulomb integral of the atom to which it becomes attached. The effect of this modification on orbital energies was discussed above. In most cases, the inductive effect is dominated by the resonance effect and does not need to be considered.

b) *The Resonance Effect.* Here, the orbital energies of the parent molecule are modified due to the interaction of the corresponding orbitals with the orbitals of the substituent. The equation which describes the energy change of an orbital i due to its interaction with other orbitals j, assuming nondegeneracy, is given below:

$$\Delta E_i = \sum_{j \neq i} \frac{H_{ij}^2}{\epsilon_i - \epsilon_j} \tag{14}$$

The equation describing the energy change of an orbital due to its interaction with a degenerate orbital is given below:

$$\Delta E_i = \pm H_{ij} \tag{15}$$

Simple, yet illuminating, discussions of these equations have been given before and, thus, further deliberations are not needed.

For illustrative purposes, we shall consider the resonance effect of substituents upon the energies of the pi MO's of the model system ethylene.

a) *First Period Electron Releasing Substituents* ($R = F, OH, NH_2$). We use as an example the molecule fluoroethylene. In this case, the major orbital interactions are between the p_z lone pair of fluorine and the ethylenic π and π^* MO's. These interactions result in raising the energies of both π and π^* by an amount inversely proportional to the energy separation of the interacting levels.

b) *Second Period Electron Releasing Substituents* ($R' = Cl, SH, PH_2$). We use as an example the molecule chloroethylene. In this case the $Clp_z - \pi$ and $Clp_z - \pi^*$ interactions contribute to the raising of the energies of the π and π^* ethylenic MOs, while the $d - \pi$ and $d - \pi^*$ interactions contribute to the lowering of the energies of π and π^*. Due to the relative energy spacing of the interacting levels, the ethylenic π MO will interact principally with the Clp_z lone pair, while the π^* MO may interact principally with the d orbitals of the heteroatom. As a result, one may expect that the ethylenic π MO will be raised in energy while the π^* MO may be lowered in energy. On the other hand, if both $\pi - d$ and $\pi^* - d$ interactions are very weak, the situation will become identical to the case discussed before.

c) *Electron Withdrawing Substituents* ($W = CN, CHO, NO_2$). We use as an example the molecule acrolein. In this case we have to consider the interactions of the ethylenic π and π^* MOs with the π' and $\pi^{*'}$ MOs of the carbonyl group. There are four orbital interactions, *i.e.* $\pi - \pi'$, $\pi - \pi^{*'}$, $\pi^* - \pi'$, and $\pi^* - \pi^{*'}$. The energy separation $\epsilon_\pi - \epsilon_\pi'$ favors greater interaction of π with π', while the matrix element of the interacting levels favors greater interaction of π with $\pi^{*'}$, *i.e.* the absolute magnitude of $H_{\pi\pi^*{}'}$ is much greater than that of $H_{\pi\pi'}$ because of the relative magnitude of the eigenvectors of π' and $\pi^{*'}$.

$$H_{\pi\pi'} = kS_{\pi\pi'} \approx kc_{12}c'_{11}S_{CC} \tag{16}$$

9

$$H_{\pi\pi*'} = kS_{\pi\pi*'} \approx kc_{12}c'_{21}S_{CC} \qquad (17)$$

and $|H_{\pi\pi'}| < |H_{\pi\pi*'}|$ because $c'_{21} > c'_{11}$

The result is that, depending upon the relative magnitude of the two interactions, the ethylenic π MO can be either raised or lowered in energy. On the other hand, the ethylenic π^* MO will interact principally with $\pi^{*'}$ and the LUMO of the composite molecule will necessarily have an energy which is lower than the ethylenic π^* MO and will be mostly localized on the substituent carbonyl group. These considerations can be understood by reference to Fig. 1.

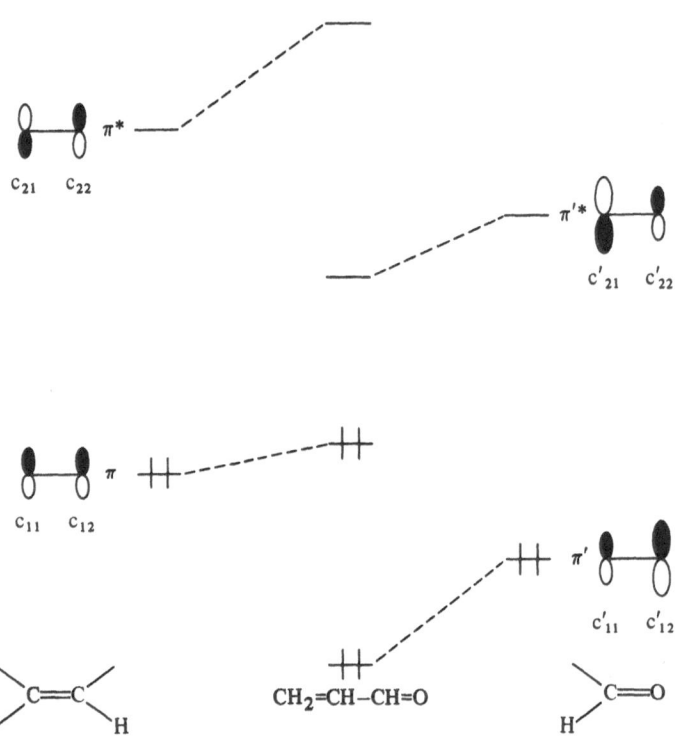

Fig. 1. Construction of the pi MO's of acrolein by the union of ethylene and formaldehyde fragments

d) Unsaturated Substituents ($V = CH = CH_2, -C_6H_5$). We use as an example butadiene. In this case, we consider the interaction of the π and π^* ethylenic MO's with the π' and $\pi^{*'}$ MO's of the unsaturated substituent. The HOMO of the composite molecule will necessarily have higher energy than that of the ethylenic π MO, while the LUMO will necessarily have lower energy than that of the ethylenic π^* MO.

The above theoretical analysis of substituent effects on orbital energies can be tested experimentally. Specifically, the energy of a given occupied MO in a closed shell system is equal to the negative of the corresponding ionization potential[28] while the energy of a given unoccupied MO can be approximated by the corresponding electron affinity[29]. The determination of the lowest ionization potential of a closed shell molecule, reflecting the HOMO energy, can easily be measured with presently available techniques and the same is true for higher ionization potentials. On the other hand, the electron affinity of a closed shell molecule, reflecting the LUMO energy, is not as simple to determine. Thus, we shall rely upon calculations of unoccupied orbital energies whenever experimental results are unavailable. This latter approach is reliable as long as our target remains the elucidation of chemical *trends*. Pertinent experimental and computational results which reveal how substitution affects orbital energies are summarized below.

a) *Pi Bonds.* The pi bond ionization potentials of representative ethylenic systems are shown in Table 2. As can be seen, the pi bond ionization potentials of $CH_2 = X$

Table 2. Ionization potentials (I.P.), electron affinities (EA), and computed orbital energies of unsaturated molecules

Molecule	IP(eV)	Ref.	EA(eV)	Ref.	π HOMO[a] Energy (eV)	π LUMO[a] Energy (eV)
$CH_2=CH_2$	10.52	30a)	−2.8	33)	−16.17	5.43
$CH_2=O$	14.09	30a)			−18.16	4.35
$CH_2=S$	11.90	30b)			−13.50	1.33
$CH_2=CHF$	10.30	30a)			−15.08	4.81
$CH_2=CHOCH_3$	8.93	31)			−13.54	5.50
$CH_2=CHN(CH_3)_2$					−12.05	5.94
$CH_2=CHCl$	10.18	30a)			−13.62	4.73
$CH_2=CHSCH_3$	8.87	32)			−12.11	5.21
$CH_2=CHP(CH_3)_2$					−17.54	4.62
$CH_2=CHCN$	10.92	30a)	.02	34)	−13.88	3.85
$CH_2=CHCHO$	10.93	30a)			−14.33	3.01
$CH_2=CHNO_2$					−16.62	0.95
$CH_2=CHCH=CH_2$	9.07	30a)			−12.49 (*trans*)	3.93 (*trans*)
					−12.46 (*cis*)	3.86 (*cis*)
$CH_2=CHC_6H_5$	8.48	30a)			−11.60	3.45

a) MO energies calculated by the CNDO/2 method using standard geometries.

decrease as the electronegativity of X decreases along a row or column of the Periodic Table. On the other hand, the π LUMO energy decreases as the electronegativity of X increases along a row or decreases along a column of the Periodic Table. The effect of substituents on the pi orbital energies of ethylene are as expected on the basis of our previous considerations.

b) *Lone Pairs.* The energy of a lone pair can be easily determined from the ionization potential of an appropriate model system. Pertinent data are shown in Table 3. As can be seen, the ionization potential increases as the electronegativity of the atom bearing the lone pair increases along a row of the Periodic Table. Also, the ionization potential decreases down a given column.

Table 3. "Lone pair" ionization potentials in CH_3X and HX model systems

Molecule	Ionization potential (eV)	Ref.
HF	16.40	35)
H_2O	12.60	35)
H_3N	10.85	35)
H_3C^-	1.12	36)
HCl	12.80	35)
H_2S	10.48	35)
H_3P	9.90	35)
CH_3F	16–18	35)
CH_3OH	11.00	37)
CH_3NH_2	9.18	35)
$CH_3CH_2\bar{C}H_2$	0.69	38)
CH_3Cl	11.30	30)
CH_3SH	9.40	30)

Table 4. Bond ionization potentials and computed orbital energies of C–X bonds

Molecule	C–X ionization potential (eV)	Ref.	$"\sigma_{CX}"^{a)}$ Energy (eV)	$"\sigma_{CX}^*"^{a)}$ Energy (eV)
CH_3-F	17.06	43a)	−21.64	9.88
CH_3-Cl	14.42	43a)	−17.04	4.96
CH_3-Br	13.49	43a)	–	–
CH_3-I	12.50	43a)	–	–
CH_3-OH	15.19	43b)	−19.78	11.13
CH_3-SH	12.00	43c)	−16.14	5.10
CH_3-NH_2	14.30	43d)	−19.42	12.17
CH_3-PH_2	12.20	43d)	−15.05	5.64
CH_3-H	12.64	43e)	b)	b)
CH_3-CH_3	11.51	43f)	b)	b)

a) MO energies were obtained from CNDO/2 calculations.
b) See Table 6.

c) C–X Bonds. Bond ionization potentials for C–X bonds are shown in Table 4. Clearly, as the electronegativity of X decreases the energy of σ_{CX} increases insofar as comparisons along a column of the Periodic Table are concerned. Additional data also show that the energy of σ_{CX} increases as the electronegativity of X decreases along a row of the Periodic Table.

Polarographic data indicates that the energy of σ_{CX}^* decreases in the order $\sigma_{C-I}^* < \sigma_{C-Br}^* < \sigma_{C-Cl}^* < \sigma_{C-F}^*$ [39].

The following pieces of experimental evidence provide indirect support for the assignment of relative energies of the σ_{CX}^* MO's:

a) In S_N2 reactions, where relative reaction rates can be predicted by reference to the energy gap between the HOMO of the nucleophile and the σ^* LUMO of the alkyl halide which is concentrated along the C—X bond, leaving group ability varies in the order $I > Br > Cl > F^{40)}$.

b) The reactivity of alkyl halides towards hydrated electrons, expected to vary in the same direction as $S_N 2$ reactivity, is also consistent with an energy variation of σ_{CX}^* in the order $\sigma_{C-I}^* < \sigma_{C-Br}^* < \sigma_{C-Cl}^* < \sigma_{C-F}^*$ [41].

The same trends are expected to be found when the atom X is varied along another column of the Periodic Table, *i.e.* decreasing electronegativity of X along a column should lead to lower σ_{CX}^* energy. Thus, for example, in the reductive C—X bond cleavage of PhCOCRR'XPh, the half wave reduction potential is less for X=S than X=O[42]. Relevant computational results are shown in Table 4.

The energy of a σ_{C-X}^* bond becomes increasingly depressed as X becomes increasingly electronegative along a row of the periodic cart.

d) *X—H Bonds.* The energy of σ_{X-H} MO's can be determined from ionization potential and computational data shown in Table 5. The energy order for first

Table 5. Bond ionization potentials and computed orbital energies of H—X bonds

Molecule	H—X ionization potential (eV)	Ref.	σ_{HX} [a] Energy (eV)	σ_{HX}^* [*a] Energy (eV)
H—F			−23.13	8.57
H—Cl	16.25	35)	−18.05	3.35
H—Br	15.28	35)		
H—I	14.00	35)		
H—OH	13.78	35, 44)	−22.07	8.43
H—SH	12.78	35, 44)	−17.59	2.44
H—NH$_2$	14.98	35)	−20.35	7.53
H—PH$_2$	13.65	35)	−15.18	.48
H—CH$_3$	13.16	43)	b)	b)
H—SiH$_3$	12.20	45)	−15.24	3.88

a) MO energies were obtained from CNDO/2 calculations except where noted.
b) See Table 6.

row elements can be seen to be $\sigma_{C-H} > \sigma_{O-H} > \sigma_{N-H}$. Again, a decrease in ionization potential is observed in going from first to second period atoms leading to the following energy orders: $\sigma_{Si-H} > \sigma_{C-H}$, $\sigma_{P-H} > \sigma_{N-H}$, $\sigma_{S-H} > \sigma_{O-H}$.

The σ_{X-H}^* energy of an X—H bond decreases with replacement of a first period atom by a second period atom. This trend is shown in Table 5. The energy of σ_{X-H}^* also decreases as the electronegativity of X increases along a row of the Periodic Table.

While the relative energies of σ_{RX} and σ_{RX}^* are well reproduced by most types of calculations when X is a heteroatom, the location of $"\sigma_{R-H}"$, $"\sigma_{R-H}^*"$, $"\sigma_{R-C}"$ and $"\sigma_{R-C}^*"$ on the range can be reliably assessed only by reference to experimental data and *ab initio* calculations. Thus, from the ionization potential data of Table 4

it is clear that a C—C bond is a better intrinsic donor than a C—H bond. In general,
this Table allows us to define an order of increasing intrinsic donor ability which
will provide the basis for discussions presented in Section 6.0. *Ab initio* computa-
tions can be used to rank $"\sigma^*_{RH}"$ and $"\sigma^*_{RC}"$ relative to other $"\sigma^*_{RX}"$ orbitals. Thus,
it is found that $"\sigma^*_{RH}"$ has the highest energy and, accordingly, the C—C bond is
a better intrinsic acceptor than the C—H bond. Typical results taken from the Salem
and Jørgensen book[20] are given in Table 6. The quotation marks used above serve

Table 6. *Ab initio* computed orbital energies of C—X model systems[a]

Model System	$"\sigma_{CX}"$ Energy (eV)	$"\sigma^*_{CX}"$ Energy (eV)
CH_3-H	−14.737 (1T$_2$)	17.519 (2T$_2$)
CH_3-F	−17.642 (σ)	−
CH_3-OH	−15.724 (σ)	13.059 (σ)
CH_3-NH_2	−13.973 (π)	14.914 (σ)
CH_3-CH_3	−13.023 (π)	15.561 (σ)

[a] All MO energies taken from Ref. [20].
Actual orbital symmetries are given in parenthesis.

the purpose of warning the reader that the corresponding orbitals do not necessarily
involve sigma overlap. Their actual symmetry is indicated in Table 6.

We now enter the discussion of interaction matrix elements, H_{ij}, and the dependence of their absolute magnitude upon the nature of the interacting fragments. We
shall consider cases which illustrate the fundamental principles.

a) Two Center H_{ij}. In this case $H_{ij} = h_{mn}$. The variation of the AO interaction
matrix element as a function of atomic constitution has already been discussed.

b) Three Center H_{ij}. In this case, our conclusions may depend upon the symmetry properties of the interacting orbitals i and j. As a first example, we consider the
matrix element H_{ij} involving an AO and a bonding MO overlapping in the manner
shown below.

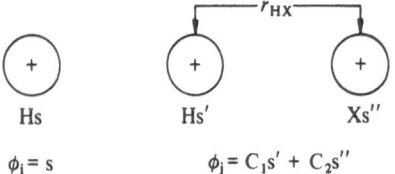

$\phi_i = s$ $\qquad\qquad$ $\phi_j = C_1 s' + C_2 s''$

The interaction matrix element H_{ij} can be expanded in terms of AO's as follows:

$$H_{ij} \propto C_1 (\beta^0_{HH}) S_{ss'} + C_2 (\beta^0_{HX}) S_{ss''} \tag{18}$$

On the basis of the above equation, the following conclusions become apparent:

a) As the electronegativity of X increases along a row of the Periodic Table the AO resonance integral and AO overlap integral terms favor an increase in the absolute magnitude of H_{ij} because β^0_{HX} becomes more negative and $S_{ss''}$ increases due to a decrease in r_{HX}. On the other hand, the ratio C_1/C_2 decreases as the electronegativity of X increases and this will tend to reduce the magnitude of the first term which makes the largest contribution to H_{ij}. The net result of the two conflicting effects can be determined by calculation. Typical results are shown in Table 7, and clearly indicate that the differences are small as a result of the opposing variations of the terms of Eq. (18).

b) As the electronegativity of X decreases along a column of the Periodic Table, β^0_{HX} becomes less negative. $S_{ss''}$ decreases due to an increase in r_{HX} but the ratio C_1/C_2 increases. Once again, calculations show that as a result of these variations, differences in H_{ij} are small (Table 7).

Table 7. Three center interaction matrix elements

Interaction i j	Geometry	$|H_{ij}|$ (eV)
$\sigma_{HF}-H1s$	F | H···H	3.203
$\sigma^*_{HF}-H1s$	F | H···H	.620
$\sigma_{HF}-H1s$	H···H−F	2.934
$\sigma^*_{HF}-H1s$	H···H−F	2.079
$\sigma_{HF}-H1s$	H···F−H	6.562
$\sigma^*_{HF}-H1s$	H···F−H	2.440
$\sigma_{HCl}-H1s$	Cl | H···H	2.339
$\sigma^*_{HCl}-H1s$	Cl | H···H	.008
$\sigma_{HCl}-H1s$	H···H−Cl	1.836
$\sigma^*_{HCl}-H1s$	H···H−Cl	1.146
$\sigma_{HCl}-H1s$	H···Cl−H	5.567
$\sigma^*_{HCl}-H1s$	H···Cl−H	2.774
$\sigma_{CF}-C2py$	F | C···C	2.401
$\sigma^*_{CF}-C2py$	F | C···C	1.654
$\sigma_{CF}-C2px$	C···C−F	8.122
$\sigma^*_{CF}-C2px$	C···C−F	3.136
$\sigma_{CF}-C2py$	C | C···F	3.09

Table 7 (continued)

Interaction i j	Geometry	$\lvert H_{ij}\rvert$ (eV)
$\sigma^*_{CF}-C2py$	C | C···F	1.041
$\sigma_{CF}-C2px$	C···F–C	1.580
$\sigma^*_{CF}-C2px$	C···F–C	5.705
$\sigma_{CCl}-C2py$	Cl | C···C	1.156
$\sigma^*_{CCl}-C2py$	Cl | C···C	.284
$\sigma_{CCl}-C2px$	C···C–Cl	3.762
$\sigma^*_{CCl}-C2px$	C···C–Cl	2.021
$\sigma_{CCl}-C2px$	C | C···Cl	1.959
$\sigma^*_{CCl}-C2py$	C | C···Cl	.9392
$\sigma_{CCl}-C2px$	C···Cl–C	2.575
$\sigma^*_{CCl}-C2px$	C···Cl–C	4.849
$\sigma_{OH}-H1s$	O | H···H	4.697
$\sigma^*_{OH}-H1s$	O | H···H	1.448
$\sigma_{OH}-H1s$	H···H–O	4.266
$\sigma^*_{OH}-H1s$	H···H–O	2.158
$\sigma_{OH}-H1s$	H···O–H	4.542
$\sigma^*_{OH}-H1s$	H···O–H	4.726
$\sigma_{SH}-H1s$	S | H···H	2.779
$\sigma^*_{SH}-H1s$	S | H···H	.300
$\sigma_{SH}-H1s$	H···H–S	1.970
$\sigma^*_{SH}-H1s$	H···H–S	1.038
$\sigma_{SH}-H1s$	H···S–H	3.681
$\sigma^*_{SH}-H1s$	H···S–H	3.687
$\sigma_{OH}-C2py$	H | C···O	2.158
$\sigma^*_{OH}-C2py$	H | C···O	.122
$\sigma_{SH}-C2py$	H | C···S	1.895
$\sigma^*_{SH}-C2py$	H | C···S	.453

At this point, the reader may wonder: what does "small" mean? The meaning of this statement becomes clear if we specify that the change of H_{ij} is measured relative to the change of $\epsilon_i - \epsilon_j$. Thus, for comparisons along a row, e.g. H···HF vs. H···H$_2$O, the change of H_{ij} is of the order of 1 eV (1.49 eV for HHF vs. HH$_2$O) while the change in $\epsilon_i - \epsilon_j$ is of the order of 3 eV (3.73 eV for HHF vs. HH$_2$O). Similarly, for comparisons along a column, e.g. H···HF vs. H···HCl, the change of H_{ij} is of the order of 1 eV (.864 eV for HHF vs. HHCl) while the change in $\epsilon_i - \epsilon_j$ is of the order of 4 eV (5.08 eV for HHF vs. HHCl).

As a second example, we consider the matrix element H_{ij} involving an AO and an antibonding MO overlapping in the same manner as before.

$\phi_i = s$ $\phi_j = C_1 s' - C_2 s''$

The interaction matrix element H_{ij} can be expanded in terms of AO's as follows:

$$H_{ij} \propto C_1 (\beta^0_{HH}) S_{ss'} - C_2 (\beta^0_{HX}) S_{ss''} \tag{19}$$

By following the same arguments as before, we can show that as the electronegativity of X increases along a row of the Periodic Table, the variations of β^0_{HX} and $S_{ss''}$ tend to decrease the absolute magnitude of H_{ij} but the variation of C_1/C_2 has the opposite effect. As a result, the difference between the H_{ij}'s is small. Similarly, as the electronegativity of X decreases along a column of the Periodic Table, a conflicting variation of β^0_{HX} and $S_{ss''}$, on one hand, and C_1/C_2, on the other, results in small changes of H_{ij}. The same type of treatment can be applied to any combination of molecular fragments in any geometry. In the following sections, we shall assume that the variations of the size of a given matrix element as a function of atomic constitution can be understood in terms of the analysis presented above.

1.2. The Concept of Matrix Element and Energy Gap Controlled Orbital Interactions

In comparing two different stabilizing interactions within a single system or one in one system and another in a second system, we shall be interested in how a change in the energy gap, $\Delta\delta\epsilon_{ij}$, and/or interaction matrix element, ΔH_{ij}, creates a change in the stabilization energy, ΔSE, where SE is set equal to $-\Delta E^2$.

$$\Delta SE = \frac{\partial SE}{\partial H_{ij}} \, \Delta H_{ij} + \frac{\partial SE}{\partial(\delta\epsilon_{ij})} \, \Delta\delta\epsilon_{ij} \tag{20}$$

The above equation allows us to distinguish the following cases:

a) $|\Delta H_{ij}| > 0$ and $\Delta(\delta\epsilon_{ij}) \simeq 0$ in which case SE increases as $|H_{ij}|$ increases.

17

b) $|\Delta(\delta\epsilon_{ij})| > 0$ and $\Delta H_{ij} \simeq 0$ in which case SE increases as $|\delta\epsilon_{ij}|$ decreases.

c) $|\Delta H_{ij}| \simeq |\Delta(\delta\epsilon_{ij})| \neq 0$. In this case, the behavior of SE depends upon the ratio of the partial derivatives $\dfrac{\partial SE}{\partial H_{ij}} \Big/ \dfrac{\partial SE}{\partial(\delta\epsilon_{ij})}$ which is given approximately by the equation shown below.

$$\left| \frac{\partial SE}{\partial H_{ij}} \Big/ \frac{\partial SE}{\partial(\delta\epsilon_{ij})} \right| \propto \left| \frac{\delta\epsilon_{ij}}{H_{ij}} \right| \tag{21}$$

In most cases of interest, $|H_{ij}/\delta\epsilon_{ij}| < 1$. This implies that for comparable $|\Delta H_{ij}|$ and $|\Delta\delta\epsilon_{ij}|$ the variations of the interaction matrix element H_{ij} will set the pattern of the variation of SE.

The above analysis clearly points to a class of highly interesting cases where $|\Delta\delta\epsilon_{ij}| > |\Delta H_{ij}| \neq 0$ and depending upon the value of $\left|\dfrac{\delta\epsilon_{ij}}{H_{ij}}\right|$, the variations of SE can be set either by $\Delta(\delta\epsilon_{ij})$ or ΔH_{ij}. The former case will obtain if $\left|\dfrac{\delta\epsilon_{ij}}{H_{ij}}\right|$ is small, i.e. if $|\delta\epsilon_{ij}|$ is small, and the latter case will obtain if $|\delta\epsilon_{ij}/H_{ij}|$ is large, i.e. if $|\delta\epsilon_{ij}|$ is large.

We are now prepared to specify a recipe for the comparison of any two stabilizing interactions A and B. The quantities ΔH_{ij} (A,B) and $\Delta\delta\epsilon_{ij}$ (A,B) are defined as follows:

$$|H_{ij} (A)| - |H_{ij} (B)| = \Delta H_{ij} (A,B) \tag{22}$$

$$|\delta\epsilon_{ij} (A)| - |\delta\epsilon_{ij} (B)| = \Delta\delta\epsilon_{ij} (A,B) \tag{23}$$

$$SE(A) - SE(B) = \Delta SE(A,B) \tag{24}$$

We now distinguish the following cases:

Case 1. ΔH_{ij} (A,B) is positive, $\Delta\delta\epsilon_{ij}$ (A,B) is negative. In this case, $\Delta SE(A,B)$ will be positive regardless of which of the two quantities $\delta\epsilon_{ij}$ or H_{ij} changes to a greater extent.

Case 2. ΔH_{ij} (A,B) is positive and $\Delta\delta\epsilon_{ij}$ (A,B) is zero or vice versa. Here, the sign of $\Delta SE(A,B)$ is self explanatory.

Case 3. ΔH_{ij} (A,B) is positive and $\Delta\delta\epsilon_{ij}$ (A,B) is positive. If $\Delta\delta\epsilon_{ij}$ (A,B) exceeds ΔH_{ij} (A,B) the variation of SE will depend upon the system(s) where the two interactions A and B obtain. In systems where $\left|\dfrac{\delta\epsilon_{ij}}{H_{ij}}\right|$ is small, $\Delta\delta\epsilon_{ij}$ (A,B) will be the controlling factor while in systems where $|\delta\epsilon_{ij}/H_{ij}|$ is large, ΔH_{ij} (A,B) will be the determining influence. These two situations are exemplified by the diagrams of Fig. 2. On the other hand, if ΔH_{ij} (A,B) exceeds $\Delta\delta\epsilon_{ij}(A,B)$ the variation of SE will be system independent and will be set by the variation of H_{ij}. For reasons discussed before, this last situation is expected to manifest itself very rarely.

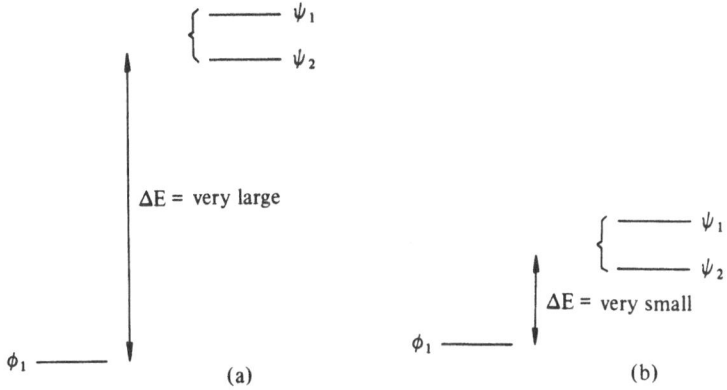

Fig. 2. (a) Orbital pattern which enforces matrix element control, *i.e.* the $\phi_1 - \psi_1$ interaction is stronger. (b) Orbital pattern which enforces energy gap control, *i.e.* the $\phi_1 - \psi_2$ interaction is stronger. In both cases, it is assumed that the $\phi_1 - \psi_1$ interaction matrix element is greater in absolute magnitude than the $\phi_1 - \psi_2$ one

The above considerations set up the stage for the introduction of the concept of energy gap controlled and matrix element controlled orbital interactions. In systems where $\delta\epsilon_{ij}$ is large, matrix element control and associated chemistry will be expected while, when $\delta\epsilon_{ij}$ is small, energy gap control and associated chemistry will obtain.

1.3. Examples of Matrix Element Control of Orbital Interactions

As a first example, we consider the pi donating ability of F and O in the model systems CH_2F and $\overset{+}{C}H_2OH$. In this case, the resonance integral will favor greater interaction of F with the vacant C2p orbital while the energy separation of the interacting levels will favor the opposite trend due to the lower ionization potential of the O2p lone pair as compared with the F2p lone pair. We can easily extend these ideas to the isoelectronic species BH_2F and BH_2OH. However, as the central atom is varied along the series C, B, the pi orbital interaction between the substituent and the central atom will tend to become matrix element controlled because the energy of the vacant 2p AO increases in the same direction, *i.e.* it is minimal for C and maximal for B.

Another interesting situation arises in the comparison of the pi donating ability of OH and SH with respect to $CH_2=CH-$ and $\overset{+}{C}H_2-$. In this case, the resonance integral will favor greater pi donation by OH, not necessarily because the S_{CO} AO overlap is greater than S_{CS} overlap as many workers seem to think, but, rather, because the $(\beta_A + \beta_B)$ term in Eq. (8) favors OH to a greater extent than the AO overlap integral favors SH. On the other hand, the energy separation of the interacting levels will favor SH over OH due to the lower ionization potential of the S3p lone pair as compared with the O2p lone pair. Once again, we distinguish two extreme cases:

a) When the LUMO of the adjacent pi system lies high in energy, the variation in the resonance integral will dominate that of the energy gap and OH will be a better donor. A system where this order can be found is $CH_2=CHX$, where $X = OH$, SH.

b) When the LUMO of the adjacent pi system lies low in energy, the variation in the energy gap will dominate and, thus, SH will be the better donor. This order can be found in $CH_2-X \cdot (X = OH, SH)$.

Typical lone pair ionization potential data have been presented before and AO overlap integral data in support of these ideas are given in Table 8. *Ab initio* results are shown in Table 9 and support the notions that the pi donating ability of heteroatoms is energy gap controlled in CH_2X but matrix element controlled in $CH_2=CHX$.

Table 8. Two center $C2p^{\pi} - Xp^{\pi}$ overlap integrals (S) computed at the optimum bond distance of $\overset{+}{C}H_2X$ cations

	S_1 a)	S_2 b)	S_3 c, d)
C–N	0.2596	0.2596	0.2454
C–P	0.2442	0.2434	0.2355
C–O	0.2169	0.2169	0.2031
C–S	0.2120	0.2106	0.2017
C–F	0.1680	0.1680	0.1571
C–Cl	0.1802	0.1786	0.1703

a) Minimal SCF atomic orbitals.
b) Single STO with exponential parameters from Clementi and Raimondi[409].
c) Single STO with exponential parameters from Slater rules[378].
d) Recent computations indicate that the trends of the overlap integrals are basis set dependent. However, those of the corresponding resonance integrals are not.

Table 9. Relative Pi donor ability of OH *vs.* SH in R–X systems[a]

Molecule	Pi charge transfer
$\overset{+}{C}H_2-OH$ b)	0.38
$\overset{+}{C}H_2-SH$ b)	0.53
$H_2C=CH-OH$ c)	0.08
$H_2C=CH-SH$ c)	0.07

a) The results were obtained from SCF-MO *ab initio* calculations at the 4-31G level.
b) Bernardi, F., Csizmadia, I. G., Epiotis, N. D.: Tetrahedron *31*, 3085 (1975).
c) Bernardi, F., Epiotis, N. D., Mangini, A.: submitted for publication.

1.4. Computational Tests of the OEMO Approach

The relative stabilization and/or destabilization of two molecular forms is best conveyed by means of an interaction diagram which depicts the interacting MO's involved in the union of two fragments. Once such a diagram has been drawn, one can predict the following:

a) Relative energies.

b) Relative bonding of atom pairs.

c) Relative atomic gross charges and AO occupation numbers.

Accordingly, the predictions arrived at on the basis of the interaction diagram can be tested by means of an explicit calculation. As we shall see in a later section, the predictive power of the interaction diagram can be exploited in formulating specific indices for the various types of interactions and effects which we shall be discussing. An explicit computation can then inform us whether the proposed interaction is indeed present and, if so, whether it is strong or weak. The computational results reported in this work should be given significance primarily in that sense.

A good computational test of the validity of OEMO theory is a high quality *ab initio* calculation[b]. However, other computational methods which are popular with chemists and which avoid the expense of *ab initio* approaches can also be useful, *if judiciously applied,* in testing the predictions of OEMO theory. These methods fall into two groups:

a) Empirical Methods. Calculations of this type are represented by the Extended Hückel (EH) theory[46-48]. The EH method utilizes an effective one electron hamiltonian and overlap is explicitly included. The off diagonal matrix elements are empirically evaluated by use of the Wolfsberg-Helmholtz approximation but interelectronic and internuclear repulsions are not treated explicitly. Calculations of the EH type, therefore, are closely tied to OEMO theory as outlined previously.

b) Zero Differential Overlap (ZDO) Methods. Representative computational methods performed within the ZDO approximation are the Complete Neglect of Differential Overlap (CNDO)[24], Intermediate Neglect of Differential Overlap (INDO)[24] and the Modified Intermediate Neglect of Differential Overlap (MINDO) methods[49-51], to mention only a few.

At this point, we shall introduce the concept of simulation and how it can be used to develop general predictive notions. For example, suppose one is interested in how a certain molecular property of Y–X will be affected due solely to the inductive effects of X as it varies along a row of the Periodic Table. In particular, suppose we wish to study the series Y–F, Y–OH, Y–NH$_2$ and Y–CH$_3$. Instead of proceeding with a MO analysis of the actual systems, something which can become very complicated due to the number of basis set AO's, we choose to replace group or atom X by hydrogen having an artificial nuclear charge which makes it as electronegative as X. Thus, instead of performing a calculation or an analysis using the actual Y–F, Y–OH, Y–NH$_2$ and Y–CH$_3$ bonds, we replace them by Y–H', Y–H'', Y–H''' and Y–H'''', where H' is a pseudohydrogen having an effective electronegativity equal to that of F, etc. It should be noted here that the bond distance of Y–H', Y–H'', etc., is kept constant. The reader will realize that this simulation technique

---

[b] The *ab initio* SCF-MO calculations reported in this work were performed using the Gaussian 70 series of programs[26]. The basis sets employed are the STO-3G, STO-4G and 4-31G basis sets developed by Pople and co-workers.

will be successful since the variation of the orbital energies along the series Y–H',
Y–H'', etc., will resemble closely the variation in the Y–F, Y–OH, etc., series
because the Y–F, Y–O, etc., bond lengths do not differ substantially.

Now, suppose we wish to study the series Y–F, Y–Cl, Y–Br and Y–I. Here,
we have to replace the X atom by hydrogen of appropriate artificial electronegativity
but we must also pick an appropriate bond length so that the variation of the orbital
energies along the series Y–F, Y–Cl, etc., is correctly reproduced.

This approach described above, henceforth named the simulation approach[52],
is useful in simplifying an otherwise bewildering analysis and providing information
about the sigma inductive effects of heteroatoms. Upon these effects, one should
superimpose any conjugative effects present. As we shall see, in many cases, sigma
inductive effects and conjugative effects are both present and operate in the same
direction. In such a situation, the relative importance of the two effects may be
simply an academic question.

Part II. Nonbonded Interactions

2. Theory of Nonbonded Interactions

While physical chemists have focused attention primarily upon van der Waals interactions in their attempts to understand why molecules or molecular fragments attract or repel each other, we have taken the position that much can be learned with
regards to the role of nonbonded interactions in chemistry within the framework
of one determinental MO theory. In other words, we have tried to convey the
message that the answer to why chemical entities attract or repel each other may
be obtained by reference to a fundamental bonding theory like OEMO theory. In
this section, we develop a theory of nonbonded interactions from that standpoint.

Our discussion in this and certain subsequent sections will involve torsional
isomerism. In such cases, we can use Eqs. (1) to (6) where H_{ij} is approximated
by kS_{ij}. Accordingly, the aforementioned equations become:

$$\Delta E_i^2 = \frac{2 S_{ij}^2 (k - \epsilon_i)^2}{\epsilon_i - \epsilon_j} \tag{1'}$$

$$\Delta E_i^2 = \frac{2 k^2 S_{ij}^2}{\epsilon_i - \epsilon_j} \tag{2'}$$

$$\Delta E_i^2 = \frac{2 S_{ij} (k - \epsilon_i)}{1 + S_{ij}} \tag{3'}$$

$$\Delta E_i^2 = 2 k S_{ij} \tag{4'}$$

$$\Delta E_{ij}^4 = \frac{4 S_{ij}^2 (\epsilon_0 - k)}{1 - S_{ij}^2} \tag{5'}$$

$$\Delta E_{ij}^4 = \frac{4 S_{ij}^2 (\epsilon_i - k)}{1 - S_{ij}^2} \tag{6'}$$

2.1. Pi Nonbonded Interactions and Pi Aromatic, Nonaromatic, and Antiaromatic Geometries

Because of the conventional upbringing of organic chemists under the auspices of
Hückel theory[53] and the concepts of aromaticity and antiaromaticity[54–58], it is
convenient to examine how the nature of pi nonbonded interactions is dependent

23

upon the number of pi electrons of a given system. Hence, we shall consider in detail model 4 N and 4 N + 2 pi electron molecules.

A. 4 N pi Electron Molecules

We first consider the case of conformational isomerism of 1,3-butadiene, the simplest 4 N pi electron molecule. We will discuss the relative stabilities of three important points on the rotational surface: the *cis* isomer ($\theta = 0°$), the *gauche* isomer ($\theta = 45°$) and the *trans* isomer ($\theta = 180°$). We shall focus our attention entirely on the effect of pi interactions on conformational preference.

$\theta=0°$ $\theta=45°$ $\theta=180°$

The pi MO's of the three conformations of 1,3-butadiene can be derived from union of the pi MO's of two ethylene molecules in the appropriate geometry. The dissection of 1,3-butadiene into two ethylenic fragments is illustrated below for the *cis* conformer:

A   B

We first consider the union of the ethylenic fragments A and B in a *cis* and *trans* geometry. The interaction diagrams for these unions are shown in Fig. 3. We distinguish two types of interactions:

a) A four electron destabilizing interaction between π and π'. Since ϵ_0 is identical for both *cis* and *trans* conformers, it is readily apparent from Eq. (6') that the relative magnitude of the four electron destabilizing interaction will be determined solely by the relative magnitude of the MO overlap integral. The overlap integral $S_{\pi\pi'}$ is greater for the *cis* conformation than the *trans* since the 1,4-pi-overlap is greater in the former than the latter case as shown below:

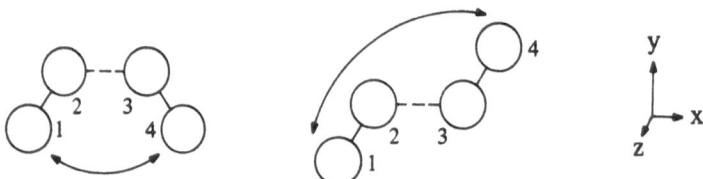

b) Two electron stabilizing interactions between π and $\pi^{*'}$ and between π' and π^*. Again, as can be seen in Fig. 3, the magnitude of the two electron stabilizing interaction is dependent only on $S_{\pi\pi^{*'}}$ ($= S_{\pi'\pi^*}$). This quantity is larger for the *trans* isomer as illustrated below:

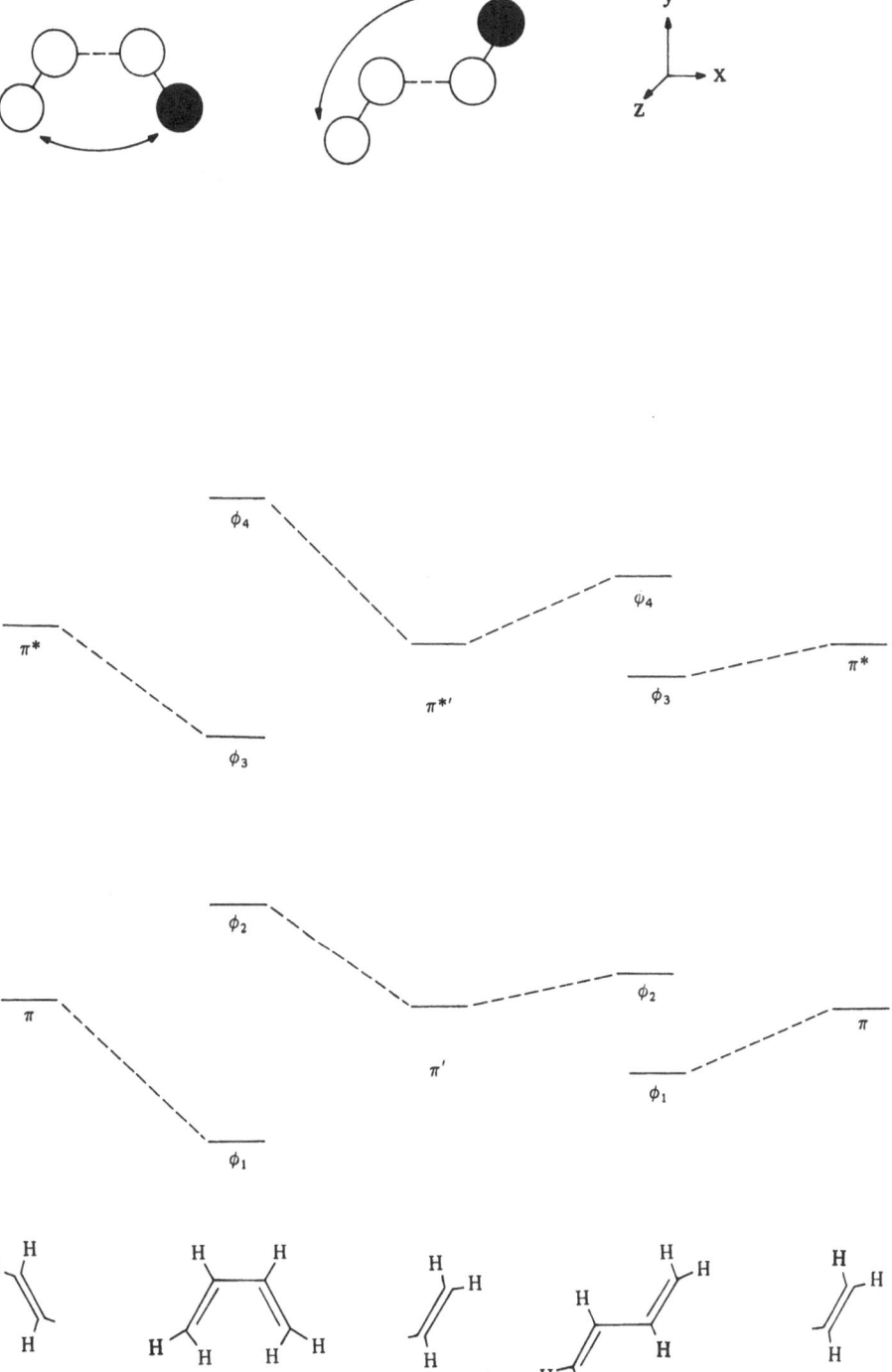

Fig. 3. The pi MO's of both *cis* and *trans* butadiene as constructed from two ethylenic fragments

We now consider the variation of the interaction energies discussed above as 1,3-butadiene is distorted from a planar *cis* geometry to a *gauche* geometry ($\theta = 45°$). Any distortion from planarity will result in a decrease of the four electron destabilization energy arising from the $\pi - \pi'$ interaction since the overlap integral $S_{\pi\pi'}$ will progressively decrease, becoming zero when the dihedral angle between the two ethylenic fragments reaches approximately the value of 45°. This is illustrated below.

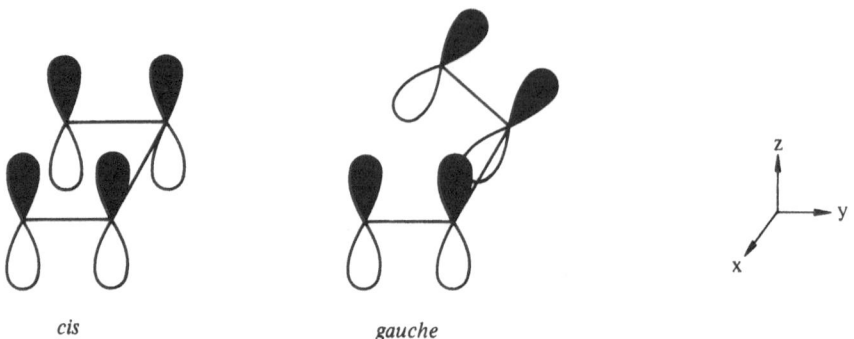

cis   *gauche*

Accordingly, the four electron destabilization energy will vary in the following order:

cis > *trans* > *gauche.*

The two electron stabilization energy will increase as the *cis* conformer is distorted from planarity since the antibonding $C_1 p_z - C_4 p_z$ overlap will decrease and eventually become bonding at $\theta = 45°$

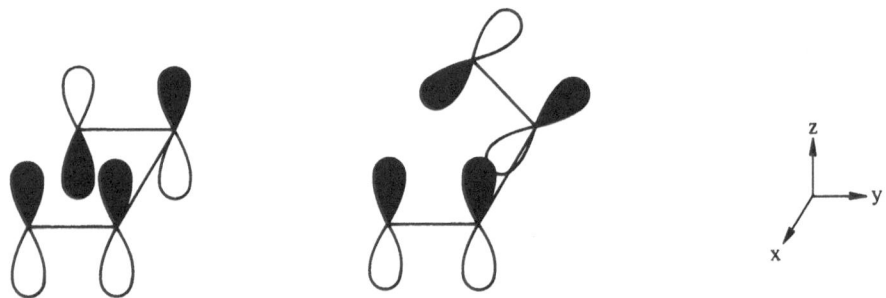

However, the increased stabilization gained by distortion due to decreased $C_1 p_z - C_4 p_z$ antibonding overlap will tend to be attenuated by the decrease in $C_2 p_z - C_3 p_z$ bonding overlap. Hence, we expect that the two electron stabilization energy will vary in the order *trans* $\simeq$ *gauche* > *cis*. Note that the $\pi' - \pi^*$ (or $\pi - \pi^{*'}$) stabilization of the *gauche* is expected to be comparable to that of the *trans* form.

On the basis of the above analysis, we predict that the relative stability of the various conformations of 1,3-butadiene will vary in the order *gauche* > *trans* > *cis*.

It should be emphasized that the previous analysis is based on the assumption of some appreciable 1—4-overlap in *cis* and *gauche* 1,3-butadiene. Actually, as this overlap goes to zero all conformations examined will be stabilized to a comparable extent and steric effects alone will dictate their relative stability. Finally, it should be pointed out that our analysis of the conformational preference of 1,3-butadiene is aimed at revealing electronic trends. As we shall see in a subsequent section, steric effects are also extremely important.

The above discussion highlights many important points:

a) *Cis*-1,3-butadiene can be thought of as a Hückel antiaromatic system, *gauche*-1,3-butadiene as a Möbius aromatic system [59] and *trans*-1,3-butadiene as a nonaromatic system.

b) The greater stability of the nonaromatic *trans* conformation relative to the antiaromatic *cis* conformation is due to both smaller four electron destabilization and greater two electron stabilization of the nonaromatic form. The greater stability of the aromatic *gauche* conformation relative to the nonaromatic *trans* conformation is due to smaller four electron destabilization and comparable two electron stabilization of the aromatic form.

c) The relative energy of the nonaromatic and Möbius aromatic structures, *i.e.* *trans* vs. *gauche* 1,3-butadiene, is a very sensitive function of spatial overlap. Accordingly, the preference for the *gauche* over the *trans* conformation is not expected to be great on strictly electronic grounds.

As a second example of a 4N-pi-electron system, we shall consider the pi dication of 1,2-difluoroethylene which can exist in a *cis* and *trans* geometry. The dissec-

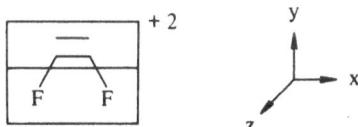

tion employed is shown above and the interaction diagrams of Fig. 4 provide the basis for the following discussion. Specifically, the effect of the various orbital interactions upon relative isomer stability is as follows:

a) The $F\,2p_z - F\,2p_z$ interaction is destabilizing and favors a *trans* geometry.

b) The $n_S-\pi$ interactions is stabilizing and favors the *trans* geometry. This result is obtained because all the terms in Eq. (1′) favor the *trans* molecule. Specifically, the overlap integral $S_{n_S\pi}$ is greater in the case of the *trans* isomer than in the *cis* isomer because the normalization constant for the n_S MO has the form $(2 + 2S_{FF})^{-\frac{1}{2}}$ and will be *smaller* in the *cis* case since S_{FF} (*cis*) $> S_{FF}$ (*trans*).

c) The $n_A - \pi^*$ interaction is stabilizing and favors the *cis* geometry for the following reasons:

1. The quantities $\epsilon_{n_A} - \epsilon_{\pi*}$ and $(k - \epsilon_{n_A})$ are smaller and greater for the *cis* isomer relative to the *trans* isomer, respectively [see Eq. (1′)].

2. The overlap integral $S_{n_A\pi*}$ is greater for the *cis* isomer because the normalization factor of the n_A MO, given by the expression $(2-2S_{FF})^{-1/2}$, is *greater* for the *cis* isomer.

Part II. Nonbonded Interactions

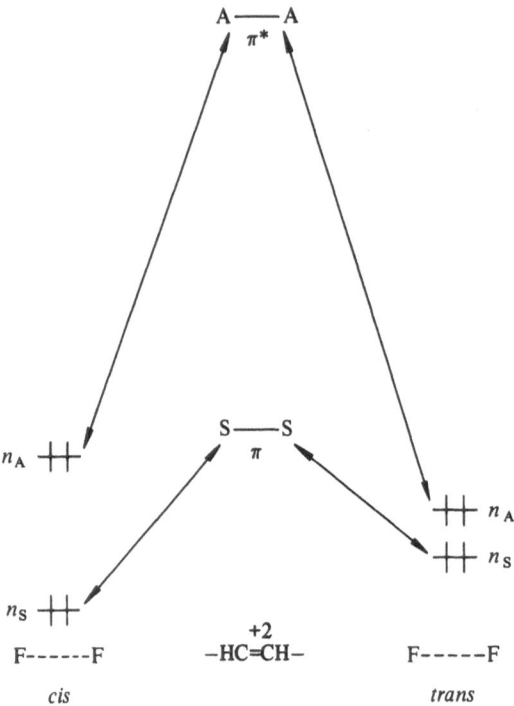

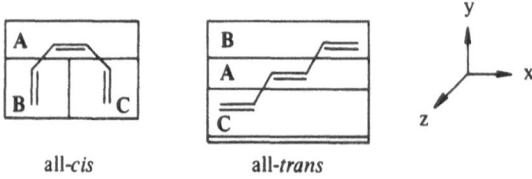

all-*cis*   all-*trans*

Fig. 4. Pi group orbital interactions in the *cis* and *trans* forms of the difluoroethylene dication. The symmetry labels are assigned with respect to a mirror plane (*cis* isomer) or a rotational axis (*trans* isomer)

Obviously, the $n_S-\pi$ stabilizing interaction is the dominant one due to the much smaller energy gap between the interacting levels. Hence, we expect that $FCH=CHF^{+2}$ will be more stable in the *trans* geometry.

Once again, we may identify the *cis*-1,2-difluoroethylene pi dication as a 4N pi electron Hückel antiaromatic system and the *trans* isomer as a 4N pi electron non-aromatic system.

B. 4N + 2π Electron Molecules

We now consider the relative stability of 1,3,5-hexatriene in its all-*trans* and all-*cis* planar conformations. 1,3,5-hexatriene can be viewed, theoretically, as the result of the union of a central ethylenic fragment A and two ethylenic fragments B and C. This dissection is shown below for the two conformations under consideration.

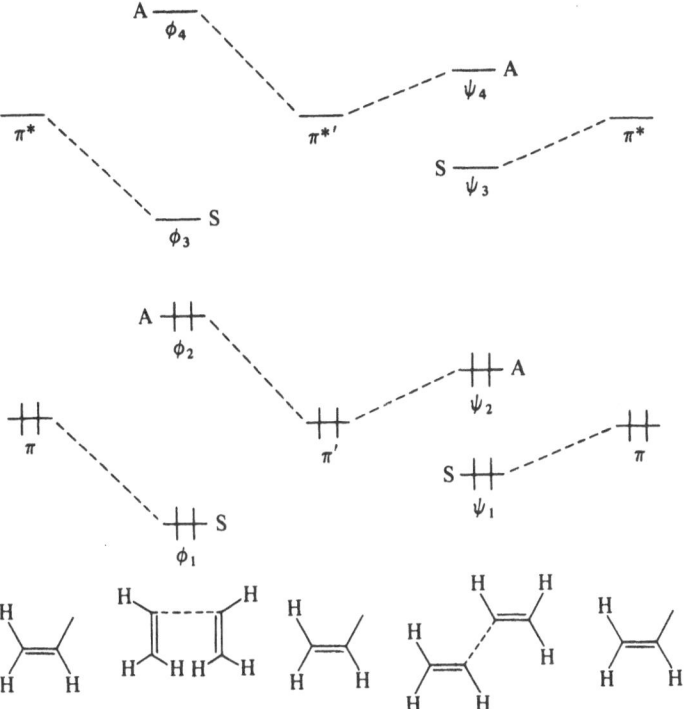

Fig. 5. The pi group MO's for the *cis* and *trans* geometries of the fragment (B + C) as constructed from the MO's of B and C. The resulting group MO's resemble the pi MO's of *cis* and *trans* butadiene, respectively. The symmetry labels are assigned with respect to a mirror plane (*cis* isomer) or a rotational axis (*trans* isomer)

We focus our attention entirely upon the influence of pi interactions on conformational preference.

The interaction diagrams of Fig. 5 show the orbital interactions which obtain in the union B + C. In the case of the all-*cis* geometry there is overlap and the two pi bonds interact appreciably. The two electron stabilizing interactions $\pi' - \pi^*$ and $\pi - \pi^{*'}$ compete with the four electron destabilizing interaction $\pi - \pi'$. In the case of the extended geometry, S_{ij} is approximately zero and the two pi bonds do not interact to any significant extent. Therefore, the stabilization and destabilization energies are near zero. We conclude that there is greater stabilization as well as greater destabilization of the cyclic form relative to the noncyclic one. The destabilization outweighs the stabilization and the all-*trans* form is favored over the all-*cis* form, at this stage of the analysis.

As can be seen from Fig. 5 the energies of group MO's resulting from the B + C union vary in the following way:

$$\epsilon(\phi_1) < \epsilon(\psi_1)$$
$$\epsilon(\phi_2) > \epsilon(\psi_2)$$
$$\epsilon(\phi_3) < \epsilon(\psi_3)$$
$$\epsilon(\phi_4) > \epsilon(\psi_4)$$

This pattern is familiar to the theoretical organic chemist since it reflects nothing more than the fact that the B + C union in the all-*cis* geometry resembles an antiaromatic structure while the same union in the all-*trans* geometry resembles a nonaromatic structure. We have already encountered such situations in our discussion of the conformational isomerism of 1,3-butadiene.

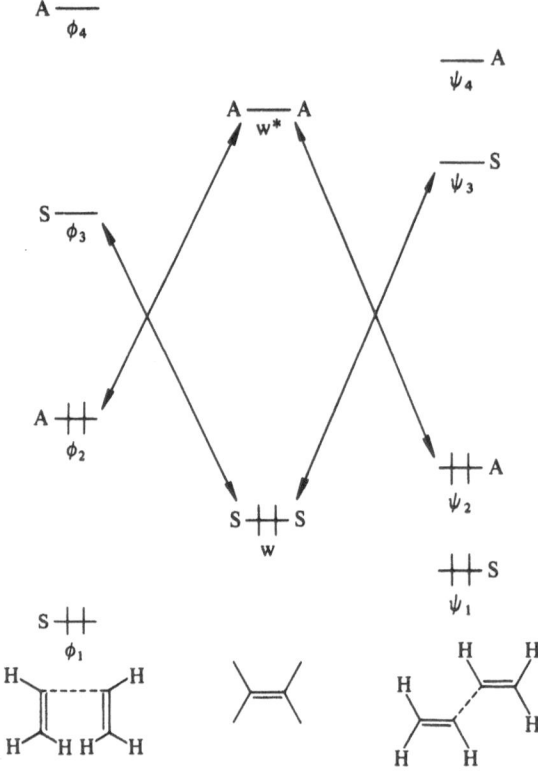

Fig. 6. Pi orbital interactions in the A + (B + C) union to form *cis* and *trans* 1, 3, 5 hexatriene. See Fig. 5 for the construction of the group MO's spanning B and C. The symmetry labels are assigned with respect to a mirror plane (*cis* isomer) or a rotational axis (*trans* isomer)

Fig. 6 shows the orbital interactions for the union A + (B + C). The diagram allows us to decide unequivocally, and without the need of a calculation, the relative degree of stabilization and destabilization of the two geometries. The following conclusions are readily apparent from considering Fig. 6 and Eq. (1'):

a) The $w - \phi_3$ will be stronger than the $w - \psi_3$ stabilizing interactions because $S_{w, \phi_3} > S_{w, \psi_3}$, ϵ_w is common for both cases and $| \epsilon_w - \epsilon_{\phi_3} |$ is smaller than $| \epsilon_w - \epsilon_{\psi_3} |$.

b) The $\phi_2 - w^*$ will be stronger than the $\psi_2 - w^*$ stabilizing interaction because $S_{\phi_2, w^*} > S_{\psi_2, w^*}$, ϵ_{ϕ_2} is less negative than ϵ_{ψ_2}, i.e. $(k - \epsilon_{\phi_2})^2$ is greater than $(k - \epsilon_{\psi_2})^2$, and $| \epsilon_{\phi_2} - \epsilon_{w^*} |$ is smaller than $| \epsilon_{\psi_2} - \epsilon_{w^*} |$.

Hence, the A + (B + C) union is more stabilizing in the case of the all-*cis* geometry relative to the all-*trans* geometry.

We now focus on the four electron destabilizing interactions. A comparison of the destabilizing interaction $w - \phi_1$ and $w - \psi_1$ is very simple. The destabilizing interaction $w - \phi_1$ will be *smaller* than the destabilizing interaction $w - \psi_1$ because $S_{w,\phi_1} < S_{w,\psi_1}$ and

$$\left| \frac{\epsilon_w + \epsilon_{\phi_1}}{2} \right| > \left| \frac{\epsilon_w + \epsilon_{\psi_1}}{2} \right|.$$ The latter inequality arises from the fact that ϵ_{ϕ_1} is

more negative than ϵ_{ψ_1}. We conclude, therefore, *that an OEMO analysis, including overlap, predicts that the all-cis geometry, for A + (B + C) union will be less destabilized relative to the all-trans geometry,* a conclusion which is in no way intuitively obvious.

We expect that the stabilizing and destabilizing energies in the A + (B + C) union will dominate these same energies in the (B + C) union since the overlap integrals in the former case will be larger than those in the latter.

Let us now, as we did in the case of 1,3-butadiene, consider the variation of the interaction energies discussed above as 1,3,5-hexatriene is distorted from an all-*cis* geometry to a *gauche* geometry ($\theta \approx 45°$) by rotation about the $C_4 - C_5$ single bond.

all-cis   *gauche* $\theta \approx 45°$

By following the same analysis as in the case of 1,3-butadiene, we can show that B + C union is most favorable in *gauche* 1,3,5-hexatriene. On the other hand, A + (B + C) union will be most unfavorable in *gauche* 1,3,5-hexatriene because four electron destabilization is maximized and two electron stabilization is minimized. The pattern set by the A + (B + C) union will dominate that set by the B + C union since the overlap integrals are larger in the former case.

On the basis of the above analysis, we predict that the relative electronic stabilization of the various conformations of 1,3,5-hexatriene will vary in the order

all-*cis* > all-*trans* > cis-*gauche*.

It should be emphasized that this analysis is based on the assumption of some appreciable 2–5 overlap in all-*cis* and cis-*gauche* 1,3,5-hexatriene. Furthermore, it should be pointed out that our analysis of the conformational preference of 1,3,5-hexatriene is aiming at revealing electronic patterns. In reality, the all-*cis* conformation of 1,3,5-hexatriene is unfavorable due to repulsive interactions between the two methylene groups, *i.e.* conformational preference varies in the order all-*trans* > cis-*gauche*.

As in the case of 1,3-butadiene, the above discussion highlights various important points:

31

Part II. Nonbonded Interactions

a) All-*cis* 1,3,5-hexatriene can be tought of as a Hückel aromatic system, *cis-gauche* 1,3,5-hexatriene as a Möbius antiaromatic system and all-*trans* 1,3,5-hexatriene as a nonaromatic system.

b) The greater electronic stabilization of the all-*cis* relative to the all-*trans* isomer and the all-*trans* relative to the *cis-gauche* isomer is due to *greater* two electron stabilization as well as *smaller* four electron destabilization.

c) All-*cis* 1,3,5-hexatriene can be thought of as a molecule which arises from an antiaromatic B + C union followed by a stronger aromatic A + (B + C) union.

As a second example of a 4N + 2 pi electron molecule we consider 1,2 difluoro-ethylene which can exist in a *cis* and a *trans* geometry. The dissection employed is shown below:

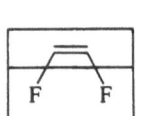 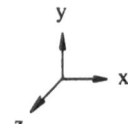

The pi frameworks of *cis* and *trans* 1,2-difluoroethylene can be constructed from the group MO's spanning the two p_z "lone pair" AO's of the fluorines and the ethylenic π MO's. In the case of the *cis* molecule, the two fluorine AO's overlap and their through space interaction lifts the degeneracy of the two "lone pair" MO's. In the trans molecule, overlap is nearly zero and the two "lone pair" MO's are degenerate. The interaction diagrams of Fig. 7 contain all the necessary information for understanding why the *cis* isomer will be predisposed to be more stable than the *trans* isomer. We distinguish three types of interactions:

a) A four electron destabilizing interaction between the fluorine lone pairs. Since the overlap integral, S_{FF}, is nonzero in the *cis* but near zero in the *trans* molecule, we know, on the basis of Eq. (6'), that four electron destabilization will be present in the *cis*, but absent in the *trans* isomer.

b) A four electron destabilizing interaction between n_S and π. The quantity $S_{n_S\pi}$ is *greater* in the case of the *trans* isomer than in the *cis* because the normalization factor for the n_S MO has the form $(2 + 2S_{FF})^{-1/2}$ and will be smaller in the *cis* case since $S_{FF}(cis) > S_{FF}(trans)$. Furthermore, ϵ_0 is more negative for the *cis* isomer and will lead to a smaller $(\epsilon_0 - k)$ value in the *cis* geometry. Hence, on the basis of Eq. (5'), we conclude that this four electron destabilization will be less for the *cis* than the *trans* isomer. Furthermore, the $n_S - \pi$ destabilizing interaction which favors the *cis* isomer will dominate the F–F destabilizing interaction which favors the *trans* isomer, the net effect being smaller overlap repulsion in the *cis* isomer.

c) A two electron stabilizing interaction between n_A and π^*. The expression for this interaction is given by Eq. (1') and the following variations obtain:
1. The energy difference $|\epsilon_{n_A} - \epsilon_{\pi*}|$ is smaller for the *cis* isomer.
2. The quantity $(k - \epsilon_{n_A})$ is greater for the *cis* isomer.
3. The overlap integral $S_{n_A\pi*}$ is greater for the *cis* isomer because the normalization factor of the n_A group MO, given by the expression $(2 - 2S_{FF})^{-1/2}$, is greater for the *cis* isomer. We conclude that the *cis* isomer of 1,2-difluoroethylene will be more stable than the *trans* isomer due to a greater two electron stabilization as well as a

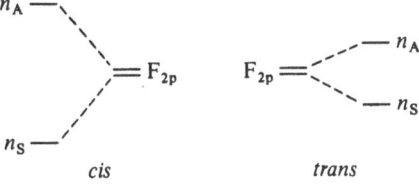

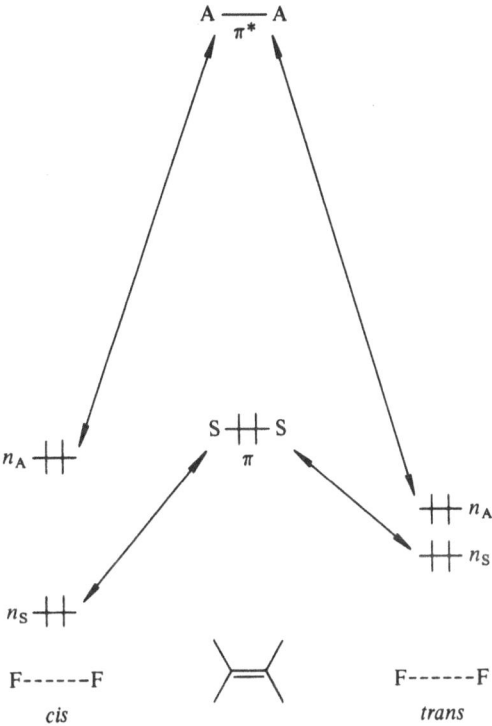

Fig. 7. Pi orbital interactions in *cis* and *trans* difluoroethylene. The symmetry labels are assigned with respect to a mirror plane (*cis* isomer) or a rotational axis (*trans* isomer)

smaller four electron destabilization. Once again, *cis*-1,2-difluoroethylene can be regarded as a $4N + 2$ pi electron Hückel aromatic system and the *trans* isomer as a $4N + 2$ pi electron nonaromatic system.

C. General Considerations

The above analysis suggests that *aromaticity or antiaromaticity in any system is the result of sequential aromatic and antiaromatic unions*. The union which dominates and, thus, determines whether the system will be aromatic or antiaromatic, is the one which involves the greatest spatial overlap between the fragments. Accordingly, one can simplify the analysis by focusing exactly on this crucial union. However, exceptions to this generalization do exist and arise in a very predictable fashion.

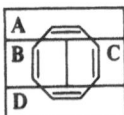

An example will serve well to further clarify the points made above. Specifically, we shall consider planar C_8H_8 as an example. The sequential unions involved here are as follows:

a) $A + B$: Antiaromatic
b) $(A + B) + C$: Aromatic
c) $(A + B + C) + D$: Antiaromatic

We can then say that planar cyclooctatetraene is antiaromatic because two anti-aromatic unions dominate a single aromatic union. Alternatively, we can say that cyclooctatetraene is antiaromatic because the crucial union, *i.e.* the union involving the greatest spatial overlap, is the $(A + B + C) + D$ union which is antiaromatic.

There are several messages hidden here:

a) In attempting to synthesize a molecule, one may deliberately accentuate or diminish the aromaticity or antiaromaticity of unions by appropriate structural modifications. For example, suppose one wished to construct a cyclic molecule having eight conjugated pi bonds.

The unions involved are as follows:

$A + B$	:	Antiaromatic
$(A + B) + C$	:	Aromatic
$(A + B + C) + D$	:	Antiaromatic
$(A + B + C + D) + E$	:	Aromatic
$(A + B + C + D + E) + F$	:	Antiaromatic
$(A + B + C + D + E + F) + G$	:	Aromatic
$(A + B + C + D + E + F + G) + H$	:	Antiaromatic

The molecule $C_{16}H_{16}$ is formally Hückel antiaromatic in a planar geometry. However, appropriate structural modifications may selectively accentuate the aromatic unions and lead to a $C_{16}H_{16}$ molecule which is *not* antiaromatic.

b) $4N$ electron systems may be more Hückel antiaromatic or Möbius aromatic than $4N + 2$ electron systems are Hückel aromatic or Möbius antiaromatic. This arises because in $4N + 2$ electron systems there are k antiaromatic and k aromatic unions while in $4N$ electron systems there are k aromatic (or antiaromatic) and $k + 1$ antiaromatic (or aromatic) unions. Hence, net aromaticity or antiaromaticity may be more pronounced in $4N$ electron systems. Thus, for example, we expect cyclobutadiene to be more antiaromatic than benzene is aromatic.

An alternative approach for determining the relative pi electronic stabilization of two torsional isomers utilizes a molecular dissection into two open shell radical fragments. This approach is illustrated by examining torsional isomerism in buta-diene and 1,3,5-hexatriene. The π MO's of the conformational isomers of 1,3,5-hexatriene can be constructed from the union of the π MO's of two formal allyl radicals. The two regiochemical modes of union of interest will be designated *cis* and *trans:*

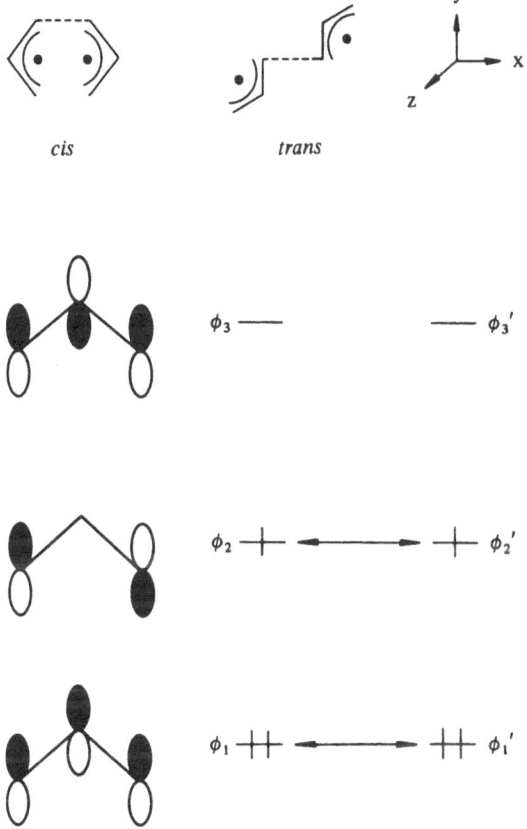

Fig. 8. Pi orbital interactions obtaining in the union of two allyl radical fragments to form 1,3,5-hexatriene

From the interaction diagram of Fig. 8, we immediately see that the two electron stabilizing interaction, $\phi_2 - \phi_2'$, will be greater for *cis* union than for *trans* since the overlap integral, $S_{\phi_2 \phi_2'}$, is larger in the former case:

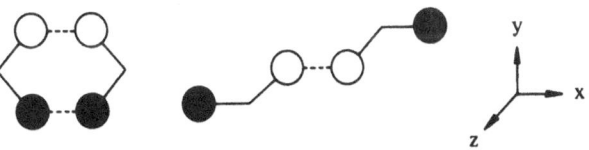

On the other hand, the four electron destabilizing interaction, $\phi_1 - \phi_1'$, is larger for *cis* union than *trans* because the overlap integral $S_{\phi_1 \phi_1'}$, is larger in the former case. However, the two electron stabilizing interaction dominates the four electron destabilizing interaction. Thus, *cis* union will be more favorable than *trans* union. In other words, here we have an application of the regiochemical rule stating that the union of two open shell fragments having a total of $4N + 2$ pi-electrons will be predisposed to be *cis*.

We can also determine the relative stability of *cis*- and *trans*-1,3-butadiene by evaluating the stabilization resulting from the union of allyl and methyl radical fragments in a *cis* and *trans* geometry. The dissection of 1,3-butadiene is shown below. It is predicted that the *trans* isomer will be more stable.

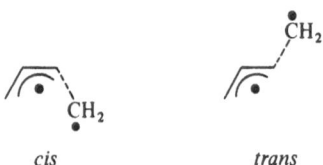

cis　　　　　　　　　*trans*

2.2. Sigma Nonbonded Interactions

The problem of sigma nonbonded interactions can be approached in exactly the same manner as the problem of pi nonbonded interactions. However, a major difference exists. Specifically, pi nonbonded interactions can be evaluated unambiguously in most systems due to the fact that the pi levels of the component system and the coupling unit are well separated in energy and their relative ordering can be assessed from first principles. On the other hand, the sigma levels of the coupling unit are closely spaced and may act competitively in determining the relative stabilization of two geometries by sigma nonbonded interactions. Thus, we have chosen to consider in detail the nonbonded interaction between the sigma lone pairs of 1,2-difluoroethylene in order to illustrate some of the problems involved.

1,2-difluoroethylene is made up of six atoms, six pi valence electrons and eighteen sigma valence electrons. The simplest approach to the problem involves the assumption that the MO's spanning the fluorine $2p_x$ lone pair AO's, *i.e.* the MO's of the component system, will interact principally with the occupied σ_g and unoccupied σ_u MO's of the sigma carbon-carbon bond. This assumption leads to the interaction diagrams shown in Fig. 9 and the conclusion, following familiar arguments, that the *cis* isomer will be favored by sigma lone pair nonbonded attraction. This type of argument has been used by us before and is based on the conclusions reached by Hoffmann and his collaborators regarding through bond coupling[60-64]. Our recent computational work has provided the basis for a more detailed analysis which provides a more realistic appreciation of sigma nonbonded interactions. The results of this analysis are not as clear cut as the ones reached on the basis of the previous more simplistic approach. However, they are important insofar as they can affect the type of reasoning employed in the unraveling of through bond and through space interactions by means of photoelectron spectroscopy.

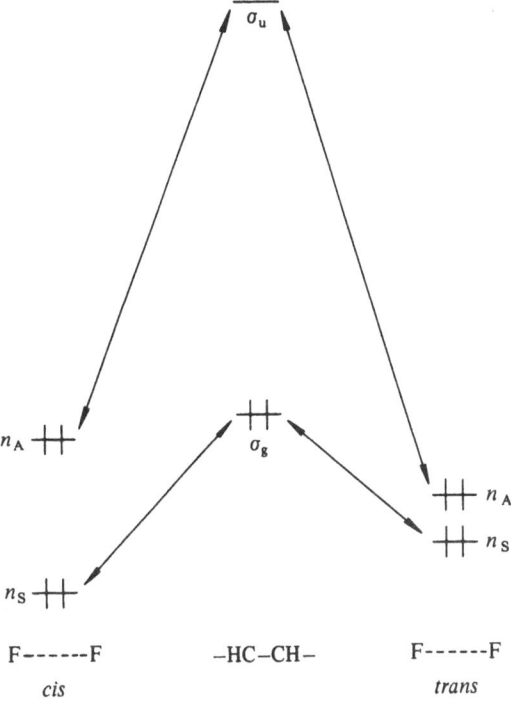

Fig. 9. Possible through bond coupling of fluorine lone pairs and the C–C sigma bond in *cis* and *trans* 1,2 difluoroethylene

As we have mentioned already, 1,2-difluoroethylene has eighteen valence sigma electrons to be distributed among fourteen sigma MO's, assuming a simple AO basis set of H1s, F2s, F$2p_x$, F$2p_y$, C2s, C$2p_x$ and C$2p_y$. The sigma orbitals of the coupling unit and component system are shown in Fig. 10 along with the electron occupancy based on the Aufbau-Principle[65].

Focusing now on the sigma lone pair interactions, we can simplify the interaction diagram as shown in Fig. 11. The relative stabilization of the *cis* and *trans* geometries due to the interaction of the lone pairs with the central C–C bond can be assessed from consideration of all orbital interactions shown in Fig. 11. These interactions and their impact upon geometrical preference are discussed below.

a) The $n_S - \sigma_g$ destabilizing interaction favors the *cis* isomer.

b) The $n_A - \pi_u$ interaction in the *trans* isomer is more stabilizing than the $n_A - \pi_g$ interaction in the *cis* isomer.

c) The $n_S - \pi_g$ interaction in the *trans* isomer is less stabilizing than the $n_S - \pi_u$ interaction in the *cis* isomer.

d) The $n_A - \sigma_u$ stabilizing interaction favors the *cis* isomer.
We now distinguish the following cases:

a) The energy difference of the π_u and π_g orbitals of the coupling unit is very small. In such a case, the sigma lone pair interactions with the pi type MO's of the coupling unit will stabilize the *cis* and *trans* geometries to a comparable extent. Accordingly, the relative stabilization of the *cis* and *trans* geometries will be determined

37

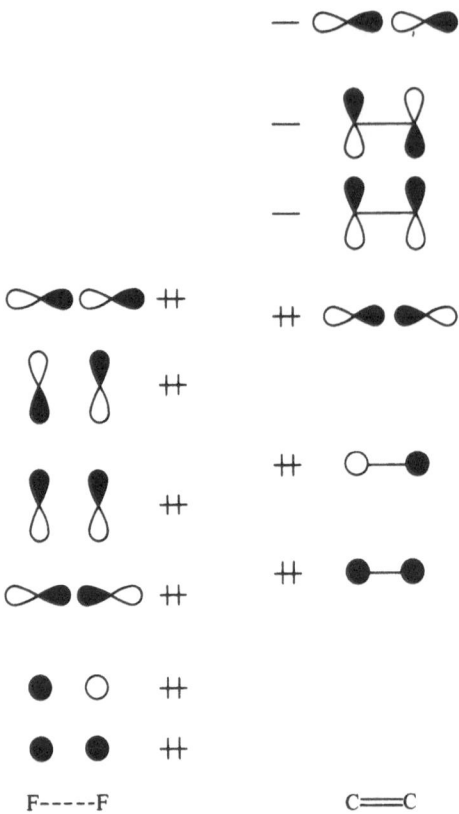

Fig. 10. Sigma orbitals of the C = C coupling unit and the F– –F component system and electron occupancy. Unfilled H– –H orbitals are omitted. Diagram is schematic

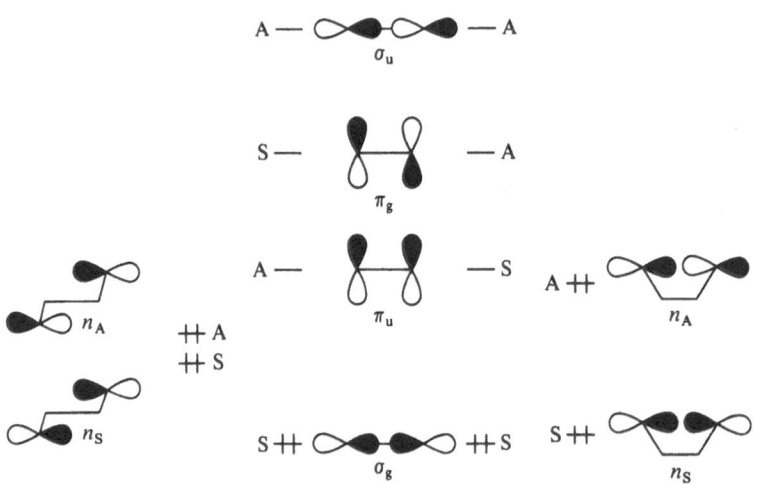

Fig. 11. The orbital pattern of sigma fluorine lone pair AO's and sigma C = C MO's in *cis* and *trans* 1,2 difluoroethylene. Orbitals q the same symmetry interact

38

by the $n_S-\sigma_g$ and $n_A-\sigma_u$ interactions which favor the *cis* isomer. In other words, this detailed analysis provides the justification for the more simplistic approach presented before.

b) The energy difference between the π_u and π_g MO's of the coupling unit is large. In this case, the dominant orbital interactions are the $n_A-\pi_u$ in the *trans* isomer and $n_S-\pi_u$ interaction in the *cis* isomer. Clearly, the former interaction is stronger and is expected to favor the *trans* geometry. Note that in this case, the σ_u orbital will not be able to tip the balance towards greater stabilization of the *cis* isomer because, in addition to the energy gap factor, the spatial overlap between the MO's of the coupling unit and component system accentuates the interaction of the pi type MO's of the C–C bond with the lone pair MO's.

The analysis provided above, leads us to anticipate trends in sigma nonbonded interactions between lone pairs. Thus, in the case of the model system FH_nX-YFH_m, the possibility of lone pair nonbonded attraction is expected to increase as the splitting of the π_u and π_g levels of the X–Y fragment gets smaller. Typical energy gaps, $\epsilon(\pi_u)-\epsilon(\pi_g)$, calculated for the isolated diatomic X–Y are given below:

Diatomic	C–C	C=C	C=N	N–N	N=N	O–O	O=O
Interatomic distance (Å)	1.54	1.34	1.32	1.45	1.25	1.48	1.21
$\epsilon(\pi_u)-\epsilon(\pi_g)$ (eV)	−17.05	−21.35	−20.85	−16.43	−21.36	−18.96	−21.98

Calculation: INDO, standard bond lengths

As can be seen, the splitting remains always substantial and does not fluctuate drastically. However, this does not constitute proof that the π_u orbital will be the one which will control the lone pair nonbonded interaction in the model systems simply because the relative orbital energies of an isolated diatomic X–Y are certainly different in a *quantitative* sense from those of a diatomic X–Y "within" a molecule FH_nX-YFH_m.

The importance of the above discussion for the evaluation of through bond and through space interactions can be understood by reference to Fig. 12. In one case, the level ordering imposed by through space coupling is reversed due to a dominant σ_g-n_S through bond interaction. In the second case, it is not reversed due to a dominant π_u-n_S through bond interaction. Accordingly, an S, A lone pair MO ordering is consistent with any of the following possibilities:

a) Strong through space interaction not overcompensated by through bond coupling involving a dominant $n_S-\sigma_g$ interaction.

b) Strong through space interaction aided by a dominant $n_S-\pi_u$ through bond interaction.

c) Zero through space interaction but strong through bond coupling involving a dominant $n_S-\pi_u$ interaction.

While the sigma nonbonded interaction of lone pairs is ambiguous, the situation improves when one considers nonbonded interactions of sigma bonds. This is illustrated by reference to the molecule 1,3-butadiene. The orbitals of the C–C coupling unit are assumed to be the same as the ones shown in Fig. 11 for the C–C coupling unit of 1,2-difluoroethylene. The sigma MO's of the CH_2 group and the

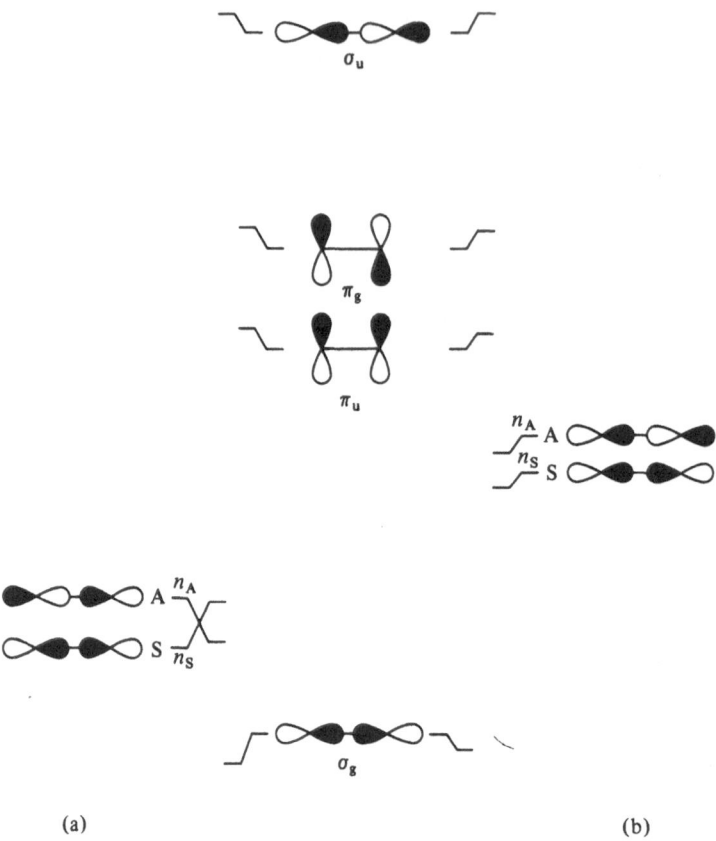

Fig. 12. Two through bond orbital interaction patterns. In (a) a dominant $n_S - \sigma_g$ interaction leads to an A, S order. In (b) a dominant $n_S - \pi_u$ interaction leads to the opposite order

MO's of the component system can be easily developed. The final simplified inter-action diagram is shown in Scheme 1. Clearly, we now have an additional inter-action, as compared to the 1,2-difluoroethylene case, which stabilizes the *cis* isomer. Hence, the sigma nonbonded interaction of bonds is expected to favor a *cis* structure when the two bonds are coupled through another sigma bond.

The above discussion is not only appropriate to 1,2-difluoroethylene and 2-butene but to other systems as well. Indeed, the sigma orbital interaction patterns discussed above obtain in diverse molecules. These and other patterns, as well as the types of molecular systems to which they apply, are pictured in Scheme 1.

2.3. Indices of Nonbonded Interactions and Steric Effects

In the previous sections, we saw that, in most cases, a nonbonded attractive or re-pulsive interaction is enforced by both four electron destabilization and two electron stabilization. Hence, in order to simplify subsequent discussions, we shall adopt the OEMO model with neglect of overlap. Consequently, in the remainder of this work

Scheme 1.

Orbital pattern A   Model system

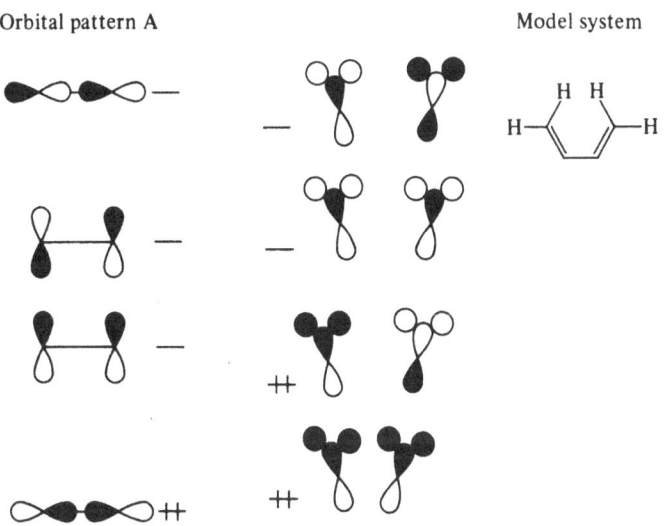

Comment: Sigma nonbonded interactions are predicted to be attractive in most cases.

Orbital pattern B   Model system

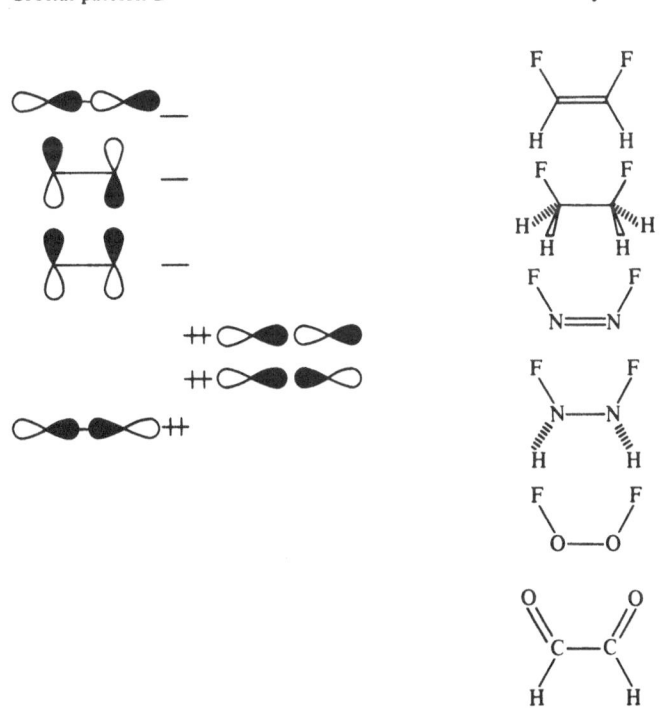

Comment: Computational test is needed to determine the sign of sigma nonbonded interaction.

Part II. Nonbonded Interactions

Scheme 1. (continued)

Orbital pattern C   Model system

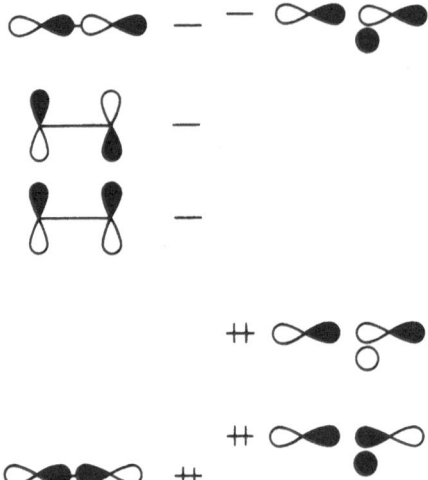

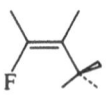

Comment: See Section 3.7.

Orbital pattern D   Model system

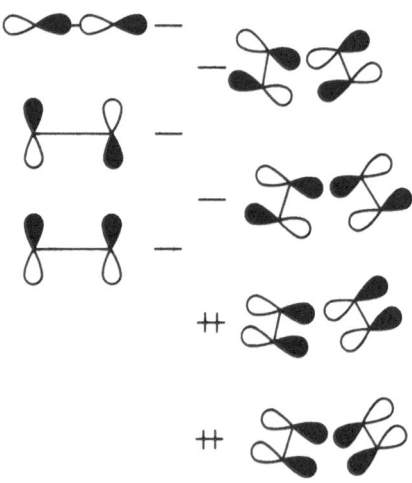

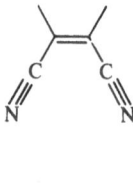

Comment: Sigma nonbonded interactions are predicted to be attractive in most cases.

Scheme 1. (continued)

Orbital pattern E   Model system

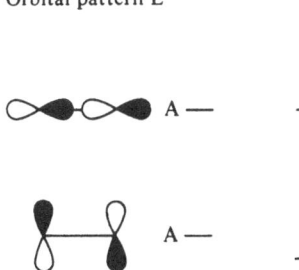

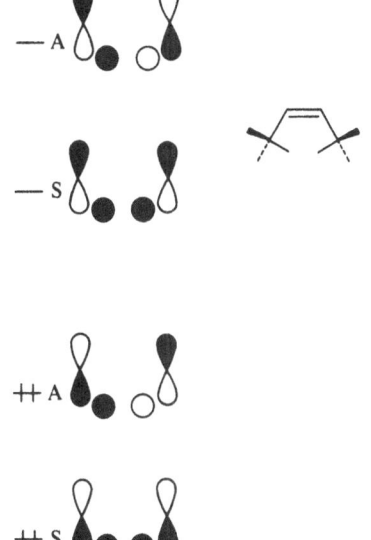

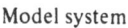

Comment: Sigma nonbonded interactions are predicted to be attractive in most cases.

Orbital pattern F   Model system

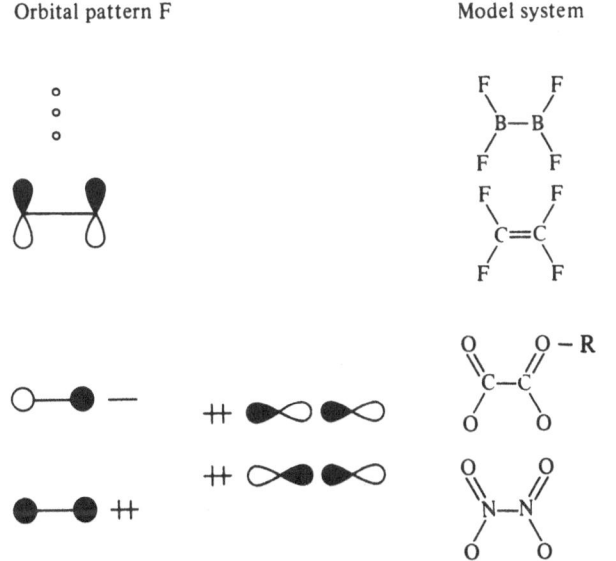

Comment: See Section 3.3.

we will focus solely on two electron stabilizing orbital interactions and will examine the effects of four electron overlap repulsion only if necessary.

So far, we have concentrated on the energetics of the union of fragment MO's in a specified geometry, *i.e.* this has been an "energy approach". Further insights can be gained by adopting a "charge transfer approach". The important points concerning charge transfer relevant to this work are:

a) The interaction between a doubly occupied orbital, ϕ_i, and an unoccupied orbital, ϕ_j, leads to charge transfer from ϕ_i to ϕ_j. The magnitude of charge transfer parallels the magnitude of the stabilizing interaction, at least in most cases.

b) The interaction between a doubly occupied orbital, ϕ_i, and an unoccupied orbital, ϕ_j, leads to an overlap distribution $\phi_i\phi_j$ which becomes increasingly bonding between the two fragments as the magnitude of the stabilizing interaction increases.

c) The interaction between two doubly filled orbitals (overlap neglected) leads to no net charge transfer.

In the charge transfer approach, we are concerned with bonding changes as a function of molecular geometry. The OEMO method can be used to formulate bonding indices which can be tested by one of the previously discussed computational methods. In general, when the interaction between two AO's becomes more bonding or antibonding as the spatial overlap of the two AO's increases, either P_{XY}, the bond order between atoms X and Y, or, P_{XY}, the overlap population between atoms X and Y, are reliable indices. However, when the aforementioned interaction becomes less bonding or antibonding as spatial overlap increases, P_{XY} becomes a misleading index. For example, consider the *cis* and *trans* isomers of 1,2-difluoroethylene and the pi bond order and overlap populations of the two fluorine atoms. We distinguish the following possibilities:

a) The pi F——F interaction is more bonding in the *cis* isomer. Here, both p^{π}_{FF} and P^{π}_{FF} will be positive and both larger in the *cis* isomer.

b) The pi F——F interaction is more antibonding in the *cis* isomer. Here, as in the previous case, either P^{π}_{FF} or p^{π}_{FF} are reliable indices.

c) The pi F——F interaction is more bonding in the *trans* isomer. Here both p^{π}_{FF} and P^{π}_{FF} will be positive but, while p^{π}_{FF} will be larger, P^{π}_{FF} will be smaller in the *trans* isomer due to the attenuating effect of S_{FF}.

d) The pi F———F interaction is more antibonding in the *trans* isomer. Here, as in the previous case, P^{π}_{FF} can be misleading.

Accordingly, we resolve that P_{XY} is the most reliable index of the interactions of two AO's.

Another reliable index is the sum of the overlap populations of AO pairs spanned by the MO's the interaction of which we focus upon. For example, consider the $n_A - \pi^*$ interaction in *cis* and *trans* 1,2-difluoroethylene (Fig. 7) and its consequences.

Interaction	*cis* isomer
F———F	more bonding
C——F (adjacent)	more bonding

C——F′ (nonadjacent)   more antibonding
C——C   more antibonding

The greater F———F pi bonding interaction in the *cis* isomer is reflected in a greater p_{FF}^{π} for the *cis* isomer but also in the greater total overlap population N_T^{π} for the *cis* isomer. The latter quantity is defined as follows:

$$N_T^{\pi} = P_{FF}^{\pi} + 2P_{CF}^{\pi} + 2P_{CF'}^{\pi} + P_{CC}^{\pi}$$

As a general rule, pi bond orders, pi overlap populations and total pi overlap populations allow a diagnosis of the type of pi interactions which obtain between two AO's. The situation is much more uncertain in the case of sigma AO interactions, and one has to rely upon bond orders and/or overlap populations.

At this point, we should call attention to the fact that the signs of the *cis* and *trans* sigma bond orders or overlap populations between the fluorine lone pairs in 1,2-difluoroethylene and related systems provide definite information about whether nonbonded lone pair interactions favor the *cis* or *trans* isomer. From Fig. 11, it is apparent that, if π_g and π_u are very close in energy, $p(F2p_x, F2p_x)$ will be more positive in the *cis* isomer. On the other hand, when π_u is the "effective" LUMO of the coupling unit, $p(F2p_x, F2p_x)$ should be positive in the *trans* isomer, negative for the *cis* isomer, and, furthermore, the *trans* bond order should be larger in absolute magnitude than the *cis* bond order.

A final useful index of sigma nonbonded interactions between lone pairs is the partial bond order $p'(Xm, Yn)$ which is evaluated over the MO's which result from the interaction of the lone pair group MO's with the sigma HOMO and vacant MO's of the coupling unit. This index is intimately connected with the type of analysis employed in this work. In our survey of a variety of problems of molecular structure we shall provide computational results pertinent to the analysis outlined, *i.e.* all or some of the following indices will be provided:

a) Total pi overlap population, N_T^{π}.
b) Long range pi bond order, p_{XY}^{π}.
c) Long range pi bond overlap population, P_{XY}^{π}.
d) Long range sigma bond order $p(Xm, Yn)$.
e) Partial long range sigma bond order $p'(Xm, Yn)$.
f) Long range sigma overlap population $P(Xm, Yn)$.

Once we have specified the indices of nonbonded interactions, we should define the models which can be used for the identification of the nature of a given nonbonded interaction and its influence upon molecular structure. Specifically, we distinguish the following two models:

a) Static Model. Here, nonbonded interactions within a molecule are evaluated for a fixed geometry and the type of the nonbonded interaction, *i.e.* attractive or repulsive, provides a basis for making a prediction of molecular geometry. The bond order or overlap population between the nonbonded atoms or groups is taken as

45

the index of the type of the nonbonded interaction, *i.e.* a positive bond order or overlap population is taken to imply nonbonded attraction and a negative bond order or overlap population is taken to imply nonbonded repulsion. This model is satisfactory in most, but not all, cases. The pitfalls awaiting the unwary practitioner will be discussed later when we examine the effect of nonbonded interactions upon bond angles.

b) Dynamic Model. Here, nonbonded interactions are evaluated by reference to the energy change which accompanies a change of a given nonbonded interaction. If the energy of the system decreases as the nonbonded interaction becomes more pronounced, the latter is identified as an attractive interaction and conversely. The change in the total pi overlap population, which includes interactions between bonded as well as nonbonded centers, is taken as the index of the type of pi nonbonded interactions. Thus, an increase in the total pi overlap population as a pi nonbonded interaction is accentuated belies nonbonded attraction, while a decrease in the total pi overlap population belies nonbonded repulsion. Of course, another index is the difference between the appropriate long range bond orders in the two geometries under consideration.

In Section 3.1., we shall show that the dynamic model leads to an unambiguous determination of the type of nonbonded interactions involved while the static model may lead to erroneous predictions as a result of an ambiguous definition of the nature of a nonbonded interaction. The superiority of the dynamic model is due to the fact that "nonbonded" interactions affect "bonded" interactions and, thus, the change in an overall overlap population rather than the change of a specific overlap population between nonbonded atoms or groups is the most appropriate index of a nonbonded interaction. Accordingly, we shall employ the dynamic model in all subsequent discussions of molecular structure, unless otherwise stated.

In order to exemplify how the various computed indices can be used in testing the predictions of OEMO theory within the framework of the dynamic model we shall consider once again the problems of conformational isomerism of 1,3-butadiene and geometrical isomerism of 1,2-difluoroethylene. We first consider the *cis* and *trans* conformers of 1,3 butadiene.

Inspection of Fig. 3 and focusing only on the stabilizing interactions, leads to the following predictions:

a) The $\pi-\pi'^*$ interaction leads to removal of electron density from a bonding π MO to an antibonding $\pi^{*'}$ MO. This will tend to reduce the C_1-C_2 and C_3-C_4 pit bond orders, increase the $C_2 - C_3$ pi bond order and decrease the $C_1 - C_4$ pi bond order. Obviously, the $\pi'-\pi^*$ interaction has the same consequence. Since charge transfer due to the MO interactions $\pi-\pi'^*$ and $\pi'-\pi^*$ is greater for the *trans* than the *cis* isomer, it is clear that the aforementioned bonding consequences of these orbital interactions will be accentuated in the *trans* relative to the *cis* conformer of 1,3-butadiene.

b) The greater stability of the *trans* conformation will be reflected in a larger total pi overlap population relative to the *cis*.

Results of INDO calculations of *cis* and *trans* 1,3-butadiene are shown below and are in complete agreement with the predictions of our theoretical model:

46

Scheme 2.

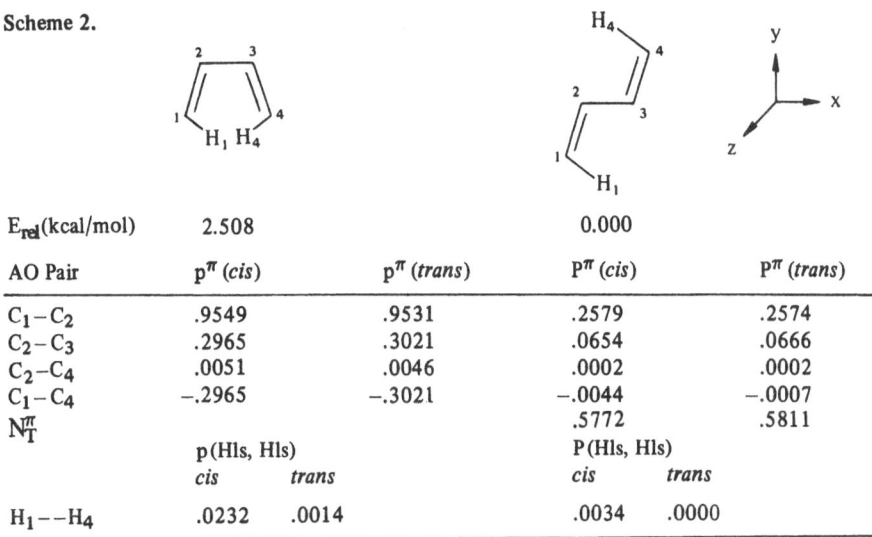

E_{rel}(kcal/mol)	2.508		0.000	
AO Pair	p^π (cis)	p^π (trans)	P^π (cis)	P^π (trans)
C_1-C_2	.9549	.9531	.2579	.2574
C_2-C_3	.2965	.3021	.0654	.0666
C_2-C_4	.0051	.0046	.0002	.0002
C_1-C_4	−.2965	−.3021	−.0044	−.0007
N_T^π			.5772	.5811
	p(HIs, HIs)		P(HIs, HIs)	
	cis	trans	cis	trans
H_1--H_4	.0232	.0014	.0034	.0000

Computation: INDO, standard geometry.

We now turn to the bonding consequences of F--F nonbonded interaction in *cis*- and *trans*-1,2-difluoroethylene. Inspection of Fig. 7, and reasoning as before, leads to the following predictions:

a) The C_1-F_1, C_2-F_2 and F--F bond orders will tend to be larger in the *cis* than in the *trans* isomer. On the other hand, charge transfer from n_A to π^* will result into a greater C_1-C_2 bond order in the *trans* relative to the *cis* isomer.

b) The greater stability of the *cis* isomer will be reflected in a larger total pi overlap population relative to the *trans*.

Results of INDO calculations of *cis*- and *trans*-1,2-difluoroethylene are shown below and are in complete agreement with the predictions of our model:

Scheme 3.

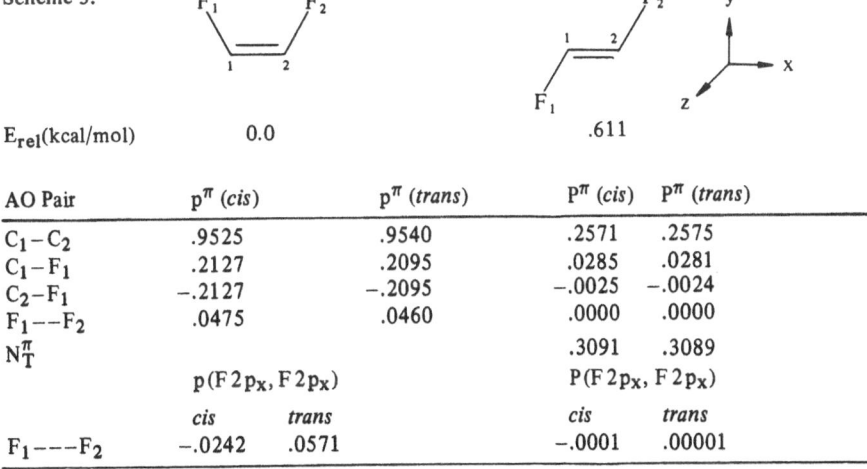

E_{rel}(kcal/mol)	0.0		.611	
AO Pair	p^π (cis)	p^π (trans)	P^π (cis)	P^π (trans)
C_1-C_2	.9525	.9540	.2571	.2575
C_1-F_1	.2127	.2095	.0285	.0281
C_2-F_1	−.2127	−.2095	−.0025	−.0024
F_1--F_2	.0475	.0460	.0000	.0000
N_T^π			.3091	.3089
	$p(F\,2p_x, F\,2p_x)$		$P(F\,2p_x, F\,2p_x)$	
	cis	trans	cis	trans
F_1---F_2	−.0242	.0571	−.0001	.00001

Computation: INDO, standard geometry.

Finally, we should establish some index of steric repulsion since in certain instances it may not be immediately obvious which of two isomers is more destabilized by steric effects. The most convenient index is the nuclear repulsion energy, E_N, which can be calculated readily for any molecular system[c].

3. The Effect of Nonbonded Interactions on Molecular Structure

We are now prepared to examine how attractive or repulsive nonbonded interactions determine molecular geometry. We shall discuss representative examples where pi and/or sigma nonbonded interactions obtain. In each case, we provide computational data in support of general theoretical arguments as well as pertinent experimental results. It should be mentioned that only crucial indices of nonbonded interactions are provided and the survey of the experimental work is by necessity incomplete, *i.e.*, it would take volumes to consider all available data. Nonetheless, at the end of this chapter, the reader should be able to apply the key ideas to problems of direct interest to him.

Now, several cautionary remarks are in order:

a) In reporting relative energies of torsional isomers calculated by different quantum mechanical procedures, it should be recognized that these quantities represent the balance of different effects as sensed by the calculation employed. Accordingly, the dominant effect can only be ascertained by Hartree-Fock calculations and these are not currently possible for most of the systems discussed in this work. On the other hand, a given electronic effect which plays a *major* role should be always detectable regardless of the quality of the calculation employed. In this sense, calculated indices of nonbonded interactions are more meaningful than calculated energies since they are testimonials of an important effect rather than a resultant of effects[d]. Thus, results

---

[c] It should be emphasized that OEMO theory does not treat *explicitly* interelectronic repulsions, which are reproduced by the two electron part of a complete hamiltonian operator, as well as internuclear repulsions. These effects are partially accounted for by virtue of the empirical evaluation of matrix elements in the OEMO method and will be grouped under the heading "steric effects". It is obvious that steric effects will tend to favor uncongested structures. It is then apparent that the OEMO theory will lead to incorrect predictions when steric effects become a *dominant* influence.

[d] These indices depend on MO symmetry properties which are well reproduced by any type of calculation. On the other hand, orbital energies and total energies do depend crucially on the type of calculation employed. For example, the three occupied pi MO's of *cis* 1,2-difluoroethylene are calculated as

$\phi_1 = .4311 (C_{Pz} + C'_{Pz}) + .5605 (F_{Pz} + F'_{Pz})$, $\epsilon = -.8654$ a.u.;
$\phi_2 = .1542 (C_{Pz} - C'_{Pz}) + .6901 (F_{Pz} - F'_{Pz})$, $\epsilon = -.7632$ a.u.;
$\phi_3 = .5605 (C_{Pz} + C'_{Pz}) - .4311 (F_{Pz} + F'_{Pz})$, $\epsilon = -.4987$ a.u.;

by the INDO method and as

$\phi_1 = .3308 (C_{Pz} + C'_{Pz}) + .5612 (F_{Pz} + F'_{Pz})$, $\epsilon = -.5678$ a.u.;
$\phi_2 = .1659 (C_{Pz} - C'_{Pz}) + .6734 (F_{Pz} - F'_{Pz})$, $\epsilon = -.5205$ a.u.;
$\phi_3 = .5463 (C_{Pz} + C'_{Pz}) - .4382 (F_{Pz} + F'_{Pz})$, $\epsilon = -.2944$ a.u.,

by an *initio* method (STO–3 G basis set). It can be seen that the eigenvectors are nearly identical while the energies differ substantially.

of extended Hückel, CNDO/2 and INDO calculations should be given significance only insofar as computed indices are concerned since relative energies determined by these methods are not as reliable as those determined by the *ab initio* procedures, at least in most cases. Unless a comparison of computational methods is made, we shall report only *ab initio* relative energies of torsional isomers.

b) Sigma-lone pair nonbonded interactions are expected to discriminate between torsional isomers to a smaller degree than pi-lone pair nonbonded interactions. Accordingly, when both types of interactions are present, as in molecules like CHF=CHF, a prediction will be made based upon consideration of pi nonbonded interactions.

c) Differential sigma lone pair nonbonded interactions, such as those which obtain in *cis* and *trans* CFCl=CFCl, are expected to play only a small role in determining geometrical preference and will not be considered.

3.1. Nonbonded Interaction Control of the XCX Angle in X_2CH_2 and $X_2C=Y$ Molecules

We shall first consider how nonbonded interactions influence bond angles in molecules. Our approach will be illustrated by reference to the model systems difluoromethane and 1,1-difluoroethylene. In these problems, we shall consider not only stabilizing orbital interactions but also overlap repulsion in order to demonstrate some interesting trends which obtain in these angle problems.

We first consider pi nonbonded interactions in difluoromethane and employ the dissection shown below. The appropriate interaction diagram for the pi system only is shown in Fig. 13. From this diagram it is obvious that only the symmetric

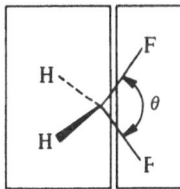 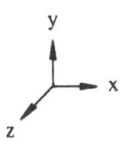

fluorine group MO, n_S, can interact with the π and π^* MO's of the methylene group. Now, shrinkage of the FCF angle has the following consequences:

a) Overlap repulsion increases due to the interaction of the fluorine $2p_z$ lone pairs.

b) The splitting of the F group MO's increases and, hence, the energy of n_S decreases. This leads to a decrease of ϵ_0 [Eq. (5')] and tends to reduce the $n_S - \pi$ overlap repulsion. On the other hand, the same effect leads to an increase in the energy separation of n_S and π^* and tends to reduce the corresponding stabilizing interaction.

c) The normalization constant of n_S, which has the form $(2 + 2S_{FF})^{-1/2}$, decreases since the spatial pi overlap between the two fluorines, S_{FF}, increases. In turn, the decrease in the n_S normalization constant will tend to reduce both the

49

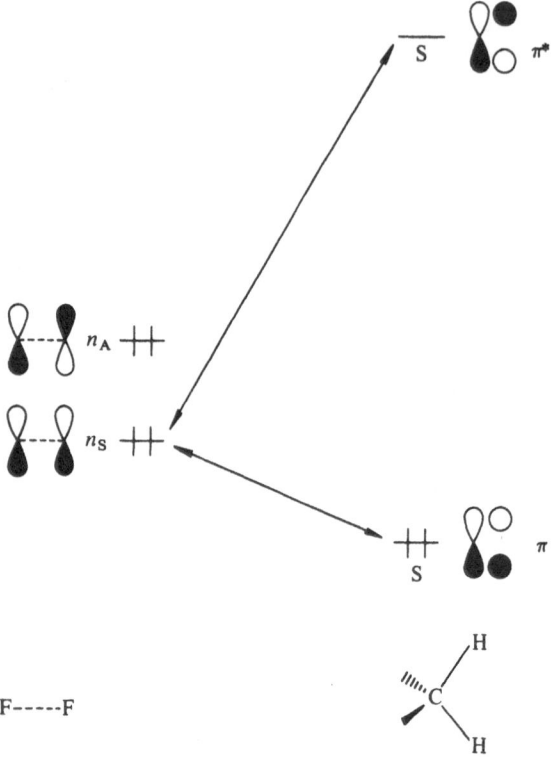

Fig. 13. Interaction of the pi fluorine lone pairs with the methylene pi system in CH_2F_2

$n_S - \pi$ overlap integral and the corresponding four electron destabilizing interaction as well as the $n_S - \tilde{\pi}^*$ overlap integral and the corresponding two electron stabilizing interaction.

d) The spatial overlap between the fluorine $2p_z$ AO's and the methylene hydrogens decreases. As a result, the $S_{n_S\pi}$ overlap integral will tend to decrease and the $S_{n_S\pi^*}$ overlap integral will tend to increase, i.e. this effect will tend to reduce overlap repulsion and increase the stabilizing $n_S - \pi^*$ interaction. These considerations can be understood by reference to the drawings shown below:

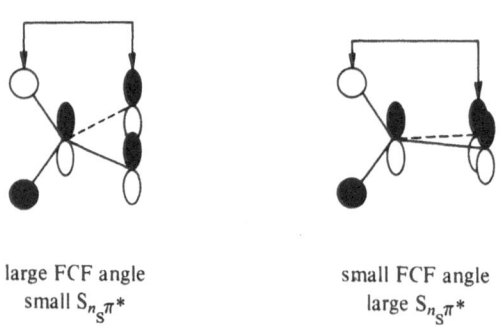

large FCF angle
small $S_{n_S\pi^*}$

small FCF angle
large $S_{n_S\pi^*}$

In short, angle shrinkage has the following effects:

a) Increase in fluorine lone pair-lone pair repulsion. This effect is favored by the variation of all quantities involved in Eq. (6').

b) Decrease in the $n_S-\pi$ overlap repulsion. This effect is favored by the variation of all quantities involved in Eq. (5').

c) Small variation of the $n_S-\pi^*$ stabilizing interaction due to conflicting variations of the quantities involved in Eq. (1').

In general, $F2p_z-F2p_z$ overlap repulsion will be outweighed by $n_S-\pi$ overlap repulsion, the net effect favoring angle shrinkage.

In addition, the question arises as to whether there is any way of testing whether the two electron stabilizing interaction will increase or decrease as the FCF angle shrinks. The answer is affirmative and in this case the information is conveyed by the variation in the bond order of the $2p_z$ fluorine AO's. We distinguish two cases:

a) The variation of the $n_S-\pi^*$ stabilization energy is dominated by the variation of the corresponding overlap integral. Thus, as the FCF angle shrinks, this interaction increases and the F———F pi bond order or pi overlap population become increasingly more negative.

b) The variation of the $n_S-\pi^*$ stabilization energy is dominated by the variation of the corresponding energy gap. Thus, as the FCF angle shrinks, this interaction decreases and the F———F pi bond order, but not necessarily the corresponding overlap population, becomes less negative.

Accordingly, the variation of the appropriate bond order as revealed by an explicit calculation can tell us something about the variation of a given interaction which is controlled by two opposing effects. On the basis of the discussions in Section 1.2, we expect that due to a small change of both H_{ij} and ϵ_{ij} as a result of angle shrinkage, the variation of the H_{ij}, which is proportional to the overlap integral S_{ij}, will dominate. We shall return to this point shortly.

The example of F_2CH_2 has pedagogical significance for it illustrates how the static model can lead to erroneous predictions unless care is exercised in the interpretation of long range bond orders and overlap populations. For example, in the case of difluoromethane, a mere inspection of the sign of the pi overlap population (vide infra) between the two fluorines might have led somebody to infer that this non-bonded interaction is repulsive. This would arise because the $n_S-\pi^*$ interaction leads to charge transfer away from a bonding F——F pi group MO and, thus, creates a

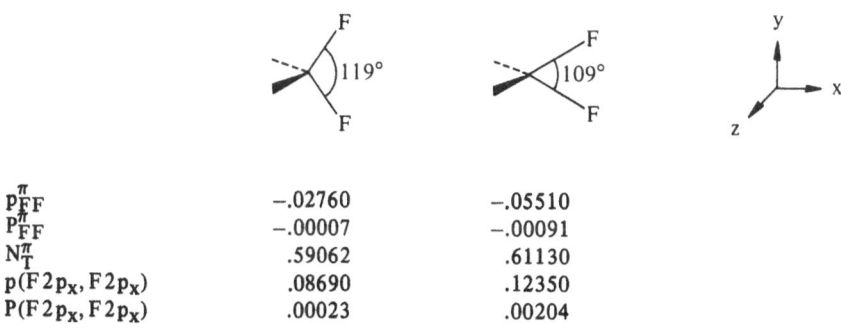

	$119°$	$109°$
p_{FF}^{π}	−.02760	−.05510
P_{FF}^{π}	−.00007	−.00091
N_T^{π}	.59062	.61130
$p(F2p_x, F2p_x)$	.08690	.12350
$P(F2p_x, F2p_x)$	.00023	.00204

Computation: INDO, standard geometry

net antibonding situation, *i.e.* negative F————F pi overlap population. As we pointed out in our original paper[2]), the artificiality of such an inference can be immediately realized by noting that a stabilizing interaction has to involve an antibonding situation *overcompensated* by bonding situations.

The problems facing the static model can be best understood by reference to the various computed indices for difluoromethane at various FCF angles. Comparing the two structures shown above, we note that the total pi overlap population increases as the angle shrinks. Thus, according to the dynamic model, there is F——F pi nonbonded attraction since the difference in the total pi overlap populations is positive. On the other hand, the F————F pi overlap population is always negative. Clearly, the FCF angle tends to shrink despite F————F antibonding character as the total energies indicate. The F————F pi bond order becomes more negative as the FCF angle shrinks, belying a dominance of the variation of the $n_S-\pi^*$ overlap integral and indicating that the $n_S-\pi^*$ interaction becomes increasingly stabilizing, a fact reflected in the variation of the total pi overlap population. In short, the static model leads to either erroneous or ambiguous conclusions depending upon how one chooses to interpret a long range overlap population.

It should be noted that in this example the sign of ΔN_T^π is positive while the sign of the F——F pi overlap population is negative and the static model fails or becomes ambiguous. Obviously, both models will lead to indentical predictions when ΔN_T^π and a given pi overlap population between nonbonded atoms or groups have the same sign.

Angle θ	P_{FF}^π	P_{FF}^π	N_T^π	$p(F2p_x, F2p_x)$	$P(F2p_x, F2p_x)$	Relative energy (kcal/mol)
120°	−.0507	−.00011	.32585	.0761	.00016	7.83
116°	−.0517	−.00013	.32642	.0809	.00020	4.07
112°	−.0526	−.00016	.32686	.0845	.00025	1.39
108°	−.0535	−.00019	.32738	.0871	.00031	0.00

Computation: CNDO/2.

Angle θ	P_{FF}^π	N_T^π	$P(F2p_x, F2p_x)$	Relative energy (kcal/mol)
120°	−.00013	.21222	.00019	3.110
116°	−.00017	.21314	.00025	1.003
112°	−.00022	.21398	.00031	0.000
108°	−.00028	.21472	.00039	0.194

Computation: *ab initio* – STO–3G basis set.

The angle problem in 1,1-difluoroethylene can be treated in a similar manner. The pi nonbonded interaction between the $F\,2p_z$ lone pairs is identical to that encountered in the case of difluoromethane due to the fact that 1,1-difluoroethylene and difluoromethane are pi isoconjugate. A confirmation of our analysis is provided by the results of computations shown above.

Once we have considered pi nonbonded interaction control of the FCF angles in F_2CH_2 and $CF_2=CH_2$, we turn our attention to sigma nonbonded interaction control in all molecules or fragments of the type FAF. Here, we focus upon the interaction of the p_x lone pairs of fluorine with the p_y AO or π_y type MO's of A.

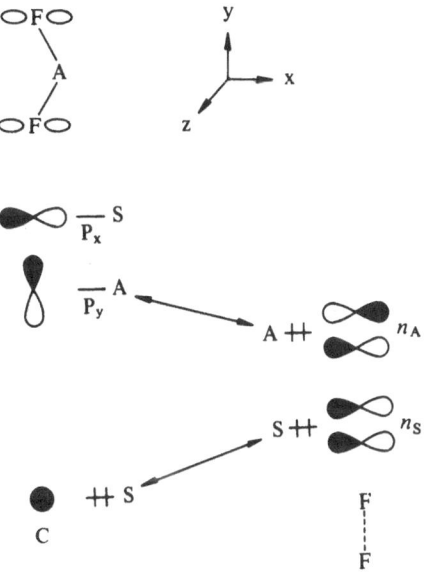

Fig. 14. Sigma fluorine lone pair interactions with the AO's of the carbon atom in F_2CH_2. A similar diagram can be drawn if the sigma MO's of the CH_2 are considered

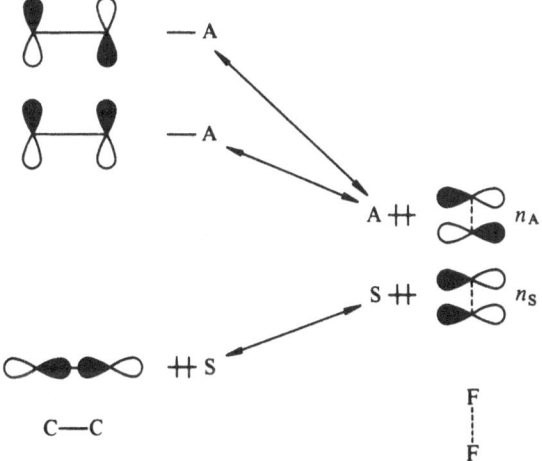

Fig. 15. Sigma fluorine lone pair interactions with the sigma MO's of the C=C fragment

The diagram of Fig. 14, shows how the nonbonded interaction between the fluorine $2p_x$ lone pairs and the $2p_y$ AO of carbon favors FCF angle shrinkage in CF_2H_2. The n_A-p_y interaction is dominant since it involves maximal spatial orbital overlap and increases as the two lone pairs are brought to closer proximity. Accordingly, $p(F2p_x, F2p_x)$ and $P(F2p_x, F2p_x)$ are both expected to be positive and increase as the FCF angle shrinks in F_2CH_2. The results of calculations shown above are in agreement with these predictions.

The sigma lone pair nonbonded interaction in $F_2C=CH_2$ can be treated in a similar fashion. The appropriate interaction diagram is shown in Fig. 15 and our conclusions are identical to the ones reached in the case of F_2CH_2, $i.e.$ nonbonded attraction between the fluorine $2p_x$ lone pairs favors angle shrinkage in $CF_2=CH_2$. The results of calculations shown above are, once again, in agreement with these predictions.

The above conclusions are general for all Y_2A molecules where Y has available sigma np lone pairs. In subsequent sections, we shall see that the nonbonded interaction between sigma lone pairs is of paramount importance in controlling bond angles in molecules.

A discussion of the experimental results relevant to the above discussion is postponed until we examine other important factors which may also contribute towards angle shrinkage in AY_2 fragments or molecules.

3.2. Conformational Isomerism of CH_3CH_3 and CH_2XCH_2X Molecules

We shall first consider rotational isomerism in ethane. In particular, we shall compare the staggered and eclipsed conformations shown below.

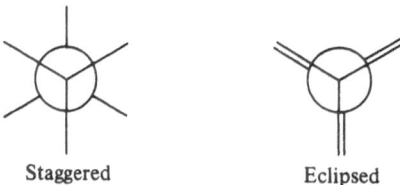

Staggered   Eclipsed

In general the reason for the greater stability of the staggered conformation relative to the eclipsed conformation has been a matter of controversy. Here, an interpretation based on electronic effects is given.

The ethane molecule can be constructed by union of two pyramidal methyl radical fragments. The interaction diagram is shown in Fig. 16 and the key stabilizing orbital interactions are depicted below.

Since the energies of the unperturbed orbitals are assumed to be independent of the rotational angle, only trends in overlap integrals need be considered in order to determine the relative stabilization of the staggered and eclipsed conformations. We now consider in detail the various MO interactions and their impact on conformational preference:

1. $\phi_4-\phi'_4$. The bonding overlap between the hydrogens will be greater, in the eclipsed geometry relative to staggered geometry. However, due to the small hydro-

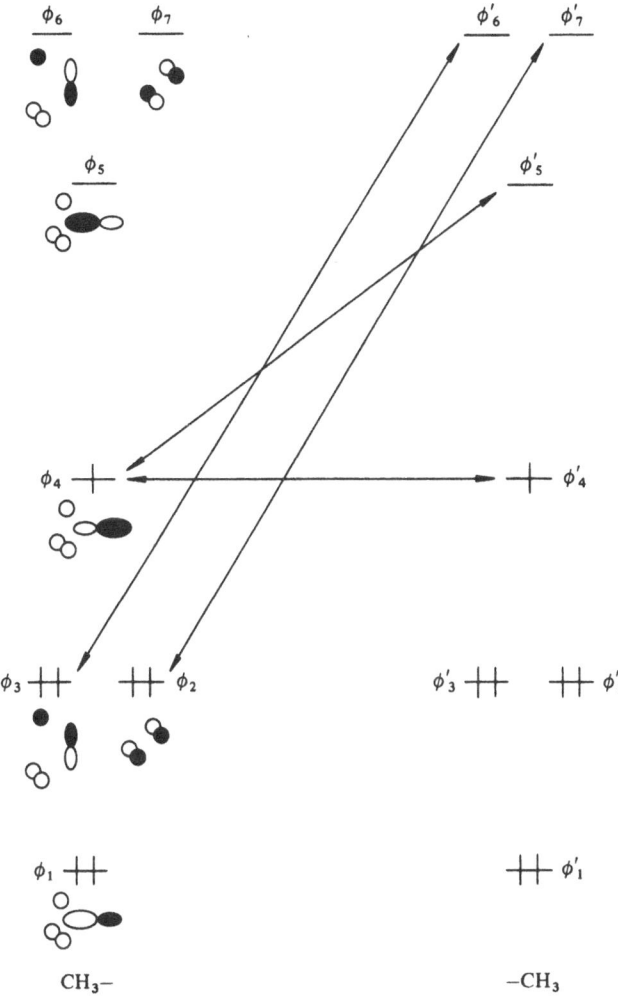

Fig. 16. Key orbital interactions obtaining in the union of two methyl radicals to form ethane. Arrows indicate the key stabilizing interactions

gen coefficients in these MO's, the preference for the eclipsed geometry will be small.

2. $\phi_4-\phi'_5$. The antibonding overlap between the hydrogens will be greater in the eclipsed geometry relative to the staggered geometry. Consequently, the staggered geometry is favored but, due to the small hydrogen coefficients, only to a small extent. This interaction will tend to offset the previous interaction.

3. $\phi_3-\phi'_6$ and $\phi_2-\phi'_7$. The situation here is analogous to the situation which obtained in the case of the $\phi_4-\phi'_5$ interaction and the staggered geometry will be stabilized relative to the eclipsed. Also, since the hydrogens have large coefficients, the degree of stabilization will be appreciable. That is, the non-bonded interaction between the hydrogen atoms will favor the staggered geometry. The rationale outlined above is basically similar to that suggested by Hoffmann[46] and Lowe[66]. It is noted that overlap repulsion also dictates a preference for the staggered form.

Staggered   Eclipsed

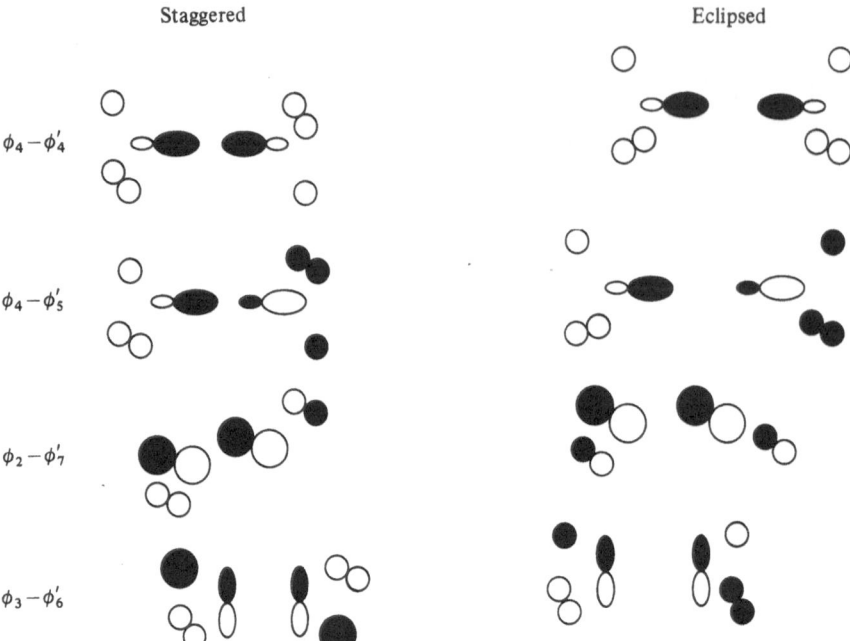

$\phi_4 - \phi_4'$

$\phi_4 - \phi_5'$

$\phi_2 - \phi_7'$

$\phi_3 - \phi_6'$

An equivalent manner of accounting for the preference of the staggered geometry will be discussed in a later section.

Turning now to disubstituted ethanes, we will consider the case where both substituents are identical and bear lone pair electrons. We shall compare the typical conformations shown below.

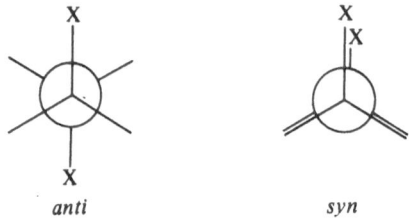

anti   syn

Our approach will be illustrated by using 1,2-difluoroethane as the model system. This molecule can be dissected as shown below. The sigma nonbonded interaction of the fluorine $2p_x$ lone pairs and its impact upon conformational preference can be

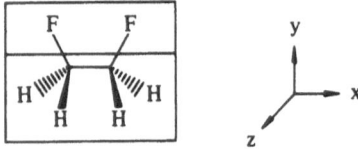

appreciated by reference to the interaction diagram shown in Fig. 17. This type of lone pair sigma nonbonded interaction has been discussed before and constitutes a typical examples of pattern b in Scheme 1. In the case where the π_u and π_g

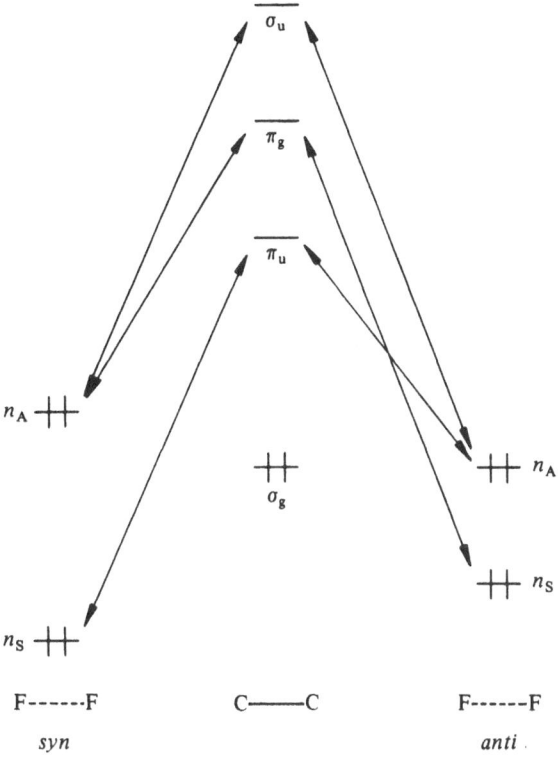

Fig. 17. Sigma type stabilizing orbital interactions in *syn* and *anti* 1,2-difluoroethane

vacant orbitals are nearly degenerate, the σ_u vacant MO becomes the "active"· MO of the coupling unit and enforces *syn* preference. On the other hand, in the case where there is substantial splitting of the π_u and π_g vacant orbitals, the π_u vacant MO becomes the "active" MO and may enforce *anti* preference.

The analysis of the $2p_z$ lone pair interaction proceeds in a similar manner. The appropriate interaction diagram is shown in Fig. 18. The key interactions will be $n_S-\phi_3$ and $n_A-\phi_4$. The former interaction will be larger in the *anti* geometry while the latter will be larger in the *syn* geometry, *i.e.*, they tend to cancel. Furthermore, this effect is expected to be secondary to the previous one because spatial overlap dictates a much greater splitting of the $2p_x$ than the $2p_z$ lone pairs in the *syn* geometry.

The nonbonded interactions of the fluorine 2s lone pairs can be neglected because F2s–F2s spatial overlap is poor. Even more important, the energy of the F2s orbital is extremely low so that interaction with the sigma vacant MO's of the coupling unit is negligible.

Good quality *ab initio* calculations have not yet been carried out in order to test whether lone pair nonbonded attraction obtains in 1,2-difluoroethane. Results of INDO calculations shown below indicate the presence of a nonbonded attractive interaction on the basis of the "partial" bond order p' ($F2p_x$, $F2p_x$). By contrast, the presence of a nonbonded repulsive interaction favoring the *anti* conformation is indicated on the basis of the bond order $p(F2p_x, F2p_x)$.

57

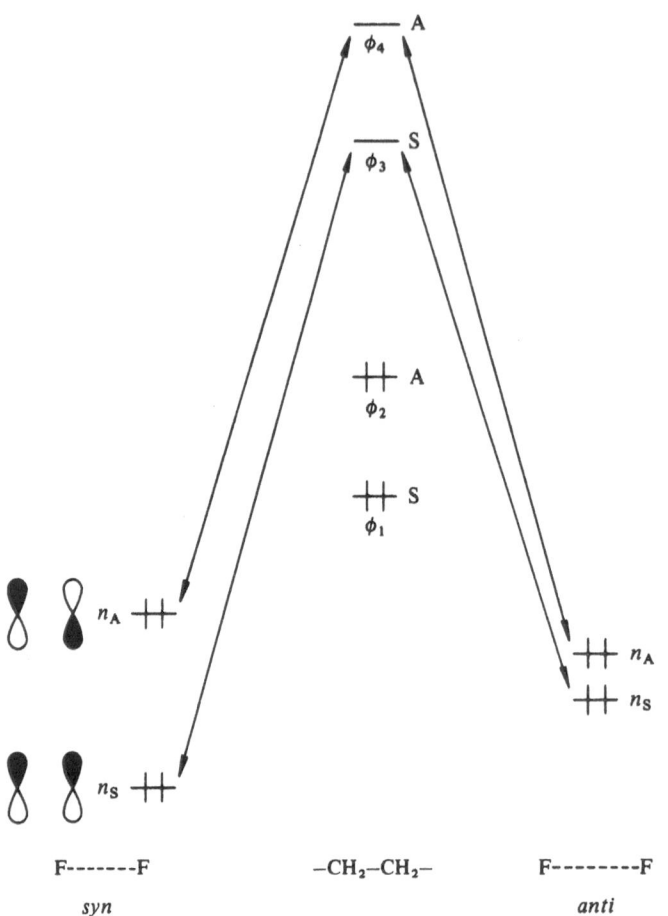

F-------F   −CH$_2$−CH$_2$−   F--------F

syn   *anti*

Fig. 18. Pi type stabilizing orbital interactions in *syn* and *anti* 1,2-difluoroethane. The symmetry labels are assigned with respect to a mirror plane (*syn* conformer) or a rotational axis (*anti* conformer)

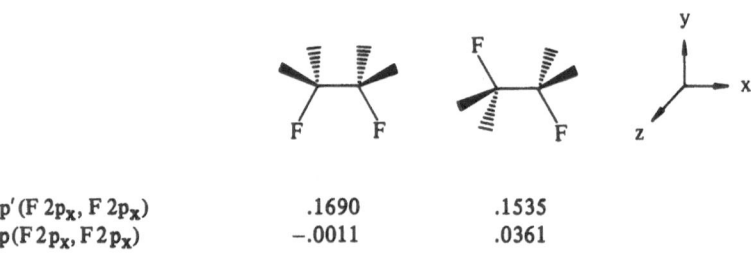

p′(F 2p$_x$, F 2p$_x$)	.1690	.1535
p(F 2p$_x$, F 2p$_x$)	−.0011	.0361

Computation: INDO, standard geometry

    The discussion of experimental results pertinent to rotational isomerism in 1,2-dihaloethanes will be deferred to Section 11.0, where we discuss additional important electronic effects determining the preferred conformation of these molecules.

The analysis outlined above can be extended to ethanes where the substituent does not bear lone pairs but rather a pi system. As an illustration, we will consider 1,2-dicyanoethane.

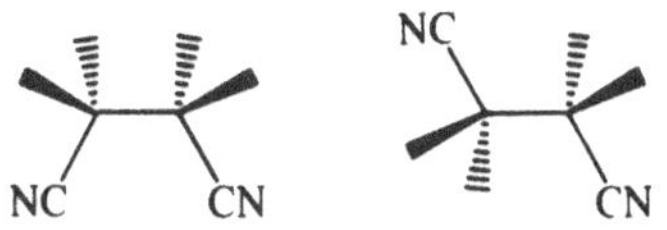

$P_{(CN, CN)}$	.00913	−.00006

Computation: INDO, standard geometry

The sigma nonbonded interaction between the two substituents fall into pattern d of Scheme 1. Here, unlike the case of 1,2-difluoroethane, we conclude that there will be a preference for the *syn* conformation due to the sigma nonbonded interaction of the pi systems of the substituents. This will be counteracted by the inherent preference of any ethane molecule for the staggered geometry and a compromise is expected to be reached in the *gauche* conformation, barring adverse steric effects.

Experimental data for 1,2-dicyanoethane are available. In the gas phase, the *anti* conformer is more stable while in the liquid phase, where dipole-dipole repulsive effects are deemphasized, the *gauche* is more stable[67].

If the two substituents are not identical, one may still focus on the interaction of the group MO's of the substituents and those of the $-CH_2-CH_2-$ fragment. As an example, we will consider 1-fluoropropane. This molecule constitutes system where a "hydrogen bond" determines conformational preference and merits special attention. This "hydrogen bond" represents a bonding situation which can be classified under the heading of nonbonded attractive interactions.

Consider the molecule 1-fluoropropane dissected in the manner shown below. First, we consider the sigma nonbonded interaction between methyl and fluorine and we construct the group MO's of the $CH_3\cdots F$ fragment as shown by the interaction diagram of Fig. 19.

Clearly, in the *syn* conformation there is a stabilizing interaction between the fluorine $2p_x$ lone pair and the LUMO of the methyl group. This type of interaction

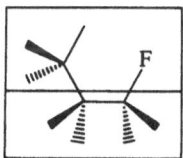

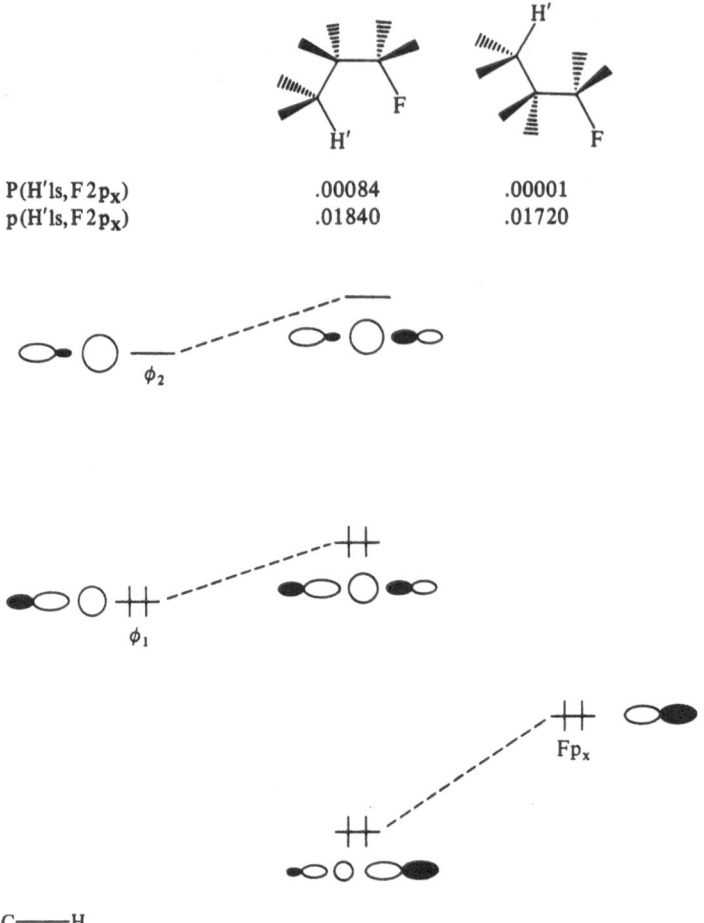

$P(H'ls, F 2p_x)$	.00084	.00001
$p(H'ls, F 2p_x)$	.01840	.01720

C——H

Fig. 19. The sigma group MO's spanning the substituents in 1 fluoropropane as constructed from the sigma MO's of a CH bond and the fluorine sigma lone pair AO. Net stabilization results due to the $Fp_x - \phi_2$ interaction

can be regarded as responsible for normal intermolecular hydrogen bond formation. As a result, the bond order between the fluorine $2p_x$ lone pair and the in-plane methyl hydrogen becomes positive, *i.e.*, there is a weak covalent bond already formed due to the through space interaction of methyl and fluorine in the *syn* conformation. Obviously, the stabilizing interaction $Fp_x - \phi_2$ is zero in the *anti* geometry since spatial overlap is negligible.

The union of the $CH_3 \cdots F$ component system with the $-CH_2-CH_2-$ coupling unit may attenuate or enhance the preference for the *syn* conformation due to hydrogen bonding depending on the relative energies of the π_u and π_g MO's of the coupling unit. The exciting possibility exists that, if systems can be found where the active vacant MO of the coupling unit is the σ_u MO, a super hydrogen bond may favor the *syn* conformer to an extent far and above that expected from a conventional hydrogen bond.

A similar analysis can be given for the pi nonbonded interaction between the methylene group and the fluorine $2p_z$ lone pair. However, due to much poorer spatial overlap, this interaction will be of limited significance compared to the sigma nonbonded interaction.

On the basis of the above discussion, we are led to the conclusion that sigma nonbonded attractive interaction in the form of a hydrogen bond will tend to favor a *syn* conformation opposing the inherent preference of ethane molecules for a staggered conformation. A compromise is expected to be reached in the *gauche* conformation. However, severe steric effects may force an *anti* conformational preference.

The importance of hydrogen bonding in determining the preferred conformation of 1,2-disubstituted ethane can be appreciated by reference to the calculations of Pople and his collaborators[68]. Representative systems were examined and in all cases the most stable conformer was calculated to be the one involving hydrogen bonding between the two vicinal substituents of the ethane molecule. Two typical examples of such structures are shown below.

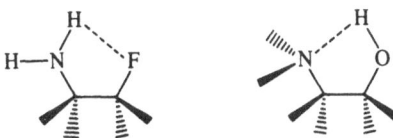

Experimental results shown in Table 10 suggest that a hydrogen bond, *i.e.*, a strong nonbonded attractive interaction, is instrumental in dictating *gauche* preference in $CH_3CH_2CH_2X$ and XCH_2CH_2OH systems where X carries at least one lone pair.

The approach used in this section can be used to treat any 1,2-disubstituted ethane. In general, sigma nonbonded interactions may favor the *syn* geometry, the

Table 10. Relative stability of the torsional isomers of YCH_2CH_2X molecules

X	Y	$E_{gauche} - E_{anti}$	Ref.
CH_3	F	$-.5 \quad \pm .3$	69)
CH_3	Cl	$-(.05 \quad -.3)$	70, 71)
CH_3	Br	$-.28 \quad \pm .1$	72, 73)
CH_3	OH	$-.29 \quad \pm .15$	74)
F	OH	$-2.07 \pm .53$	75–77)
		$-(> 2.8)$	
Cl	OH	$-(1.8 - 4.5)$	75, 78)
		$-1.2 \quad \pm .09$	
		$-.95 \quad \pm .02$	
Br	OH	$-1.25 \pm .08$	75)
I	OH	$-.81 \quad \pm .09$	75)

inherent nonbonded interaction present in any ethane system will tend to favor a staggered conformation, and steric effects will tend to favor an *anti* conformation. Clearly, then, a *gauche* preference is most likely a sign of the importance of non-bonded attractive interactions. Another factor favoring *gauche* preference will be discussed in a subsequent section.

3.3. Conformational Isomerism of X_4Y_2 and $X_2H_2Y_2$ Molecules

In this section we shall consider conformational isomerism in typical hexaatomic molecules, X_4Y_2 and $X_2H_2Y_2$. A diagram which depicts the MO's spannend by the $2p_x$ lone pairs of the heteroatoms X and the MO's of the coupling unit Y–Y in planar Y_2X_4 is shown in Fig. 20. We distinguish two cases:

  a) The energy gap between the lone pair MO's and the vacant MO's of the coupling unit is large, *i.e.* their interaction is matrix element controlled. In such an event,

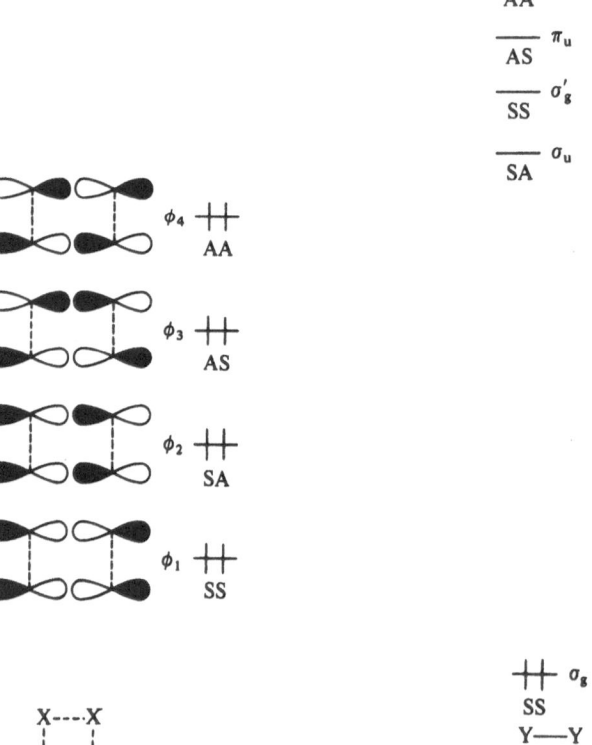

Fig. 20. The sigma lone pair group MO's of X_4 component system and the sigma group MO's of the Y–Y coupling unit

the "active" unoccupied MO's of the coupling unit will be the ones which overlap maximally with the lone pair MO's, namely the π_u and π_g MO's. If there is substantial splitting between these two MO's, the "active" MO will be the π_u MO. This occurs typically in B_2F_4. The $\phi_3 - \pi_u$ stabilizing interaction is weak and overlap repulsions can force a preference for a perpendicular geometry.

b) The energy gap between the lone pair MO's and the vacant MO's of the coupling unit is small, *i.e.* their interaction is energy gap controlled. In such an event, the "active" unoccupied MO of the coupling unit is simply the sigma LUMO. This situation obtains in N_2O_4 where the formal negative charge in the oxygens raises the energy of the lone pair MO's. The $\phi_2 - \sigma_u$ stabilizing interaction is strong and can force a preference for a planar geometry.

In B_2F_4 and N_2O_4 we note the following:

a) Experimentally, the most stable conformation of N_2O_4 in the gas phase is found to be the eclipsed conformation predicted by our analysis[79-82]. Also, the N–N bond length of N_2O_4 (1.75 Å) is substantially greater than in N_2H_4 (1.47 Å) mainly due to charge transfer from the oxygen MO's to the N–N σ_u LUMO which results into a weaker N–N bond. This interaction does not obtain in N_2H_4.

Ab initio calculations show the eclipsed conformation to be energetically more stable than the perpendicular conformation due to an attractive 1,4-interaction between the oxygen atoms[83]. Furthermore the N–N sigma overlap population is found to be less for the eclipsed conformation than the perpendicular due to greater charge transfer to the crucial N–N sigma antibonding MO. Extended Hückel calculations of N_2O_4 have been employed to complement the *ab initio* calculations. Interestingly, when the 1,4-interaction between the oxygen atoms is eliminated, there is an energy shift in favor of the perpendicular conformation[e]. This latter result strongly supports the idea that nonbonded attraction is primarily responsible for controlling conformational isomerism in N_2O_4.

We note here, that the eclipsed conformation of N_2O_4 can be viewed as being a sigma aromatic system geometrically similar to the pi aromatic system napthalene as illustrated below:

b) The experimental studies of B_2F_4 in the gas phase indicate that the molecule exhibits a staggered conformation shown below[84-86].

---

[e] The 1,4 interaction was eliminated by setting the appropriate off diagonal matrix element, H_{ij}, along with the corresponding overlap matrix element, S_{ij}, to zero.

Ab initio calculations furthermore, predict, in agreement with experiment, the staggered conformation of B_2F_4 to be more stable than the planar conformation[83].

c) Results of CNDO/2 calculations are shown below and confirm our analysis disscussed above.

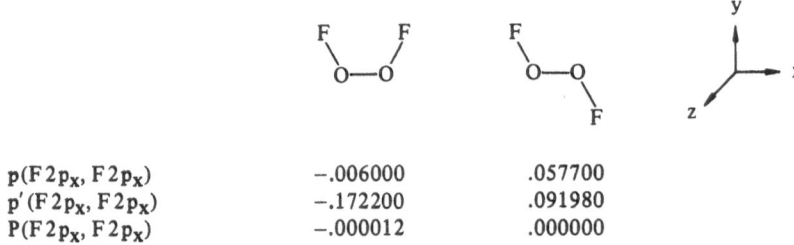

$$p(O\,2p_x, O\,2p_x) \quad = \quad .0415 \qquad p(F\,2p_x, F\,2p_x) = -.019200$$
$$P(O\,2p_x, O\,2p_x) \quad = \quad .00054 \qquad P(F\,2p_x, F\,2p_x) = -.00003$$

Calculation: CNDO/2, experimental geometry

3.4. Conformational Isomerism of X–Y–Y–X and W–X–Y–Z Molecules

We first consider the electronic factors responsible for conformational preference in tetraatomic molecules of the type X–Y–Y–X. Our model compound is O_2F_2 and the two conformations we will compare are shown below:

$p(F\,2p_x, F\,2p_x)$	−.006000	.057700
$p'(F\,2p_x, F\,2p_x)$	−.172200	.091980
$P(F\,2p_x, F\,2p_x)$	−.000012	.000000

Calculation: CNDO/2, standard geometry

The problem of sigma nonbonded fluorine lone pair interactions is identical to the one encountered in the case of 1,2-difluoroethane. In this case, CNDO/2 results indicate that fluorine lone pair interaction favors the *trans* isomer, *i.e.* the fluorine-fluorine sigma interaction is repulsive in the *cis* isomer as indicated by the negative sign of $p'(F\,2p_x, F\,2p_x)$, $p(F\,2p_x, F\,2p_x)$ and $P(F\,2p_x, F\,2p_x)$. We note here that an additional electronic factor which is important in dictating the preferred conformation of FOOF will be discussed in a subsequent section.

Another example of torsional isomerism in X–Y–Y–X tetraatomic molecules is the relative energetics of *cis* and *trans* N_2F_2.

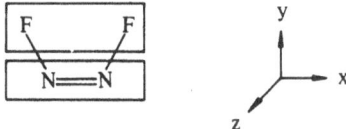

	cis	trans
E_{rel} (kcal/mol)	0.000	.634
P_{FF}^{π}	.00015	.00000
$P(F 2p_x, F 2p_x)$	−.00016	.00000
N_T^{π}	.17057	.17520

Computation: STO–3G at STO–3G optimized geometry

	cis	trans
E_{rel} (kcal/mol)	2.535	0.000
P_{FF}^{π}	.00115	.00002
$P(F 2p_x, F 2p_x)$	.00003	.00009
N_T^{π}	.18317	.17959

Computation: 4–31G at STO–3G optimized geometry

$N_2 F_2$ can be dissected as shown below:

In this case both pi and sigma nonbonded interaction between the fluorine lone pairs are important in dictating the preferred isomer of $N_2 F_2$. Arguing as before, we predict the following:

a) Pi nonbonded interactions stabilize cis $N_2 F_2$ more than *trans* $N_2 F_2$.

b) Sigma nonbonded interactions depend upon the identity of the "active" unoccupied MO of the N=N coupling unit.

Ab initio calculations of the geometric isomers of $N_2 F_2$ have revealed the following interesting trends:

a) Pi nonbonded attraction between the fluorine pi lone pairs favors the *cis* isomer only when an extended basis set is employed. At the level of a minimal basis set the total pi overlap population, the best index of pi nonbonded interaction, favors the *trans* isomer. This basis set dependence may be due to the fact that the *cis* isomer exhibits longer N–F bonds than the *trans* isomer due to an effect which will be discussed later.

b) Sigma nonbonded interaction between the fluorine sigma lone pairs appears to favor the *trans* isomer. This may also be due partially to the difference in N–F bond lengths in the two geometric isomers.

Experimentally, the *cis* isomer of $N_2 F_2$ is found to be about 3.0 kcal/mol more stable than the *trans* isomer[87, 88]. Another electronic factor which contributes to the substantial stabilization of the *cis* isomer will be discussed in a subsequent section.

Finally, we turn our attention to tetraatomic molecules of the type W–X–Y–Z. Our model system ClNSO exhibits *cis-trans* isomerism:

P^{π}_{ClO}	.09270	.08380
P^{π}_{ClO}	.00026	.00001
N^{π}_{T}	.24808	.24660
$p(Cl3p_x, O2p_x)$	−.01700	.07050
$p'(Cl3p_x, O2p_x)$	−.05200	.01044
$P(Cl3p_x, O2p_x)$	−.00036	.00004

Computation: CNDO, experimental geometry

Arguing as before we predict that *cis* ClNSO is stabilized more than *trans* ClNSO by pi lone pair nonbonded attraction. The results of CNDO/2 calculations are shown above and it can be readily seen that pi nonbonded attraction favors the *cis* isomer. On the other hand, the indices of sigma nonbonded interaction imply that the *trans* geometry is stabilized more by sigma lone pair nonbonded attraction than the *cis*.

Infrared and Raman spectroscopic studies of XNSO, where X=F, Cl, Br, I, indicate that these compounds have a *cis* configuration in the gas phase[89]. This result has also been confirmed by electron diffraction study of ClNSO[90]. Once again, an additional important electronic factor is responsible for the greater stability of the *cis* isomer and this will be discussed in a later section.

3.5. Conformational Isomerism of CH₃CH=X Molecules

In this section, we shall discuss the electronic factors which determine conformational preferences in $CH_3CH=X$ systems. We shall attempt to predict the relative energy of the staggered and eclipsed conformations shown below as well as its dependence upon the nature of X.

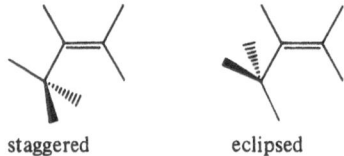

staggered   eclipsed

Our approach will be illustrated by reference to propene as the model system. The dissection employed here is depicted below.

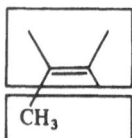

The appropriate interaction diagram is shown in Fig. 21. The energies of the unperturbed orbitals are independent of the degree of rotation about the C—C single bond. Consequently, we need only consider changes in the overlap integrals that accompany rotation.

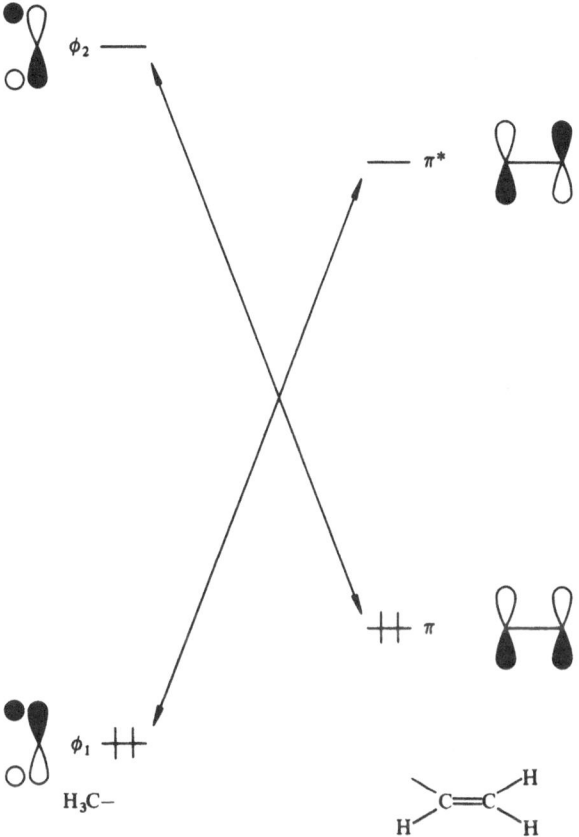

Fig. 21. Stabilizing pi orbital interactions in propene
The key stabilizing orbital interactions are sketched below

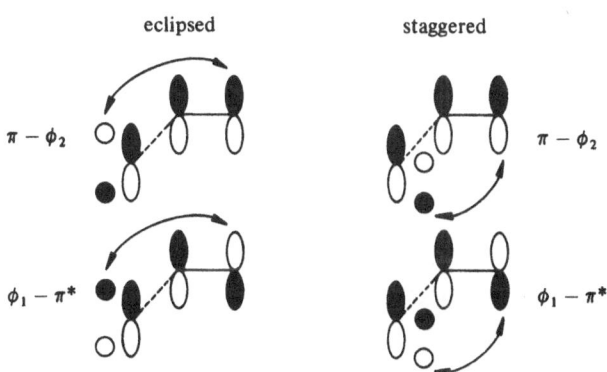

As can be seen, there is an increase in the antibonding overlap between the methyl hydrogens and the outer olefinic Cp_z AO which effectively reduces the total overlap integral in the staggered geometry. Thus, the stabilization due to both interactions will be greater in the eclipsed isomer. That is, electronic factors will favor the eclipsed conformer while steric effects favor the staggered conformer. In propene the electronic factors dominate and the molecule exists in the eclipsed geometry with a barrier to rotation of 2.00 kcal/mol[91–93].

In CH_3CH_2=X molecules, the methyl rotational barrier is the energy difference between the eclipsed and staggered conformations with the former being the energy minimum and the latter being the energy maximum.

When a CH_2 is replaced by O several things happen. First, the energies of both the π and π^* MO's decrease in energy. Secondly, the orbitals are no longer symmetrical but have the shape shown below.

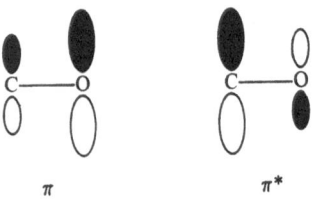

π $\qquad\qquad\qquad\qquad$ π^*

As a consequence of the energy decrease in both π and π^*, only the $\phi_1 - \pi^*$ interaction need be considered when dealing with the rotational barrier in CH_3CH=O. As in the propene case, the eclipsed conformer will be favored. However, as can be seen from the diagrams below, the eclipsed form will be favored to a lesser extent in acetaldehyde relative to propene.

eclipsed $\qquad\qquad\qquad$ staggered

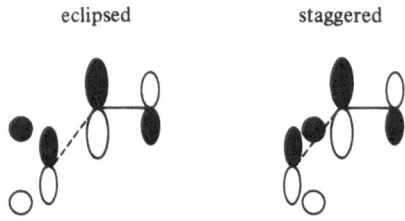

This is due to the small coefficient on the oxygen atom in the π^* MO. The methyl rotational barrier for CH_3CH=O should be less than for CH_3C=CH_2

Experimentally, acetaldehyde is known to exist in the eclipsed conformation. It has a methyl rotational barrier of 1.16 kcal/mol[94–96] as contrasted with a barrier of 2.00 kcal/mol in the case of propene.

We note here that similar reasoning has been employed by Lowe[97] as well as Hehre and Salem[98] in their studies of conformational isomerism of propene.

3.6. Torsional Isomerism of CHX=CHX and CHX=CHY Molecules

A 1,2-disubstituted alkene can exist in either of two geometries, the *cis* and *trans*, as shown below for 1,2-difluoroethylene which we shall use to illustrate our approach:

cis   *trans*

	cis	trans
E_{rel} (kcal/mol)	0.244	0.000
P^{π}_{FF}	.00001	.00000
$P(F2p_x, F2p_x)$	−.00001	.00000
N^{π}_T	.20789	.20754

Computation: STO−3G at STO−3G optimized geometry.

	cis	trans
E_{rel} (kcal/mol)	.634	0.000
P^{π}_{FF}	.00023	.00001
$P(F2p_x, F2p_x)$	.00046	.00007

Computation: 4−31G at STO−3G optimized geometry

	cis	trans
E_{rel} (kcal/mol)	1.293	0.000
P^{π}_{FF}	.00025	.00001
$P(F2p_x, F2p_x)$	.00041	.00008

Computation: 4−31G at 4−31G optimized geometry

The nature of the nonbonded interactions between the pi and sigma fluorine lone pairs has been discussed in detail before. The results of *ab initio* calculations are shown above. They clearly demonstrate that pi nonbonded attraction obtains in 1,2-difluoroethylene and related systems. The predicted greater stability of the *trans* isomer is most likely due to an exaggeration of dipolar interaction effects, favoring the *trans* molecule, by the basis sets used.

On the other hand, sigma nonbonded interaction appears to favor the *trans* isomer in the STO−3G calculations as can be seen from the indices shown above, *i.e.* $P(F2p_x, F2p_x)$ is negative in the *cis* isomer and zero in the *trans* isomer. However, $P(F2p_x, F2p_x)$ is positive for both isomers in the computations employing a 4−31G basis set. It appears, therefore, that in this case the nature of sigma nonbonded interaction is basis set dependent and, hence, further work is needed to clarify the situation.

The experimentally determined stability of geometric isomers for several 1,2-disubstituted olefins where the substituents bear lone pairs, *i.e.* the substituents are of the $-\ddot{X}$, $-\ddot{X}-R$, or $-\ddot{X}R_1R_2$ type, are shown in Table 11. From inspection

Part II. Nonbonded Interactions

Table 11. Relative stability of the geometric isomers of XHC=CHY olefins

X, Y	ΔH° (cal/mol)[a]	$K_{eq}^{a)}$	Ref.
F, F (g)	928 ± 29	.401 – .582	99, 100)
Cl, Cl (g)	650 ± 70	.577 – .683	99, 101, 102)
Br, Br (g)	−130 ± 300	.984	103)
I, I (g)	0 ± 200	1.690	99, 104)
Br, Br (l)	320 ± 200	.66	99, 105)
I, I (l)	−1,550 ± 1,000	.681 – .792	99, 104, 106, 107)
F, Cl (g)	780 ± 20	.485 – .669	99, 108)
OMe, OMe (l)	1,549 ± 19	.150 – .338	109)
(g)	1,445 ± 54	.210 – .410	109)
OEt, OEt (l)		.25	110)
Cl, OEt (l)	660 ± 130	.221	110)
Cl, CN (l)		.449	111)
F, CH=CH₂ (l)		.613	112)
Cl, CH=CH₂ (l)		.428	112)

a) A positive enthalpy and an equilibrium constant less than unity mean that the *cis* isomer is more stable relative to the *trans*.

of these data it is obvious that the *cis* isomer is thermodynamically more stable, at least in most cases.

As a second example, we consider torsional isomerism in 1,2-dihydroxyethylene. The six conformations of dihydroxyethylene along with their labels are shown below:

	C_{ss}	C_{se}	C_{ee}	T_{ss}	T_{se}	T_{ee}
E_{rel} (kcal/mol)	4.439	0.000	3.511	3.072	2.313	.765
P_{OO}^π	.0002	.0002	.0001	.0000	.0000	.0000
N_T^π	.4164	.4162	.4201	.4142	.4168	.4194

Computation: STO–3G at STO–3G optimized geometry

	C_{ss}	C_{se}	C_{ee}	T_{ss}	T_{se}	T_{ee}
E_{rel} (kcal/mol)	7.323	0.000	5.204	4.815	5.561	4.633
P_{OO}^π	.0028	.0028	.0012	.0002	0.0001	.0001
N_T^π	.3756	.3718	.3826	.3674	0.3712	.3783

Computation: 4–31G at STO–3G optimized geometry

By following the same arguments as we did in the case of 1,2-difluoroethylene, and restricting our attention to pi nonbonded interactions, we predict that the *cis* geometry of dihydroxy ethylene will be favored over the *trans* geometry in a fixed

70

conformation on account of pi nonbonded attraction between the oxygen lone pairs.

We now focus on conformational preference within the *cis* and *trans* isomers, respectively. If steric effects play the major role in dictating the preferred conformation in the *trans* isomer, one would expect that the relative order of stability will be $T_{ss} > T_{se} > T_{ee}$. However, the results of *ab initio* calculations shown above predict that the order of stability of the *trans* isomers is $T_{ee} > T_{se} > T_{ss}$ in the calculations utilizing the STO–3G basis set and $T_{ee} > T_{ss} > T_{se}$ in the 4–31G computations. An additional electronic factor which can account for the greater stability of the more sterically crowded T_{ee} isomer will be discussed in a subsequent section.

Turning our attention to the *cis* isomer, the C_{ss} conformation involves a potential sigma nonbonded interaction between the hybrid in-plane oxygen lone pairs, the C_{se} involves a hydrogen bond, and the C_{ee} conformation involves a sigma nonbonded interaction between the hydroxyl hydrogen atoms. These interactions are pictorialized below:

| | C_{ss} | | C_{se} | | C_{ee} |

Basis Set	STO–3G	4–31G	STO–3G	4–31G	STO–3G	4–31G
$P(O2p_x, O2p_x)$	−.0001	.0038	−.0006	−.0010	−.0006	−.0028
$P(O2p_x, Hl'_s)$	.0000	−.0001	.0030	.0070	.0008	.0102
$P(Hl_s, Hl'_s)$	.0000	.0000	−.0004	−.0006	.0016	.0015

We expect the hydrogen bonded geometry C_{se} to be more stable than the C_{ss} or C_{ee} conformations because substantial stabilization of the C_{se} conformation is already obtained in the initial fragment union, *i.e.* OH + OH.

The results of *ab initio* calculations exhibit several noteworthy features:

a) Sigma long pair nonbonded interaction appears to be sensitive to choice of basis set.

b) Pi nonbonded attraction between the oxygen lone pairs always obtains as expected and favors a greater stability of the *cis* isomer. However, the calculations show that the *trans* isomer is more stable for fixed ss and ee conformations. This may be due to either a dominance of steric effects or the exaggeration of dipolar effects by the basis sets employed.

c) Hydrogen bonding, *i.e.* sigma nonbonded attraction, as well as pi nonbonded attraction, make the C_{se} geometry the preferred one.

As a final example, we consider the relative stability of the torsional isomers of 2-butene. The six conformations of 2-butene are shown below along with the definitions to be used in the discussion, *e.g.* the label C_{ss} refers to the *cis* isomer where the in plane hydrogens of the methyl groups are staggered relative to the double bond.

71

	C_{ss}	C_{se}	C_{ee}	T_{ss}	T_{se}	T_{ee}
E_{rel} (kcal/mol)	2.855	2.315	1.864	3.721	1.819	0.000
$P^\pi_{CH_3}$, CH_3	.0034	.0006	.0005	.0000	.0000	.0000
N^π_T	.9835	.9604	.9640	.9503	.9580	.9895
$P(H1s, H'1s)$	.0000	.0001	.0009	–	–	–

Computation: 4–31G calculation at STO–4G optimized geometry[5]

The relative stability of any two torsional isomers can be predicted by following the same analyses as the one presented for propene and 1,2-difluoroethylene. The following conclusions are reached:

a) Attractive pi nonbonded interactions obtain in the C_{ss} isomer and attractive sigma nonbonded interactions obtain in the C_{ee} isomer.

This has been confirmed by *ab initio* calculations as shown by the appropriate indices of nonbonded attraction shown above.

b) By following the same type of analysis as in the case of propene, it is predicted that the T_{ee} form will be more stable than the T_{ss} form.

This has been confirmed by *ab initio* calculations. The relative energies of the two conformations are shown above.

c) The relative energy of the C_{ss} and C_{ee} conformations may be understood by realizing that the quantity $E(C_{ee}) - E(C_{ss})$ is a composite of nonbonded attractive terms and steric terms. We can write:

$E(C_{ee}) - E(C_{ss}) \alpha$ (Sigma Nonbonded Attraction) –
(Pi Nonbonded Attraction) + (Steric Destabilization of C_{ee}) –
(Steric Destabilization of C_{ss})

Now, since sigma nonbonded interactions are weaker than pi nonbonded interactions, there will be an electronic bias in favor of C_{ss}. However, steric effects favor the C_{ee} conformation. Hence, steric effects may make it appear that sigma nonbonded interactions dominate pi nonbonded interactions.

Ab initio calculations reveal the order of conformational preference shown below for the *cis* conformations:

$$C_{ee} > C_{se} > C_{ss}$$

This order of stability is reflected in the nuclear repulsion energy, which constitutes an index of "steric effects", as shown below:

E_N (rel), kcal/mol

C_{ee}	0.000
C_{se}	166.770
C_{ss}	441.390

72

The above is an example, therefore, where the apparent dominance of sigma over pi nonbonded attraction is attributable to steric effects.

d) The relative energy of the T_{ee} and C_{ee} conformations may be understood by realizing that the quantity $E(C_{ee})-E(T_{ee})$ is a composite of nonbonded attractive terms and steric terms. As in the C_{ee} vs C_{ss} comparison, these two terms favor different isomers, *i.e.* nonbonded attraction favors C_{ee} and steric effects favor T_{ee}.

Ab initio calculations reveal that T_{ee} is favored over C_{ee}. This constitutes an example where steric effects dominate sigma nonbonded attractive effects.

e) The relative energy of the T_{ss} and C_{ss} conformations may be understood by realizing that pi nonbonded attraction favors the C_{ss} conformation and steric effects the T_{ss} conformation.

Ab initio calculations indicate that the C_{ss} geometry is more stable than the T_{ss} one. This constitutes an example where pi nonbonded attractive effects dominate steric effects.

f) The methyl rotational barrier in the *cis* isomer is given by the energy difference between C_{ee} and C_{se} while the same barrier in the *trans* isomer is given by the energy difference between T_{ee} and T_{se}. *Ab initio* calculations show that the smaller barrier in the *cis* isomer has a steric origin, *i.e.* C_{ee} is more destabilized relative to T_{ee} than C_{se} is relative to T_{se}. Pertinent results are shown in Fig. 22.

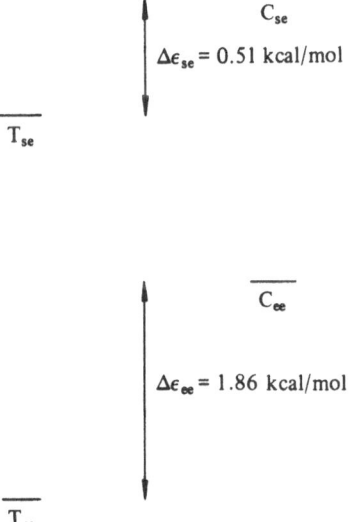

$$C_{se}$$

$$\Delta\epsilon_{se} = 0.51 \text{ kcal/mol}$$

$$T_{se}$$

$$C_{ee}$$

$$\Delta\epsilon_{ee} = 1.86 \text{ kcal/mol}$$

$$T_{ee}$$

Fig. 22. Energy difference between the T_{ee} and C_{ee} and T_{se} and C_{se} conformations, respectively, of 2-butene. Energies from ab initio calculations.

Experimentally, the methyl rotational barrier is found to be smaller for the *cis* isomer relative to the *trans* isomer. The experimental methyl rotational barrier as well as *ab initio* values are shown in Table 12.

Table 12. Methyl rotational barriers in *cis*- and *trans*-2-butene

Transformation	Exptl. kcal/mol	Ref.	4–31G kcal/mol	STO–4G kcal/mol
$C_{ee} \rightarrow C_{se}^{a)}$	0.45–0.73	113, 114)	0.450	0.319
$T_{ee} \rightarrow T_{se}^{b)}$	1.95	114)	1.814	1.495

a) Threefold rotational barrier in the *cis* isomer.
b) Threefold rotational barrier in the *trans* isomer.

3.7. Torsional Isomerism of $CH_3CH{=}CHX$ Molecules

l-substituted propenes are instriguing systems because they exhibit geometric and conformational isomerism which display unexpected trends. Our approach will be illustrated by reference to the model system l-fluoropropene. We shall examine the four possible conformations shown below and attempt to identify the electronic factors which determine the relative stability of the geometrical isomers as well as the relative magnitude of methyl rotational barriers in the *cis* and *trans* geometries.

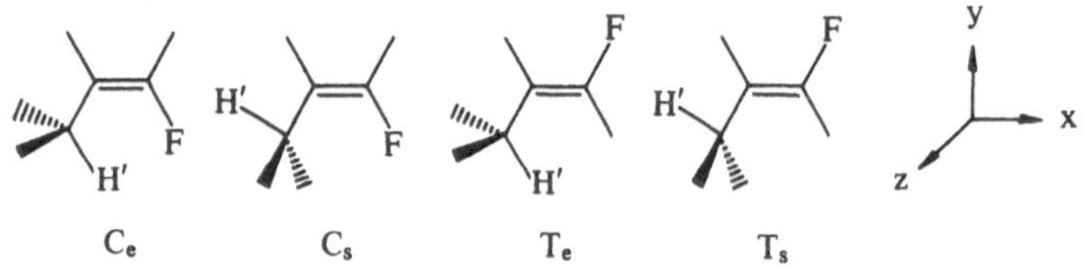

		C_e	C_s	T_e	T_s
$P^\pi_{CH_3, F}$		.00003	.00032	.00001	.00001
N^π_T		.87480	.87579	.87428	.87466
$P_{(F2p_x, H'1s)}$		.00038	.00001	.00001	–.00001

Computation: CNDO/2, standard geometry

     Geometrical isomerism in $CH_3CH{=}CHX$ molecules can be discussed in the same manner as in the case of CHX=CHX molecules. By following familiar arguments we can show that the following electronic factors operate:

     a) Pi CH_3---F nonbonded attraction favors Cs over Ts and Ce over Te.

     b) Sigma CH_3---F nonbonded attractions, *i.e.* hydrogen bonding, favor Ce over Te but does not differentiate Cs and Ts. From these considerations, it is clear that a *cis* geometry will be favored relative to a *trans* geometry by nonbonded attractive effects.

     The next question is: which of the two possible *cis* rotamers, *i. e.* Cs and Ce, is the stable conformation? Clearly, pi nonbonded attraction favors Cs more than Ce while sigma nonbonded attraction, *i. e.* hydrogen bonding, stabilizes the Ce rotamer exclusively. Due to the greater spatial overlap, the sigma nonbonded interaction will be stronger than the pi nonbonded interaction. Hence, it is reasonably expected that Ce will be the energy minimum and Cs the energy maximum on the rotational sur-face of $CH_3CH=CHF$. As will be seen later, this is a general pattern, *i. e.* hydrogen bonding appears to dominate any other pi or sigma nonbonded attractive interaction, at least in most cases.

A further insight provided by the above analysis concerns the relative magnitude of the methyl rotational barriers in the *cis* and *trans* isomers. These rotational barriers can be set equal to $E(Ce)-E(Cs)$ and $E(Te)-E(Ts)$ for the *cis* and *trans* isomers, respectively. Now in the case of the *cis* isomers, both minimum (Ce) and maximum (Cs) are dominated by nonbonded attractions and their energy difference is expected to be small. On the other hand, in the case of the *trans* isomer, the maxi-mum is destabilized relative to the minimum by nonbonded repulsion, *i. e.* the Te and Ts structures are expected to be well separated in energy. This suggests that in $CH_3CH=CHF$, and all $CH_3CH=CHX$ molecules where the *cis* isomer is more stable than the trans isomer, the methyl rotational barrier will be smaller in the *cis* geometry due to nonbonded attractive effects.

Experimentally, it is known that the *cis* isomer in l-substituted propenes is more stable and has a lower rotational barrier. Some pertinent data are shown in Tables 13–14. In most cases, the experimental results agree with our predictions. An interesting trend obtains in the alkyl vinyl ether series. Specifically, two types of nonbonded attraction can obtain in these molecules:

a) 1,4-attractive interactions between the ether oxygen atom and vicinal groups which favor the *cis* isomer.

b) 1,4-attractive interactions between the double bond and the alkyl group bonded to the ether oxygen. These interactions will be sterically possible only in the *trans* geometry which they will tend to favor.

These two types of nonbonded attractive interactions are illustrated below:

When the alkyl group attached to oxygen, R', has pi type $MO's$ capable of inter-acting with the olefinic bond, *e. g.* R=Me or Et, the *trans* isomer is enthalpically favored due to attractive R'–––vinyl nonbonded interactions. However, when this interaction is effectively "shut off", *e. g.* $R'=t-Bu$, the trend reverses.

Ab initio calculations of substituted propenes agree with our predictions. The barriers calculated by Allen and Scarzafava are 1.07 and 1.34 kcal/mol. for *cis* and *trans* 1-fluoropropene, respectively[125]. The *cis* isomer is also predicted to be more stable than the *trans* by 0.894 kcal/mol. Similarly, the work of English and Palke gives a barrier of 1.204 and 2.03 kcal/mol in the *cis* and *trans* isomers, respective-ly[126]. Once again, the *cis* isomer is found to be more stable by 1.52 kcal/mol.

Table 13. Relative stability of the geometric isomers of $XHC=CHCH_3$

X	$\Delta H°$ (cal/mol)[a]	Keq[a]	Ref.
Cl (l)		.316	111)
Br (l)		.471	115)
OMe (l)	-910 ± 50	.408	116)
		1.033	110)
		1.210	117)
OEt (l)	-370 ± 20	.234	116)
		.722	110)
OiBu (l)	-430 ± 50	.699	110)
OiPr (l)	$+560 \pm 40$	.368	110)
OtBu (l)	$+680 \pm 130$	.296	110)
OPh (l)		.538	118)
CN (l)		.754	111)

[a] A positive enthalpy and an equilibrium constant less than unity mean that the *cis* isomer is more stable relative to the *trans*.

Table 14. Rotational barriers in 1-substituted propenes

Molecule	Isomer	Barrier[a]	Ref.
1-Fluoropropene	c[b]	1057	119)
	t[c]	2150	120)
1-Chloropropene	c	620	121)
	t	2170	122)
1-Cyanopropene	c	1400	123)
	t	2100	124)
2-Butene	c	730	114)
	t	1950	113)

[a] In cal/mol.
[b] c = *cis*.
[c] t = *trans*.

Indices of nonbonded attraction calculated by the CNDO/2 method are shown above, and indicate that both pi and sigma nonbonded attraction obtain in the *cis* isomer with the former being maximized in the Cs conformation and the latter being maximized in the Ce conformation.

Finally, a few words about previous explanations of torsional isomerism in 1-substituted propene. The most common rationale for the observed trends has been steric hindrance. However, steric effects alone cannot account for the greater stability of the *cis* isomer. Hence, the steric interpretation should be rejected.

The other common explanation for the experimental trends in 1-fluoropropenes involves dipolar effects. The dipoles in both isomers are depicted below.

This effect correctly predicts the *cis* isomer to be more stable than the *trans* and *may* well contribute to the observed difference in thermodynamic stability between them.

However, one is unable to predict anything about the rotational barriers in these molecules by considering only the dipolar effect. Consequently, the only rationale capable of explaining both the geometric as well as the conformational preferences observed appears to be nonbonded attraction.

3.8. Torsional Isomerism of CXY=CXY Molecules

In this section we will consider the effect of nonbonded interactions upon the relative stability of the *cis* and *trans* isomers of symmetrically tetrasubstituted ethylenes. The dissection employed is the one shown below *i. e.* the MO's of CXY=CXY are constructed from the group MO's spanning X, Y, X, Y and the group MO's of the C=C fragment.

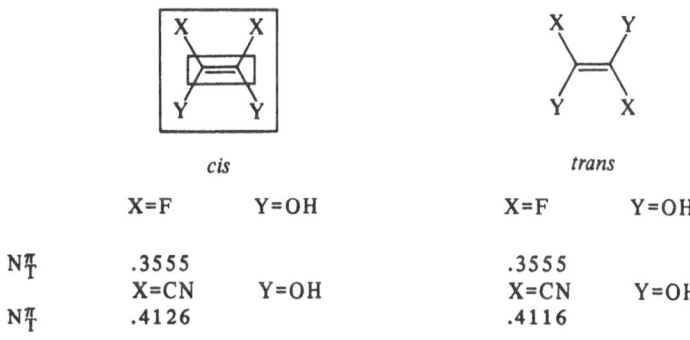

	cis		*trans*	
	X=F	Y=OH	X=F	Y=OH
N_T^π	.3555		.3555	
	X=CN	Y=OH	X=CN	Y=OH
N_T^π	.4126		.4116	

Computation: INDO, standard geometry

The XXYY group MO's can be constructed from the group MO's of the subfragments XY and XY taken in the appropriate geometry.

We begin by considering a case where X and Y are pi donor species, namely, 1,2 difluoro-1,2 dihydroxyethylene. The construction of the FFOHOH group MO's in the *trans* geometry is shown in Fig. 23. The splitting between ϕ_1 and ϕ_2 as well as between ϕ_3 and ϕ_4, shown in Fig. 24, is assumed to be comparable for both isomers. The symmetry designations are with respect to a mirror plane bisecting the C–C double bond in the *cis* isomer and a twofold rotational axis in the *trans* isomer. From inspection of this diagram, it is obvious that the dominant stabilizing interaction favors slightly the *cis* isomer. Therefore, the *cis* isomer is expected to be stabilized relative to the *trans* isomer.

Unfortunately, there is no experimental data available for molecules of this type. Furthermore, good quality *ab initio* calculations are also unavailable for these systems.

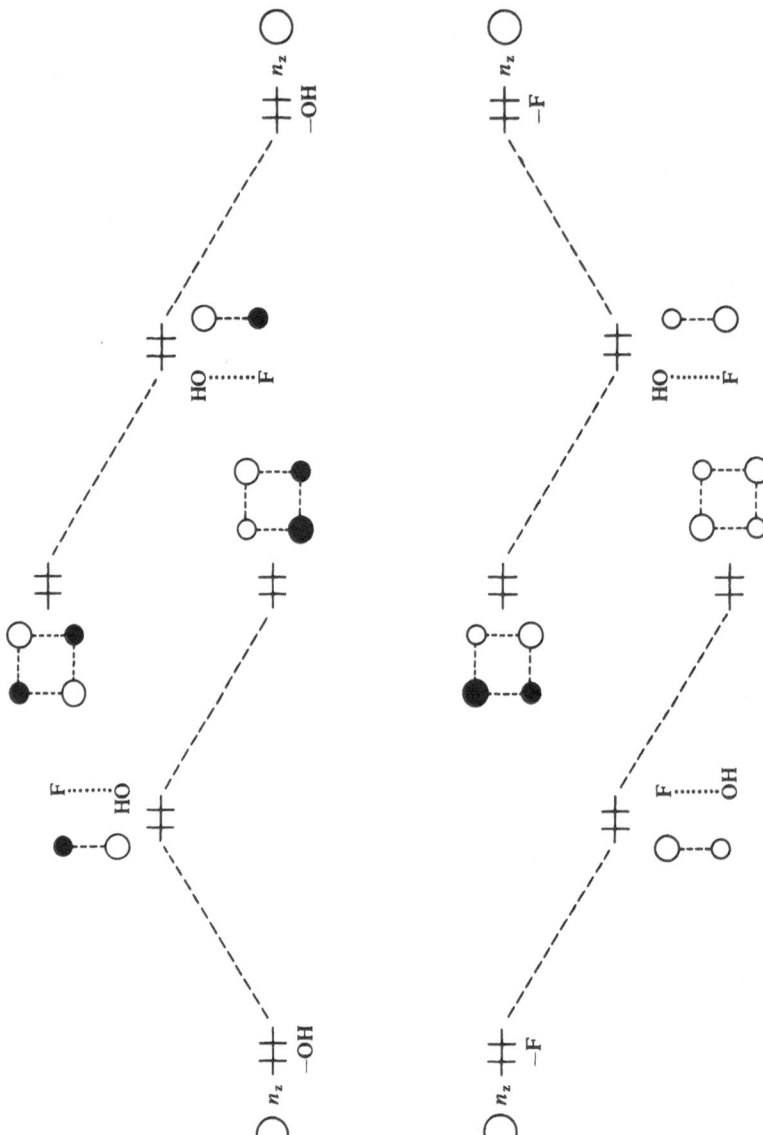

Fig. 23. Construction of the pi lone pair group MO's of FF (OH) (OH) in *trans* 1,2 difluoro-1,2 dihydroxyethylene. The same procedure is applicable to the *cis* isomer

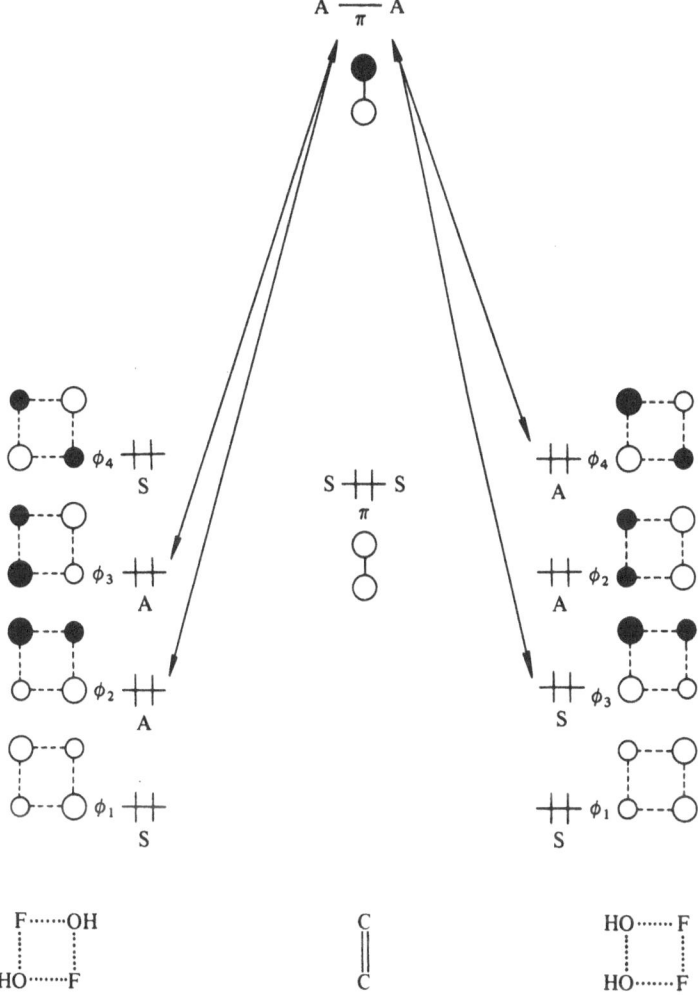

Fig. 24. Stabilizing orbital interactions in cis and trans 1,2-difluoro-1,2-dihydroxy ethylene. Symmetry labels are with respect to a rotational axis (*trans* isomer) and mirror plane (*cis* isomer)

However, INDO calculations have been performed for the *cis* and *trans* isomers of 1,2-difluoro-1,2-dihydroxyethylene and the results are shown above. As can be seen, the total overlap population is the same in both cases and, hence, the difference in the stabilities of the *cis* and *trans* isomers is expected to be very small.

We now proceed to the next case in which the substituents consist of a donor and acceptor. The specific molecule will be 1,2-difluoro-dicyanoethylene. Following the previous procedure, the group MO's of the substituents and the ethylene π and π^* MO's are used to construct the MO's of the composite system. The interaction diagrams are shown in Fig. 25. The results are analogous to the previous example except for the additional stabilizing interactions which will also favor the *cis* isomer. Consequently, the *cis* isomer is expected to be the most stable isomer.

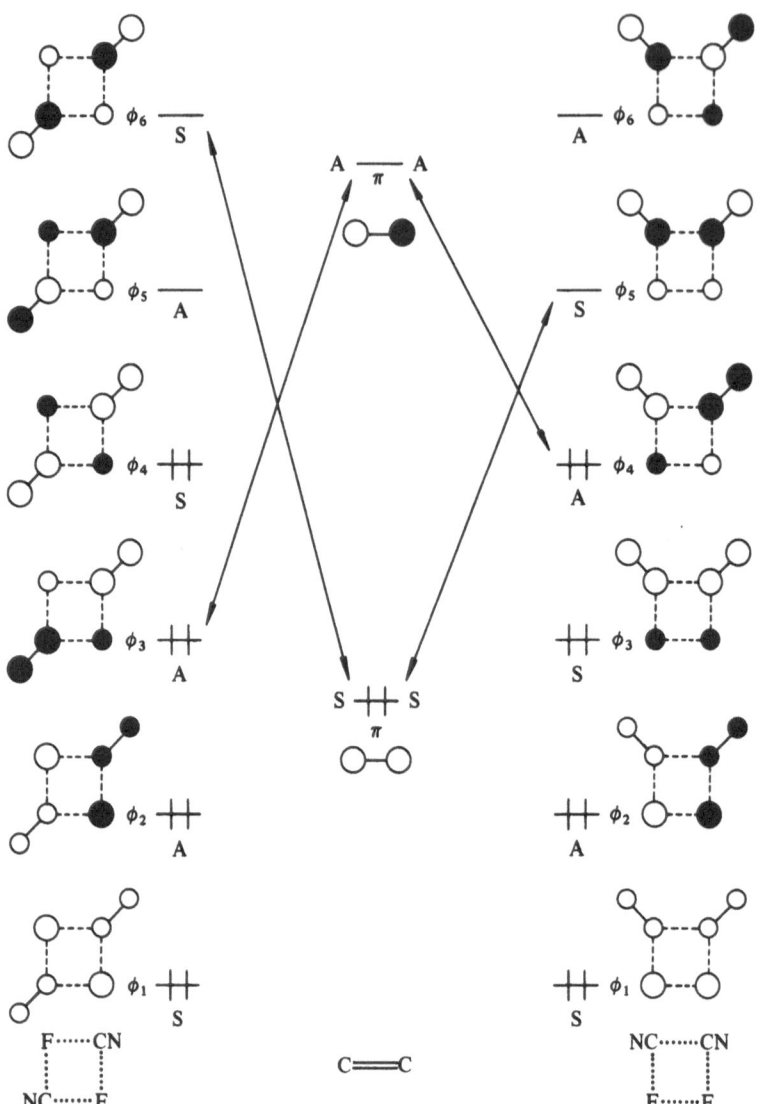

Fig. 25. Stabilizing pi orbital interactions in *cis* and *trans* 1,2-difluoro-1,2-dicyanoethylene. Symmetry labels are with respect to a rotational axis (*trans* isomer) and mirror plane (*cis* isomer)

This conclusion is further supported by an inspection of the indices of non-bonded attraction as calculated by the INDO method.

Examples of these effects can be found. In the case of the systems shown below, the *cis* isomer is known to be the more stable isomer[127]. This is especially striking for the dianion since there exists a large coulombic repulsion between the sulfur atoms. Here, however, the possibility of counterion coordination should not be disregarded as a potential source of the greater stabilization of the *cis* isomer.

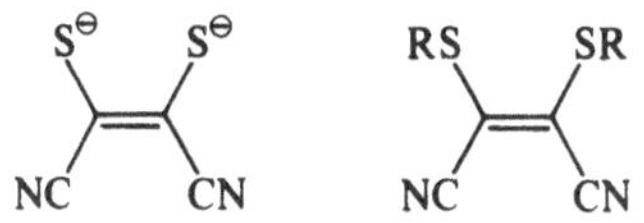

As a final example, we consider torsional isomerism in 1,2-diamino-1,2-dicyanoethylene (DADCE) which is known to be a precursor in the evolutionary synthesis of nucleic acids[128]. Some torsional isomers of DADCE are shown below:

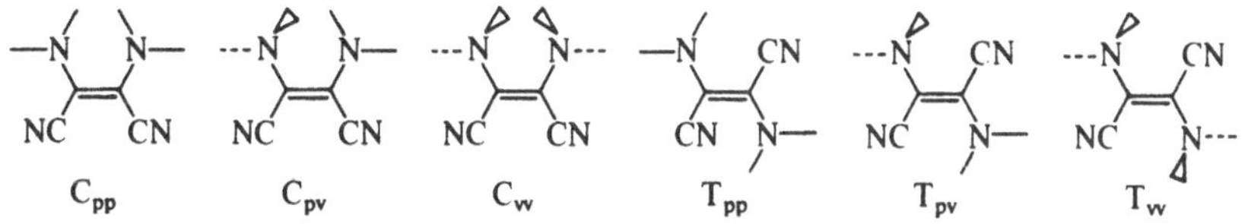

$$C_{pp} \qquad C_{pv} \qquad C_w \qquad T_{pp} \qquad T_{pv} \qquad T_w$$

Arguing as before, we expect pi nonbonded attraction to favor the *cis* geometry for a fixed conformation. Furthermore, sigma nonbonded attraction in the form of a hydrogen bond is expected to make C_{pv} the most stable conformation in the *cis* series. A complete study of all possible torsional isomers of DADCE will be reported in the future.

Experimentally, the *cis* isomer of DADCE seems to be more stable than the *trans* isomer[129]. Furthermore, in the crystalline state, the *cis* isomer is known to exist in the hydrogen bonded conformation[130].

3.9. Conformational Isomerism of CH_3COX Molecules

We shall now examine molecules of the type CH_3COX, where X can be a halide, a hydroxyl group, an alkoxy group, or an amido group. We shall first consider the conformational preference, *i. e.* staggered vs. eclipsed, imposed on such molecules by the presence of lone pairs or pi systems on the group X, *e. g.* X=F.

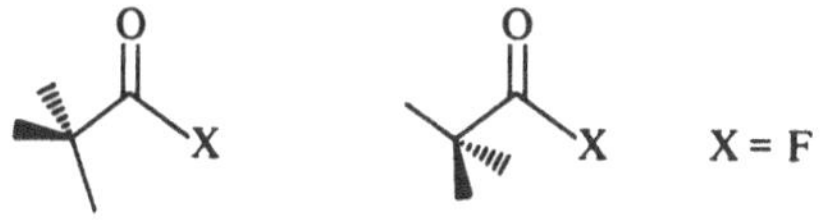

S (staggered)	E (eclipsed)	X = F

N_T^{π} .8152   .8249

Calculation: INDO, standard geometry

Our analysis is illustrated by reference to the model system CH_3COF and the dissection employed is shown below:

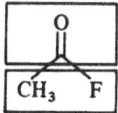

Proceeding as before, we construct the pi MO's of CH_3COF from the pi group MO's of fragment B and the pi MO's of the central carbonyl fragment as shown in Fig. 26. From Fig. 26 we can see that the two electron stabilizing interaction $\phi_2 - \pi^*$ will favor the eclipsed conformer relative to the staggered conformation. Furthermore, the preferred conformation of CH_3COF will also be influenced by the inherent conformational preference exhibited by the parent system $CH_3CH=O$, *i. e.* the eclipsed conformation. Hence, it is predicted that CH_3COF and its analogues, *e. g.* acyl halides (X=O, CL, Br, I), acetic acid (X=OH) and its corresponding esters (X=OR) and amides (X=NR_2), will be biased towards adopting the eclipsed conformation.

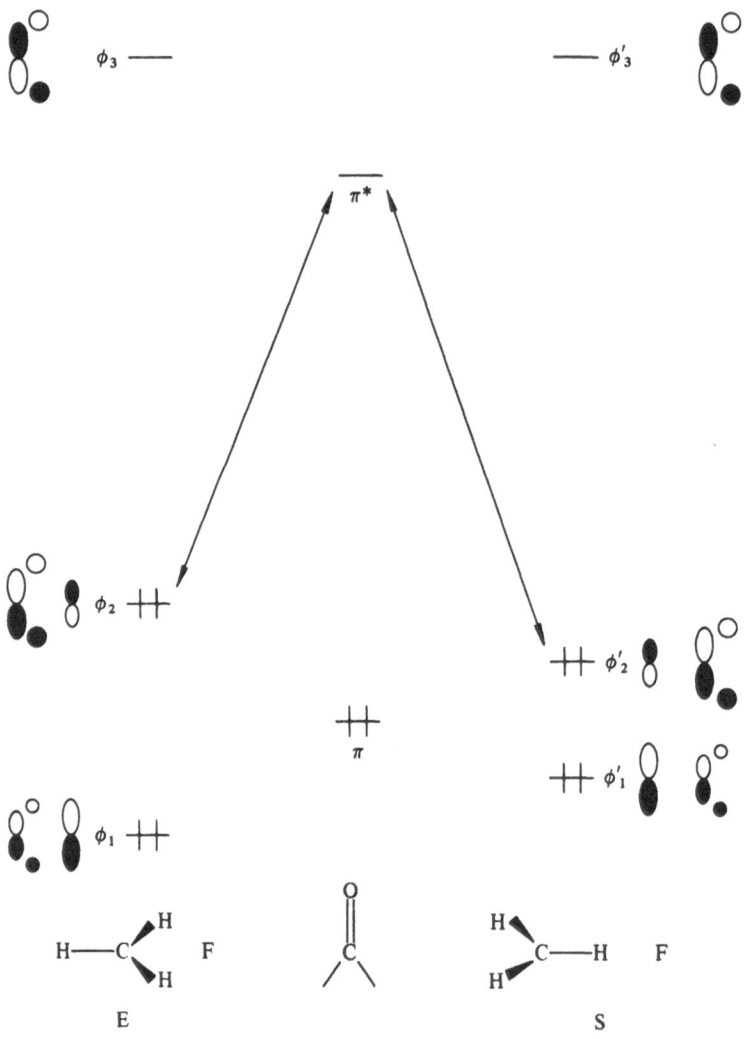

Fig. 26. Dominant stabilizing pi type orbital interactions in the S and E conformers of acetyl fluoride

Results of INDO calculations for staggered and eclipsed CH_3COX are shown above and support our analysis. Specifically, a lower energy of the eclipsed conformation can be attributed to pi nonbonded interaction since the total pi overlap population is larger in the E conformation relative to the S conformation.

The above analysis for CH_3-CO-X molecules where X bears lone pair AO's is strongly supported by the experimental data. Specifically, the preferred conformation for the case where X = halogen is the eclipsed conformation[131–136].

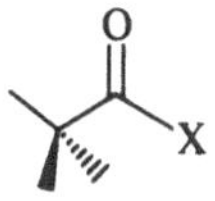

X = F, Cl, Br, I

The most stable conformation of acetic acid, as determined by microwave spectroscopy, is shown below and is again the one predicted by our theoretical approach[137]:

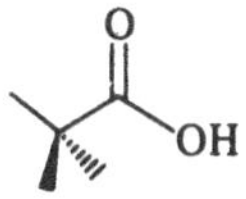

So far, we have discussed only those cases where X bears a lone pair AO but our analysis will be similar for molecules where X has π type MO's, i. e. CH_3, $-C\equiv N$, etc. For example, in the case of $X=CH_3$, the molecule can exist in three possible conformations:

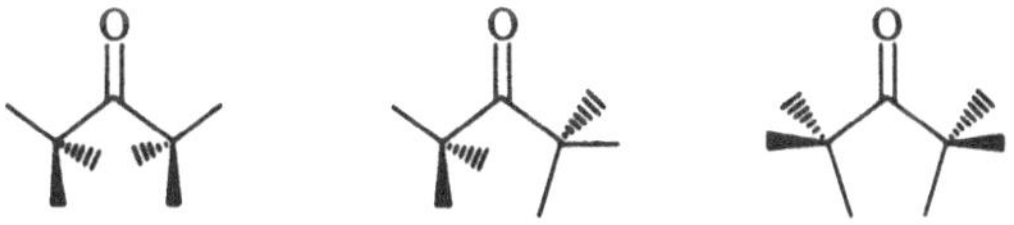

C_{ee} $\qquad$ C_{se} $\qquad$ C_{ss}

The relevant interaction diagrams for the C_{ee} and C_{ss} conformations are shown in Fig. 27. Arguing as before, we conclude that the C_{ee} conformation will be preferred over the C_{se} and C_{ss} conformations due to an attractive interaction between the out of plane methylene hydrogens.

Experimental results confirm the predictions of the above analysis. Specifically, the vibrational spectra of acetone[138] and 3-pentanone[139] are consistent with the most stable rotamer having a C_{ee} geometry:

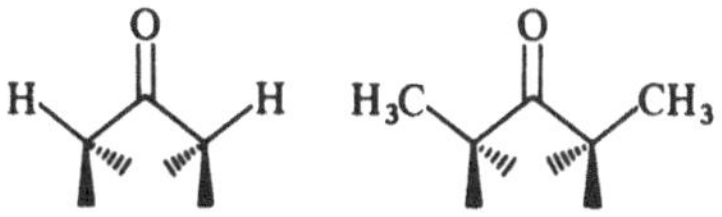

A similar analysis can be performed for the cases where X is an unsaturated group, i. e. $C\equiv CH$, $H_2C=CH$, $C\equiv N$. The microwave spectrum for the cases where X is a cyano[140, 141] or alkynyl group[142] shows that the most stable conformation is the eclipsed one as predicted by OEMO theory:

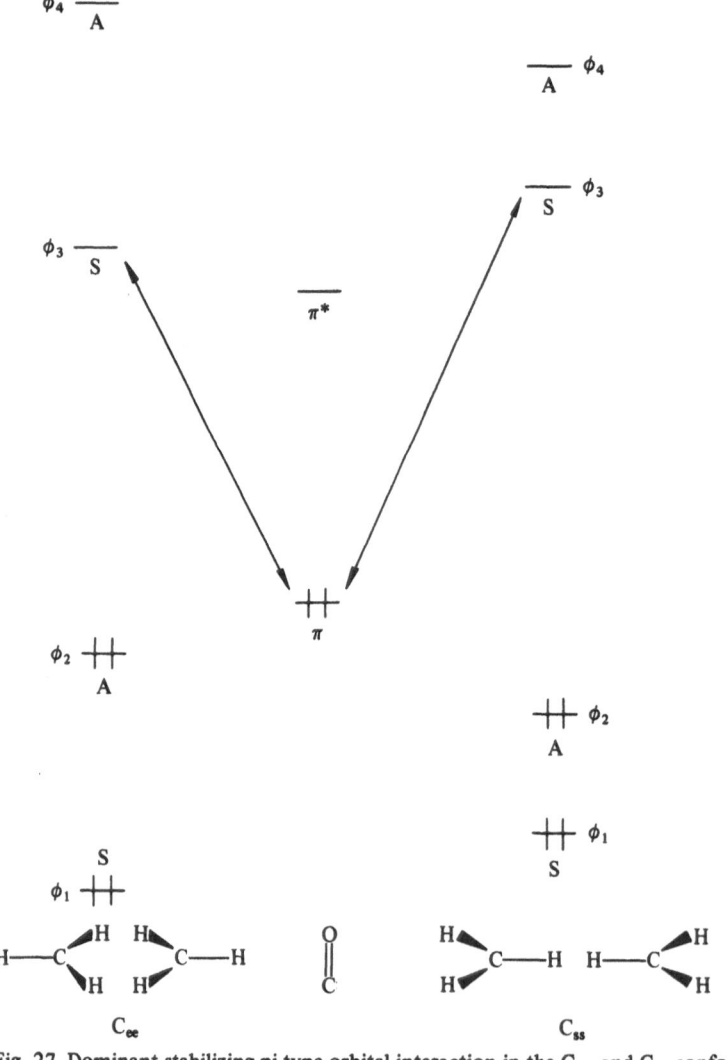

The theoretical approach used above to elucidate the conformational preferences of CH_3-CO-X molecules can also be applied to a discussion of the methyl rotational barrier in these systems. The methyl rotational barrier corresponds to the energy difference between the eclipsed and staggered conformations with the eclipsed conformation being an energy minimum and the staggered conformation being an energy

Fig. 27. Dominant stabilizing pi type orbital interaction in the C_{ee} and C_{ss} conformers of acetone. Symmetry labels are assigned with respect to a mirror plane (both isomers)

maximum. The relative magnitude of the methyl rotational barrier in a series of related compounds can be deduced by comparing the nonbonded interactive factors which raise or lower the energy of the maximum and minimum.

We shall return to CH_3COX molecules in a later section to discuss another very important sigma effect which obtains in these systems.

3.10. Conformational Isomerism of XCH_2COY Molecules

We turn now to molecules of the type XCH_2COY where X and Y may be different or identical groups or atoms which bear lone pair AO's. The two conformations we will consider are shown below:

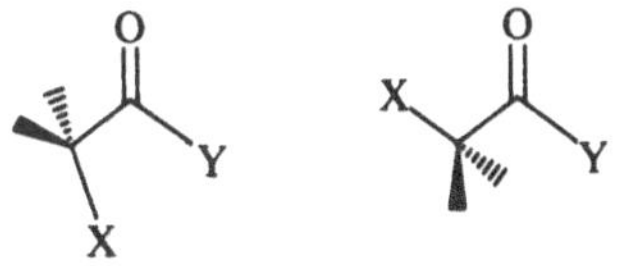

E (eclipsed)   S (staggered)

The labels E and S refer to the orientation of the X group relative to the Y group. In these systems, both conformations shown above can be affected by sigma lone pair nonbonded interactions between X and Y in one case (E) and X and O in the other case (S). However, these nonbonded interaction effects, whether they are attractive or repulsive, will not significantly alter the conformational preference due to pi effects. Hence, the S conformation is expected to be the most stable form of XCH_2COY molecules.

Experimental results indicate that the sigma nonbonded interactions introduced by the substituents X and Y do not play an important role in determining the conformational preference of XCH_2COY molecules. Specifically, the IR and Raman spectra of XCH_2COY for the case X=Y=Br, X=Y=Cl and X=Cl, Y=Br show that the most stable conformation is the S conformation[143, 144].
A particularly interesting result is that the preferred conformation for 2-bromo-2-methyl-propionyl-bromide is the one with the two bromine atoms eclipsed[145]:

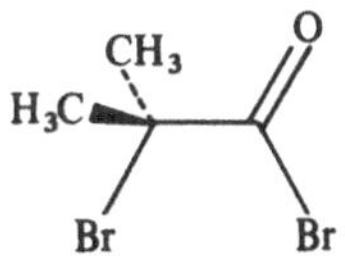

The elimination of the attractive CH_2————O interaction by methyl substitution may well be the reason for this observation.

3.11. Conformational Isomerism of R—X—R Molecules

The conformational preferences of R—X—R molecules will now be examined. The model system which we will use to illustrate our approach is dimethyl ether, *i. e.*

X=O and R=CH$_3$. The three possible conformations of·CH$_3$–O–CH$_3$ are shown below along with the labels to be used in this section, e. g. the label C$_{ee}$ refers to that conformation in which both in-plane methyl hydrogens are eclipsed by the C–X bonds:

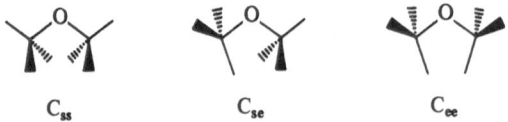

Dimethyl ether can be dissected in the manner indicated below for the C$_{ss}$ conformation and the interaction diagrams for the C$_{ss}$ and C$_{ee}$ conformations can be constructed as shown in Fig. 28:

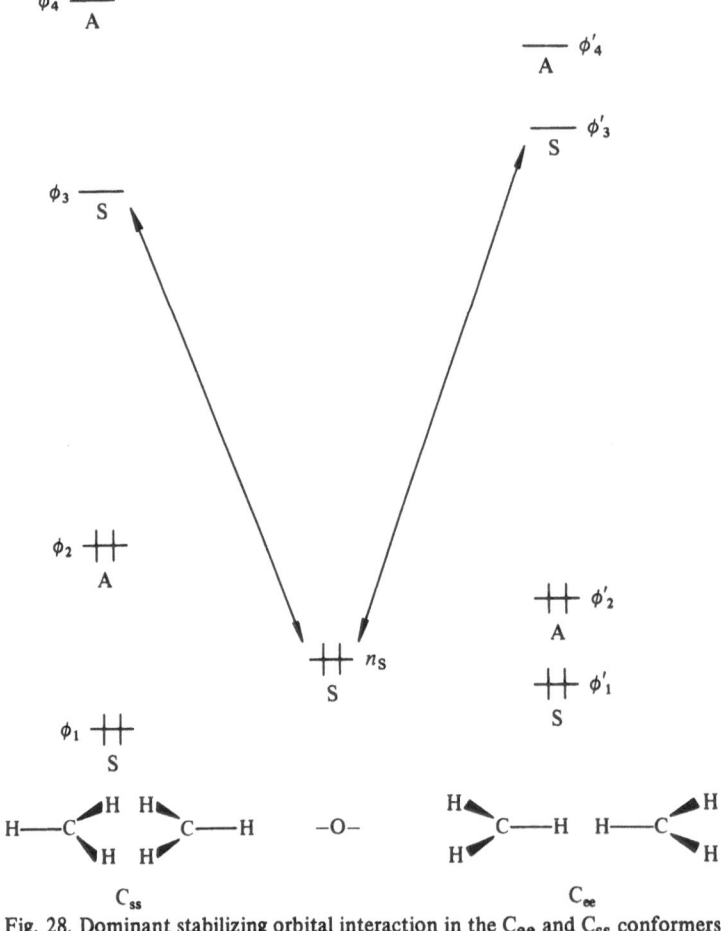

Fig. 28. Dominant stabilizing orbital interaction in the C$_{ee}$ and C$_{ss}$ conformers of dimethylether. The symmetry labels are assigned with respect to a mirror plane

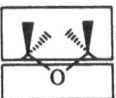

By following the same arguments as before, we conclude that the C_{ss} conformation will be lower in energy than the C_{ee} conformation. The same analysis can be used to compare the relative energetics of the C_{ss} and C_{se} conformations. In conclusion, we can say that the C_{ss} conformation will be favored over the C_{ee} and C_{se} conformations due to a larger pi attractive nonbonded interaction which obtains in the 6 pi electron aromatic geometry of the C_{ss} conformation.

Sigma nonbonded interactions in dimethyl ether can be analyzed in the same way as before and the conclusion is that sigma nonbonded attraction is greatest for the C_{ee} conformation. We expect pi nonbonded attraction to dominate sigma non-bonded attraction and the final result is that the C_{ss} conformation will be the most stable torsional isomer of R−X−R molecules.

An interesting problem arises when we examine the relative stabilization of the C_{ss} and C_{ee} conformations of the model systems dimethyl ether and isobutene. In the former case, there is only one dominant two electron stabilizing interaction which favors the C_{ss} conformation. On the other hand, in the case of isobutene there are two key two electron stabilizing interactions, one favoring the C_{ss} and the other favoring the C_{ee} conformation. These considerations can be best understood by reference to Fig. 29.

On the basis of simple considerations, we can predict that the C_{ss} conformation will still be favored in isobutene but less so than in dimethyl ehter. In the latter case, $n_S \rightarrow \phi_3$ charge transfer renders the H_2−−−−H_2 interaction attractive while in isobutene, $\psi_1 \rightarrow \phi_3$ charge transfer, rendering the H_2−−−−H_2 interaction attractive, is partly counteracted by $\phi_1 \rightarrow \psi_2$ charge transfer rendering the H_2−−−−H_2 interaction repulsive. The net result is a positive overlap population in C_{ss} dimethyl ether and a smaller but still positive overlap population in C_{ss} isobutene reflecting a greater energy difference between the C_{ss} and C_{ee} conformations of dimethyl ether relative to that between the same conformations of isobutene.

On the basis of the above analysis, substituent effects on the energy difference between the C_{ss} and C_{ee} conformation of R−X−R molecules can be readily understood:

a) Decreasing the lone pair ionization potential of the central heteroatom, X, will increase the stabilizing $n_S - \phi_3$ interactions favoring the C_{ss} conformation. Thus, the energy difference between C_{ss} and C_{ee} conformations should increase in the order Y=CH^{-1} > NH > Ö.

b) Increasing the electron acceptor ability of a central pi bond $>$X=Z will increase the $\phi_1 - \psi_2$ interaction favoring the C_{ee} conformation and decrease the $\psi_1 - \phi_3$ interaction favoring the C_{ss} conformation, thus resulting into a decrease of the energy difference between the C_{ss} and C_{ee} conformations. When $>$X=Z becomes a very good acceptor the C_{ee} conformation may be preferred.

The above analysis can also be used in connection with the problem of the methyl rotational barrier in double rotor molecules, e. g. dimethyl ether, relative to

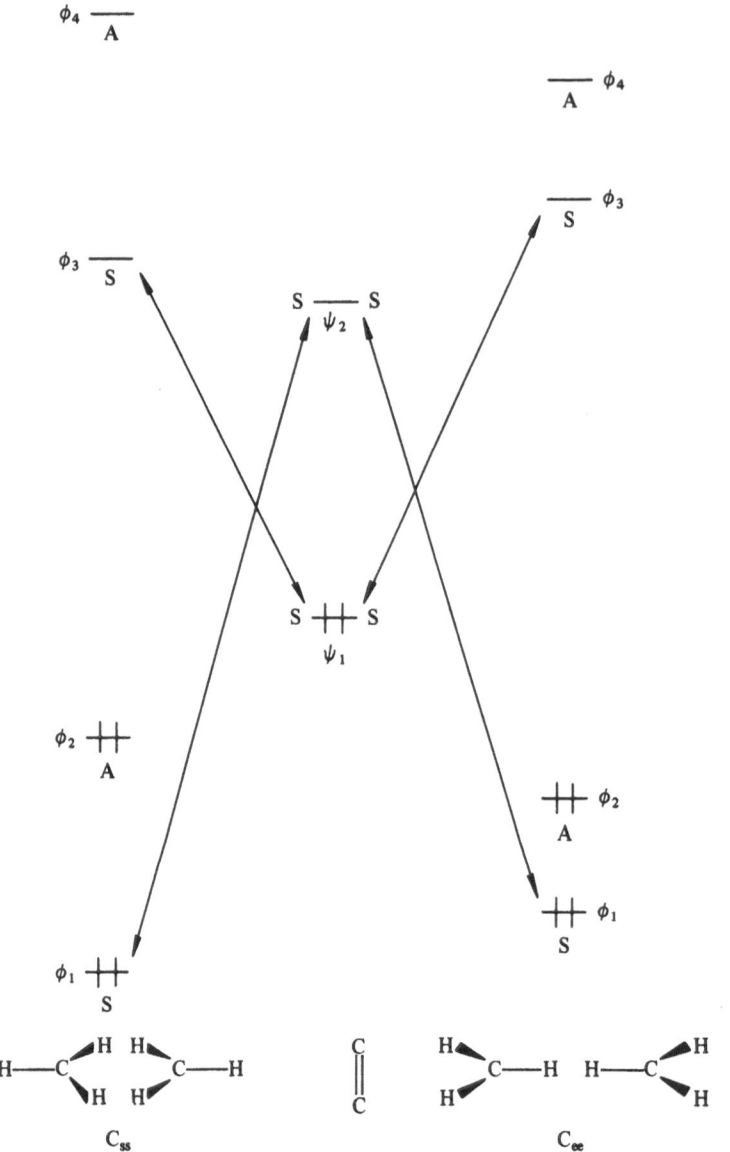

Fig. 29. Dominant stabilizing pi type orbital interactions in the C_{ee} and C_{ss} conformers of isobutene. The symmetry labels are assigned with respect to a mirror plane

single rotor molecules, *e. g.* methanol. The methyl rotational barrier in dimethyl ether is the energy difference between the C_{ss} conformation and the C_{se} conformation with the former being the energy minimum and the latter being the energy maximum. By focusing on pi nonbonded attractive interactions only, we predict that the methyl rotational barrier in dimethyl ether will be greater than the methyl rotational barrier in methanol since the energy minimum of dimethyl ether is stabilized by nonbonded attractive interactions which do not obtain in methanol. This barrier difference between a double rotor molecule and its corresponding single rotor analogue will be designated ΔB.

The analysis presented above can be extended to the systems where the $\underset{-\text{X}-}{\overset{\text{Z}}{\|}}$ fragment is replaced by a XH_2 group which acts as an effective pi bond.

A similar reasoning to the one outlined above can be employed to understand conformational preferences and barrier problems in the molecules shown below and related systems.

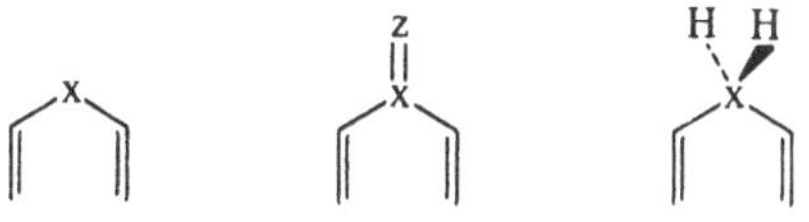

Experimental studies show that the C_{ss} conformation is, indeed, the lowest energy form for R−X−R molecules. Specifically, the vibrational spectra of diethyl ether[146], acetone[138], and diethyl ketone[139] are consistent with the most stable rotamer assuming a C_{ss} conformation:

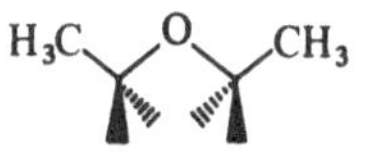

| Diethyl ether | Acetone | Diethyl ketone |

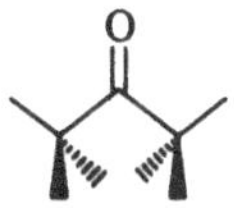

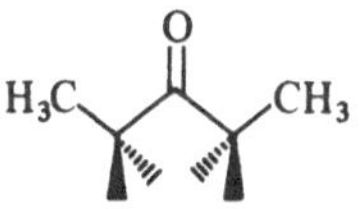

Interestingly, the next most stable form of diethyl ether and diethyl ketone may correspond to an 8 pi electron Möbius aromatic system:

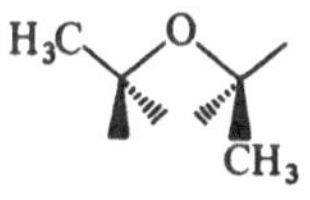

| Möbius aromatic | Möbius aromatic |

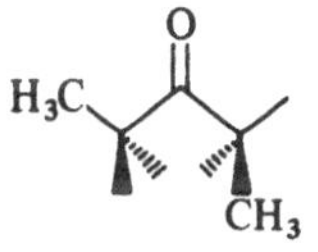

The preferred conformation of ethyl-methyl ether has also been determined and is such as to maximize pi nonbonded interactions as shown below[147]:

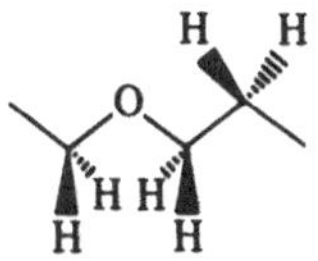

An apparent exception is the case where X=−NH− and R=−HC=O. Experimental results indicate that the lowest energy conformation involves a noncyclic geometry[148]:

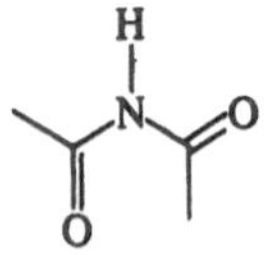

In this case, severe dipolar repulsion can destabilize the cyclic conformation relative to the noncyclic geometry.

Ab initio calculations by Pople and co-workers confirm all predictions[149].
Typical results are shown in Tables 15–16, where the following trends are obvious:

a) In all the molecules studied, the C_{ss} conformation is found to be the most stable rotamer.

b) The energy difference between the C_{ss} and C_{ee} conformation decreases as the lone pair ionization potential of the central heteroatom X increases and as the electron acceptor ability of a central pi bond $>X=Z$ increases. In addition, ΔB is

found to be greater for CH_3-Y-CH_3 than for $\underset{CH_3-X-CH_3}{\overset{Z}{\|}}$ molecules in the

calculations as well as experimentally.

Table 15. Relative energies of the three rotamers of CH_3-X-CH_3 molecules[a]

| | Relative energy (kcal/mol) | | |
X	C_{ss}	C_{se}	C_{ee}
C=O	0.00	0.75	2.22
C=CH$_2$	0.00	1.93	4.31
$-\bar{\bar{O}}-$	0.00	2.98	7.00
$-\bar{N}H-$	0.00	3.62	8.25
$-CH_2-$	0.00	3.70	8.77

a) See Ref.[149].

Table 16. Methyl rotational barrier difference between double rotor (CH_3-X-CH_3) and single rotor (CH_3-X-H) molecules

| | ΔB[a] (kcal/mol) | | |
X	Ab initio	Experiment	Ref.
$-\bar{N}H-$	1.49	1.22	150, 151)
$-\bar{\bar{O}}-$	1.86	1.65	152, 153)
$-CH_2-$	.44	.40	154, 155)
C=CH$_2$	.23	.21	156, 157)
C=O	.01	-.39	158, 159)

a) ΔB = (Double Rotor Barrier) – (Single Rotor Barrier).

90

The indices of nonbonded attractions for the C_{ss} and C_{ee} conformations of dimethyl ether are shown below and confirm that the greater stability of the C_{ss} conformation relative to the C_{ee} conformation is due to pi nonbonded attraction:

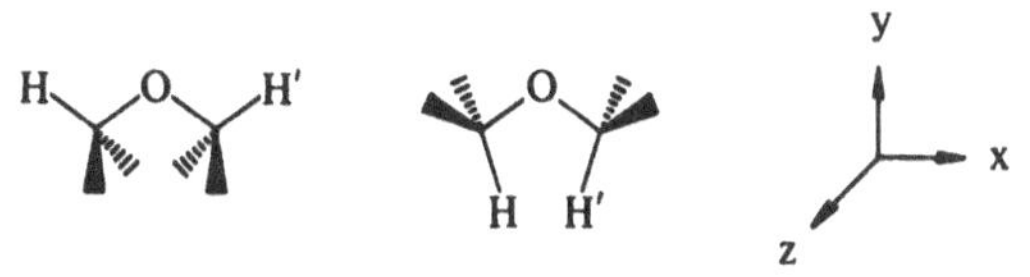

	C_{ss}	C_{ee}
$P^{\pi}_{CH_3, CH_3}$	.00051	−.00128
$P_{(Hls, H'ls)}$	−.00010	.05154
N^{π}_T	1.14444	1.12715

Computation: INDO, standard geometry

This conclusion regarding the greater stability of the C_{ss} conformation of dimethyl ether was also arrived at by Pople and co-workers on the basis of their *ab initio* calculations[149].

Finally, *ab initio* calculations of trimethylene show a tendency for adopting a "crab" conformation[160]:

This result is perfectly consistent with our discussion since the so called "crab" conformation benefits from the stabilizing 1,5-nonbonded attractive interaction which obtains in the C_{ss} conformation of $CH_3—X—CH_3$ molecules.

3.12. Conformational Isomerism of R—X—R' Molecules

We shall now trace the key factors dictating conformational preference of R—X—R' molecules where R≠R'. Our model system will be methyl vinyl ether and the four conformations of this system are shown below along with the corresponding labels which will be used in this section.

	Cs	Ce	Ts	Te
E_{rel}(kcal/mol)	0.000	4.451	1.254	3.116
$P^{\pi}_{CH_3, C(3)}$	0.0010	0.0001	0.0001	0.0000
N^{π}_T	0.5716	0.5702	0.5682	0.5675

Computation: STO–3G at STO–3G optimized geometry[8]

Part II. Nonbonded Interactions

	Cs	Ce	Ts	Te
E_{rel}(kcal/mol)	0.000	1.611	2.326	2.012
$P^\pi_{CH_3, C(3)}$	0.0039	0.0012	0.0001	0.0000
N^π_T	0.5803	0.5721	0.5704	0.5659

Computation: 4–31G at STO–3G optimized geometry[8]

The label Cs, for example, refers to the *cis* conformation in which the in-plane hydrogen is staggered relative to the O–C$_2$ bond. Since steric effects are expected to favor the *trans* conformations we shall illustrate our theoretical approach by comparing the Cs and Ts conformations.

Methyl vinyl ether can be dissected into two fragments, A and B, as shown below for the *cis* geometry.

The key difference between the two conformations Cs and Ts is that in the former case the two 1s AO's of the methyl hydrogens can overlap with the $2p_z$ AO of the outer olefinic carbon, while in the latter case they cannot:

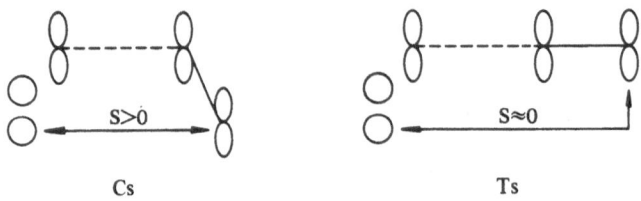

Cs	Ts

We are now prepared to consider the construction of the composite pi system of methyl vinyl ether from the pi system of fragment B and oxygen lone pair AO. The orbital interactions which obtain in this union are depicted in Fig. 30. On the basis of the principles outlined before, we can see that the p_z–ϕ_3 interaction is more stabilizing in the Cs conformation. As a result, the more "crowded" structure, Cs, will be lower in energy than the Ts conformation.

A similar approach can be used for comparing the stability of the Ce and Te conformations relative to that of the Cs conformation. Since appreciable overlap between the methyl hydrogens and the outer olefinic carbon obtains only in the Cs conformation, we would expect this conformation to have lower energy than either the Ce or Te conformation.

The MO analysis presented above can be effectively utilized in predicting the relative methyl rotational barrier in the *cis* and *trans* isomers. The barrier in the *cis* geometry will correspond to the energy difference between the Cs and Ce conformations with the Cs conformation being the energy minimum and the Ce conformation being the energy maximum. On the other hand, the barrier in the *trans* geometry

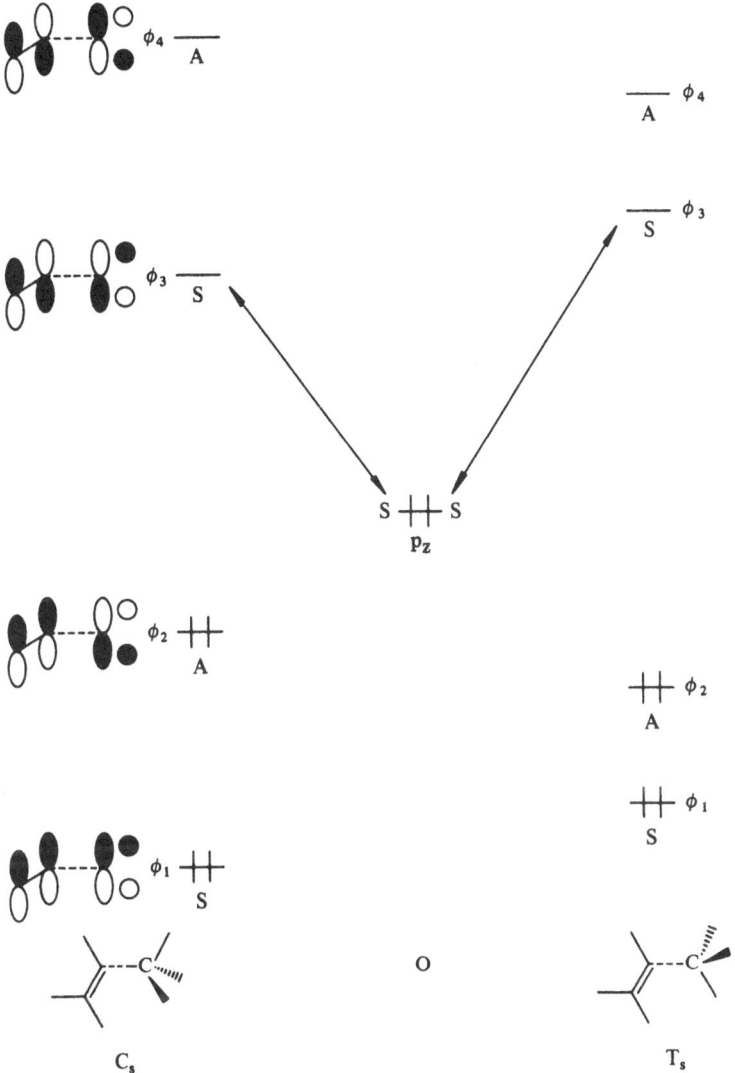

Fig. 30. Dominant stabilizing pi orbital interactions in the C_s and T_s conformers of methyl vinyl ether

will be the energy difference between the Ts and the Te conformations with the Ts conformation being the energy minimum and the Te conformer being the energy maximum. It is immediately obvious that rotation of the methyl group in the *cis* geometry will be unfavorable relative to *trans* because in the former case an attractive nonbonded interaction is destroyed in the process while such a destructive effect is absent in the *trans* geometry. In other words, we expect that the energy minimum Cs will be more stable than the energy minimum Ts and the two energy maxima, Ce and Te, will be comparable in energy resulting in a larger methyl rotational barrier in the *cis* geometry.

93

By following arguments which are familiar from Section 3.11., we can further make these predictions:

a) In $CH_3-\ddot{Y}-CH=CH_2$ molecules and related systems the energy difference between the Cs and Ts geometries will increase as the ionization potential of $\ddot{Y}$ decreases.

b) In $CH_3-\overset{Z}{\underset{\|}{C}}-CH=CH_2$ molecules and related systems the energy difference between the Cs and Ts geometries will decrease as $-\overset{Z}{\underset{\|}{C}}-$ becomes a better pi acceptor.

c) The methyl rotational barrier will be larger in the *cis* geometry.

d) The Cs preference will be greater in $CH_3-\ddot{Y}-CH=CH_2$ then in $CH_3-\overset{Z}{\underset{\|}{C}}-CH=CH_2$ type molecules.

e) In general, ΔB will be greater for $CH_3-\ddot{Y}-CH=CH_2$ type molecules than $CH_3-\overset{Z}{\underset{\|}{C}}-CH=CH_2$ molecules.

Experimental studies have suggested the existence of two rotamers of methyl vinyl ether[161, 162] and methyl vinyl thioether[163] with the more stable isomer having a planar *cis* conformation, a surprising result in view of the fact that a classically trained chemist may have dismissed it on grounds of being "sterically" unfavorable.

Additional experimental confirmation of our theoretical predictions can be found in the cases of methyl and ethyl formate. Microwave[164, 165] and IR spectroscopy[166, 167] show that the preferred geometry is the Cs conformation in both systems. Strikingly, the Cs conformation of methyl formate is 2.7 kcal/mol more stable than the most stable *trans* conformation.

Ultrasonic measurements of methyl formate, ethyl formate, methyl acetate, ethyl acetate, and n-propyl formate show that the *cis* conformation is more stable than the *trans*[168−171].

Furthermore, the entropy change for the following process was found to be positive[172].

R = CH₃, C₂H₅

R = CH_3, C_2H_5

A positive $\Delta S°$ implies that the *cis* conformation is characterized by less degrees of freedom, a fact which can be attributed to a "locking" effect of the methyl group into a Cs conformation by virtue of attractive pi nonbonded interactions.

Finally, IR[173], NMR[174], UV[175], and Raman spectroscopy[175] show that the *cis* conformation of N-methyl formamide and N-methyl acetamide is more stable than the *trans*.

Recent *ab initio* calculations have shown, in agreement with the conclusions reached by OEMO theory, that the most stable conformation of methyl vinyl ether is the Cs conformer[176]. The calculated relative energies of the four conformations of methyl vinyl ether at the STO–3G and 4–31G levels are shown above along with the appropriate indices of nonbonded attraction. As can be seen, the greater stability of the Cs conformation can be attributed to an attractive 1,5 pi nonbonded interaction.

Finally, experimental measurement as well as *ab initio* computation show that the methyl rotational barrier is also higher in the *cis* than the *trans* conformation. These results are shown in Table 17.

Table 17. *cis* – *trans* energy differences and methyl rotational barriers for methyl-vinyl-ether

Transformation	Experimental (kcal/mol)	STO–3G (kcal/mol)	4–31G (kcal/mol)
Ts – Cs	1.15 ± 0.25[a, b]	0.96	2.47
Ce – Te		1.34	– 0.40
Ce – Cs	3.83 ± 0.1[b]	4.14	2.97
Te – Ts		1.84	0.89

[a] Experimental energy difference between the two isomers.
[b] See Ref.[177].

3.13. Torsional Isomerism of Cations and Anions

The theoretical approach we have developed for determining the preferred conformation of uncharged molecules is also applicable to those cases where the molecule bears a net positive or negative charge. In this section we will examine the importance of nonbonded interactions in torsional isomerism of selected cations and anions.

We first consider the electronic factors which influence torsional isomerism in cations. Our first model system will be the 2-propyl cation and the two conformations which we will compare are shown below:

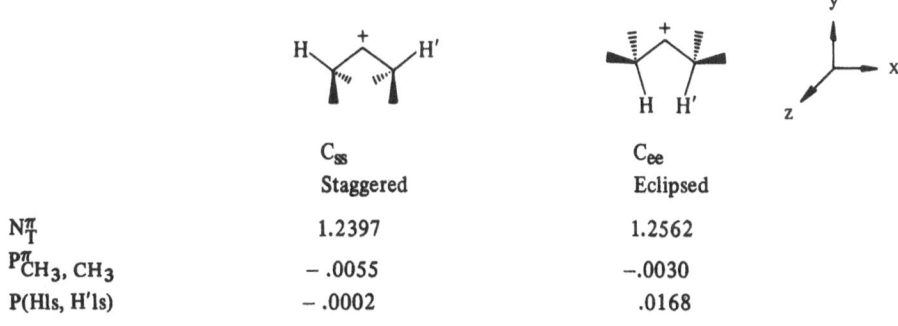

	C_{ss} Staggered	C_{ee} Eclipsed
N_T^π	1.2397	1.2562
P_{CH_3, CH_3}^π	– .0055	–.0030
P(Hls, H'ls)	– .0002	.0168

Computation: INDO, standard geometry

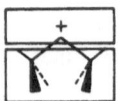

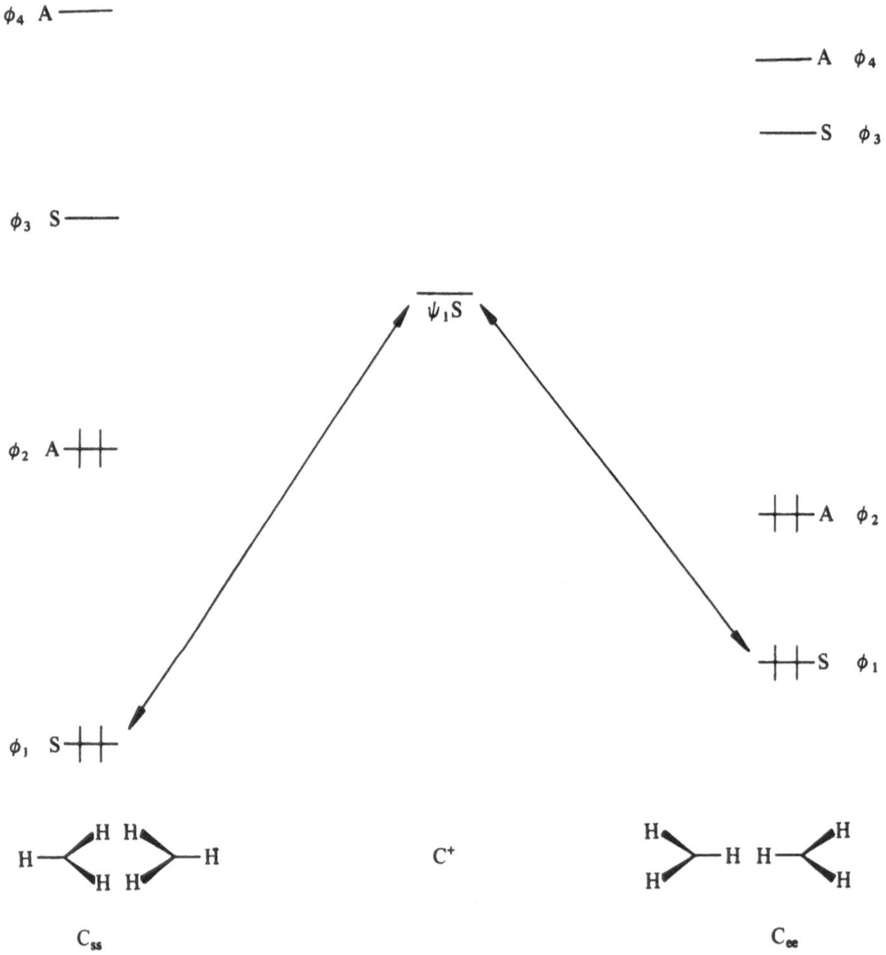

ϕ_4 A——

——A ϕ_4

——S ϕ_3

ϕ_3 S——

$\overline{\psi_1 S}$

ϕ_2 A++

++A ϕ_2

++S ϕ_1

ϕ_1 S++

C_{ss} C⁺ C_{ee}

Fig. 31. Dominant stabilizing pi orbital interaction in the C_{ee} and C_{ss} conformers of 2-propyl cation. The symmetry labels are assigned with respect to a mirror plane

The dissection of this cation, illustrated for the staggered conformation, is shown above and the interaction diagrams for the staggered and eclipsed conformations are shown in Fig. 31. It is concluded that the eclipsed conformation will be favored over the staggered conformation since the two electron stabilizing $\phi_1 - \psi_1$ interaction is greater for the former case. Furthermore, change transfer from ϕ_1 to ψ_1 results into a net antibonding CH_3––CH_3 pi bond order. Since charge transfer is greatest in the C_{ee} conformation, the resultant CH_3––CH_3 pi bond order will be more negative than in the C_{ss} conformation. However, since the spatial overlap between the methylene hydrogens is smaller for the C_{ee} conformation than for the C_{ss} conformation, we conclude that the antibonding CH_3––CH_3 pi overlap population will be smaller in the former case.

Sigma nonbonded attraction also obtains in the isopropyl cation and, arguing as before, we conclude that sigma nonbonded attractive interactions are maximized in the C_{ee} conformation. The 2-propyl cation constitutes a good example where pi and sigma nonbonded interactions reinforce each other.

Results of *ab initio* calculations of the 2-propyl cation, as well as the isoelectronic molecule, dimethyl borane, are shown in Table 18. As can be seen, the C_{ee} conformation is predicted to be more stable than the C_{ss} conformation in both the 2-propyl cation and dimethylborane. However, the relative energies of the C_{se} and C_{ee} conformations are reversed when the basis set is changed from STO–3G to 4–31G.

Table 18. Calculated relative conformational energies of CH_3–Z–CH_3 molecules

Z	Basis Set[a]	Relative conformational energy (kcal/mol)		
		C_{ss}	C_{se}	C_{ee}
$\overset{\oplus}{C}$–H	STO–3G	–––	0.08	0.00
	4–31G	1.02	0.00	0.09
B–H	4–31G	0.47	0.00	0.14

[a] See Ref.[149, 178].

Indices of pi and sigma nonbonded interactions in the C_{ss} and C_{ee} conformations of the 2-propyl cation as calculated by the INDO method are shown above. As can be seen, both pi and sigma nonbonded interactions favor the C_{ee} relative to the C_{ss} conformation.

Another interesting case is the preferred conformation of 1-butenyl cation for which the two conformations, Cs and Ts, are shown below:

Cs   Ts

The interaction diagrams for the above conformations are identical with that of methyl vinyl ether (Fig. 30) except that the oxygen lone pair AO is replaced by an unoccupied carbon 2p AO. With this in mind we conclude that the transoid conformation of the cation, Ts, will be more stable than the cisoid conformation, Cs, since the ϕ_1–p_z two electron stabilizing interaction is greater for the Ts conformation.

Experimental evidence concerning the relative stabilities of cation conformations is limited. The few examples known, however, strongly support our analysis. For example, the hydrolysis of α-methyl allyl chloride under $S_N 1$ conditions affords exclusively the *trans* cotyl alcohol, presumably via a transoid butenyl cation[179]:

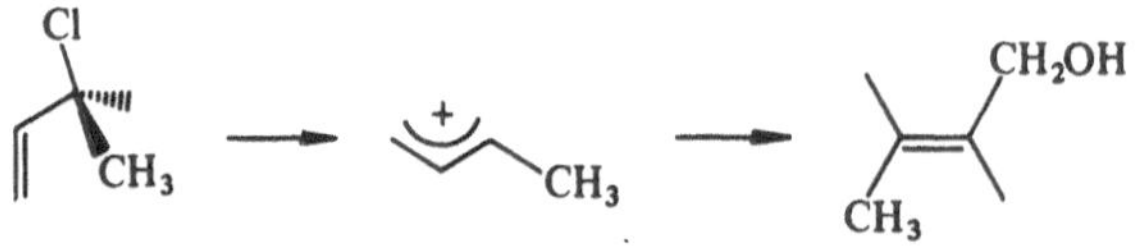

Furthermore, under the same conditions the *trans* cotyl chloride hydrolyzes faster than the *cis* isomer.

Focusing our attention on anionic molecules, we turn first to the problem of conformational isomerism in the 2-propyl carbanion, $CH_3\overline{C}HCH_3$. The two geometries which we will compare are again the staggered and eclipsed conformations shown below:

C_{ss} C_{ee}

The appropriate interaction diagrams are similar to that of Fig. 28. Reasoning as before, we conclude that pi nonbonded interactions favor the staggered conformation. Sigma nonbonded interactions, on the other hand, favor the eclipsed conformation. We expect pi nonbonded attractive interactions to dominate sigma interactions and the resulting order of stability is predicted to be $C_{ss} > C_{ee}$.

We now turn to the problem of torsional isomerism in the 1-butenyl anion. By following arguments similar to those for methyl vinyl ether, we conclude that the C_s conformation will be lower in energy than the T_s conformation due to an attractive 1,5 nonbonded interaction.

The experimental probe of the conformational preferences of allyl anions is based primarily on the kinetics of base catalyzed isomerization reactions of olefins and many results have been summarized by Bank[180]. Kinetic data of base catalyzed isomerization of 1-butene[181] show that the less stable *cis* isomer of 2-butene is formed through a thermodynamically more stable allylic anion:

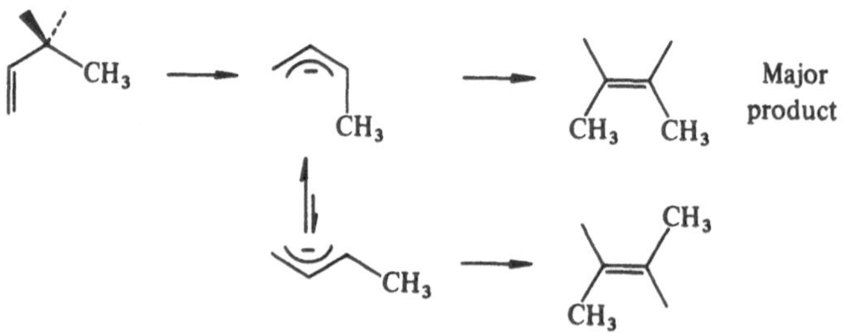

Additional evidence for the greater stability of the *cis* conformation of allylic anions is provided by other base catalyzed isomerization studies of 1-butene and 1-pentene. It was found that the thermodynamically less stable *cis* isomers of 2-butene and 2-pentene were the major products of the reaction[182-186]. Furthermore, *cis*-2-butene isomerizes, under the same conditions, faster than the *trans* isomer to give 1-butene.

3.14. Torsional Isomerism of Substituted Benzenes

Nonbonded interactions may obtain in halosubstituted benzenes. For example, consider the model compound *ortho*-difluoro-benzene which can exist in either a planar geometry or a distorted geometry in which the C–F bonds are bent, alternately above and below the mean plane as a result of severe dipolar and steric repulsion.

P^{π}_{FF}	.00001	$-.00003$
$P(F\,2p_y, F\,2p_y)$	$-.00005$	.00000

Computation: INDO, standard geometry

The dissection of this molecule is shown below for the planar geometry:

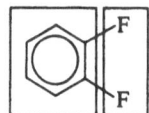

The interaction diagram for the two geometries can be constructed from the fluorine group MO's and the pi MO manifold of benzene as shown in Fig. 32. By focusing on the n_A-ϕ_5 stabilizing interaction, we can immediately conclude that planar *ortho* difluorobenzene is stabilized more by nonbonded attractive interactions than the distorted geometry. We note that the n_S-ϕ_4 two electron stabilizing interaction favors the distorted geometry. However, the n_A-ϕ_5 interaction dominates the n_S-ϕ_4 interaction because of a smaller energy gap and a larger MO overlap integral, *i. e.* $S_{n_A\phi_5} > S_{n_S\phi_4}$. The preceeding inequality obtains because the C_2 and C_3 coefficient for ϕ_4 are larger than for ϕ_5:

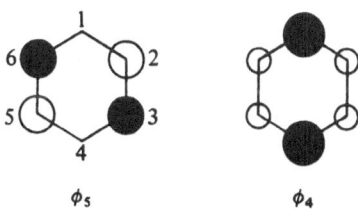

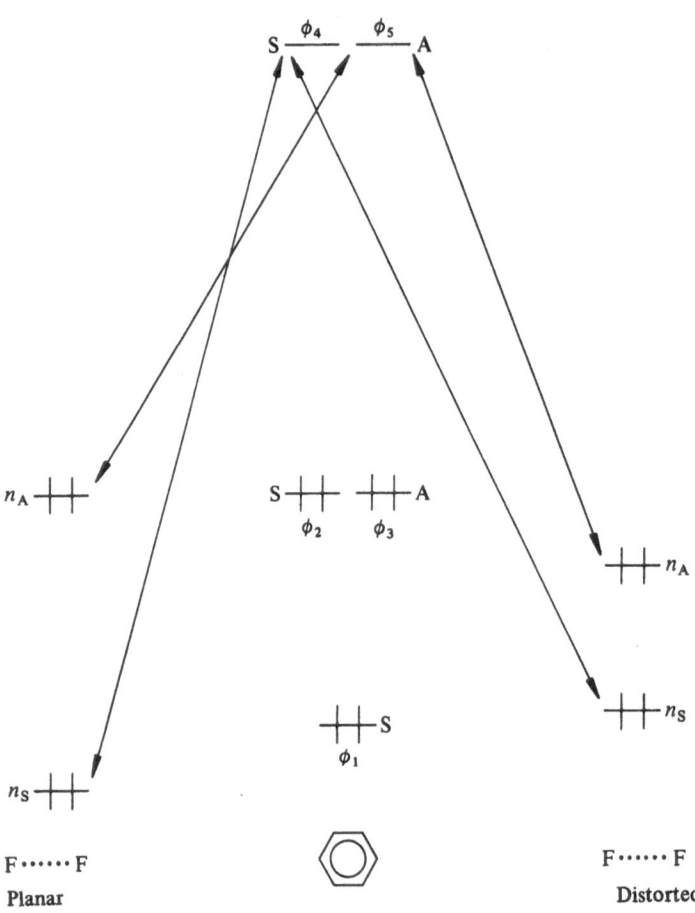

Fig. 32. Dominant stabilizing pi MO interactions in the distorted and planar geometries of *ortho* difluorobenzene

The results of INDO calculations of planar and distorted difluorobenzene in which the C–F bonds have been displaced a few degrees above and below the mean plane, are shown above. As can be seen, the planar geometry maintains pi nonbonded attractive interaction. Sigma lone pair interactions will be similar to those which obtain in 1,2-difluoroethane and appear to be repulsive in nature.

The above approach can be extended to tri-, tetra-, penta-, and hexasubstituted fluorobenzenes. Arguing as before, we conclude that the planar geometry of these systems will be more favorable since pi attractive nonbonded interactions between the fluorine atoms will be maximized in the former case. The pi F––F overlap populations for the tri-, tetra-, penta-, and hexasubstituted fluorobenzenes are shown below.

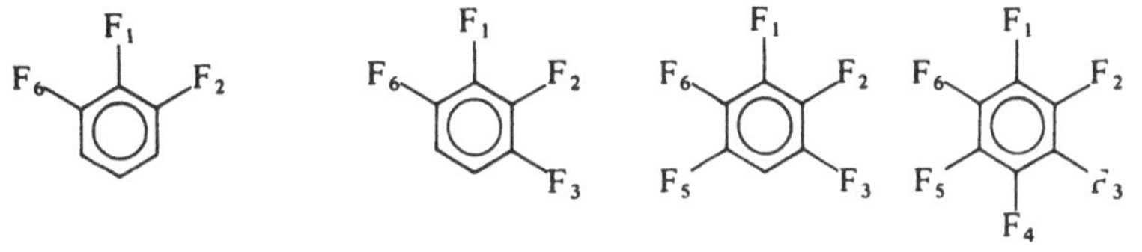

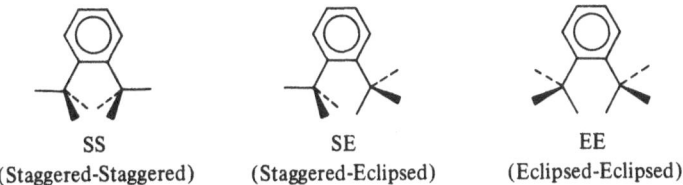

$P_{F_1--F_2}^{\pi}$	.00001	.00001	.00001	.00001
$P_{F_2--F_3}^{\pi}$		.00001	.00001	.00001
$P_{F_3--F_4}^{\pi}$				.00001

These results mean that polyhalobenzenes may resist the intuitively expected tendency toward deformations in order to relieve the "steric" repulsion of bulky vicinal halogens such as Cl, Br, and I, pi nonbonded attraction being the reason.

Electron diffraction studies of hexachlorobenzene[187], 1, 2, 4, 5-tetrachlorobenzene[187] and *ortho*-dibromobenzene[188] show that these molecules are planar:

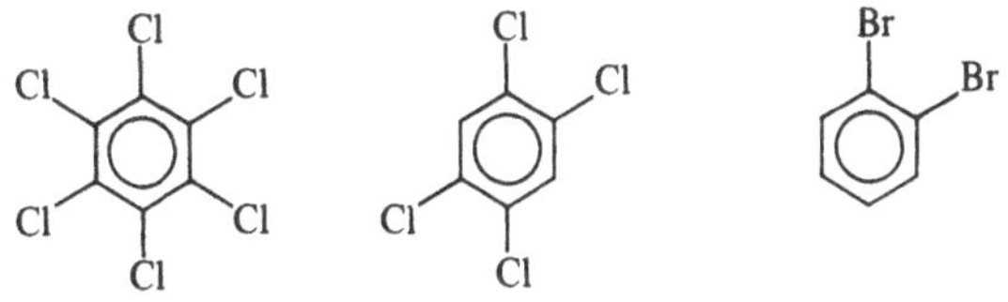

Hexachlorobenzene is also planar in its crystalline state as shown by an X-ray diffraction study[189]. It is of interest to point out that the distance between the two vicinal chlorine nuclei (3.12Å) in hexachlorobenzene is shorter than the sum of the van der Waals radii (3.60Å). This fact prompted one author to comment "... it is reasonable to assume that the ring of perchlorobenzene is puckered with the C–Cl-bonds being bent alternatively upwards and downwards from its mean plane"[190]. In view of our theoretical approach and the available data, this type of reasoning seems unjustified.

Interesting conformational effects may be encountered in alkyl substituted benzenes such as the three rotamers of *ortho* xylene shown below:

SS
(Staggered-Staggered)

SE
(Staggered-Eclipsed)

EE
(Eclipsed-Eclipsed)

Arguing as before, we conclude that pi nonbonded attractive interactions are maximized in the SS conformation while sigma nonbonded attractive interactions are maximized in the EE conformation. This conflict between pi and sigma nonbonded attractive interactions is similar to the conflict observed in *cis* 2-butene in which steric effects resulted in an *apparent* dominance of sigma over pi aromaticity.

101

The differences in the two cases are:

a) The C—C bonds in benzene are longer than the olefinic C—C bond in 2-butene resulting in a deemphasis of steric effects.

b) The pi LUMO of benzene is lower in energy than the pi LUMO of the ethylenic fragment of 2-butene as indicated by the electron Affinities (A) of benzene (A = −.54)[191] and ethylene (A = −2.8)[25]. Hence, pi nonbonded attractive interactions are enhanced in *ortho* xylene relative to *cis* 2-butene.

We conclude from the above considerations that pi nonbonded attractive interactions may dominate sigma nonbonded attractive interactions and, consequently, conformational stability in *ortho* xylene could vary in the following way SS > EE > SE.

The preferred conformation of *ortho* xylene has been determined by C^{13} NMR and the most stable rotamer was found to be the SS conformation[192].

Substituted phenols are also interesting since a hydrogen bond is possible in the *ortho* isomer. For example, consider the three conformations of catechol:

By using the arguments presented for dihydroxyethylene, we conclude that the hydrogen bonded conformation will be the preferred conformation for catechol.

3.15. Conformational Isomerism of Diene Systems

Conjugated olefins have played a key role in the development of theoretical models in organic chemistry. In this section we will examine in detail conformational isomerism in 1,3 dienes and related molecules.

On the basis of a previous discussion, we predicted that pi nonbonded interactions will produce an order of stability of *gauche* > *trans* > *cis*, assuming appreciable C_1–C_4 pi overlap in 1,3-butadiene. However, if sigma nonbonded interactions dominate pi interactions the order of stability will be *cis* > *gauche* > *trans*. Finally, in the event that steric effects are the controlling factor the order of stability will become *trans* > *gauche* > *cis*.

Experimentally, 1,3-butadiene is known to exist in a planar *trans* form but the identity of the second energy minimum has been a matter of controversy[193, 194].

Ab initio results are summarized in Table 19[f]. They seem to suggest the existence of two energy minima, a *trans* and a *gauche*, the *trans* being the more stable. In other words, the apparent order of stability of 1,3-butadiene is *trans* > *gauche* > *cis*. We conclude, therefore, that the preferred conformation of 1,3-butadiene is dictated by steric effects which overwhelm attractive pi and sigma nonbonded interactions.

---

[f] For earlier *ab initio* calculations of the electronic structure of 1,3-butadiene see Refs. [195] and [196].

Table 19. Computed relative energies of the conformers of 1,3-butadiene

	E_{rel}(kcal/mol)		
trans	gauche	cis	Ref.
0.0	3.1(θ=40°)	3.50	197)
0.0	2.6(θ=45°)	3.40	198)
0.0	3.0(θ=30°)	3.44	199)

We can extend the above theoretical approach to isoconjugate molecules such as acrolein and glyoxal. Arguing as before, we conclude that the relative order of conformational stability will be *gauche > trans > cis* if pi nonbonded interactions dictate the preferred conformation, *cis > gauche > trans* if sigma interactions are dominant, and *trans > gauche > cis* if steric effects are the most important factor.

The experimental situation is much the same as in butadiene. It has been established that the *trans*-conformation is favored for both acrolein[200] and glyoxal[201]. In the case of glyoxal, the *cis* conformer has been detected but, as of now, no stable *gauche*-conformation has been found.

The results of *ab initio* calculations of glyoxal and acrolein are shown in Table 20. The low *cis-trans* energy difference in acrolein may be a result of hydrogen bonding which obtains only in the *cis* conformation as illustrated below:

This effect may be exaggerated by the *ab initio* calculation since the experimental *cis-trans* energy difference in glyoxal and acrolein is 3.2 kcal/mol[200] and 2.0 kcal/mol[201], respectively.

Once we have established the factors responsible for conformational isomerism in 1,3-butadiene, we can attempt to put the insight that theory affords into use by designing dienic systems which will have a maximum chance of existing in a preferred crowded conformation, *i. e.* a *gauche* or *cis* conformation.

As was discussed earlier, *cis* 1,3-butadiene is favored by attractive sigma nonbonded interactions. However, this conformation is destabilized relative to the *trans*

Table 20. Computed relative energies of the *cis* and *trans* conformations of glyoxal and acrolein

Molecule	E(*cis*)-E(*trans*) (kcal/mol)	Ref.
Glyoxal	2.99	202)
Glyoxal	5.02	203)
Glyoxal	5.90	204)
Glyoxal	5.40	199)
Acrolein	0.80	199)

conformation by severe steric repulsions between the methylene hydrogen atoms. These facts immediately give us an insight into the possible design of a stable *cis* 1,3-butadienic system. Specifically, we wish to design a system where steric effects are deemphasized and sigma nonbonded attraction is accentuated relative to the case of 1,3-butadiene. A typical system is shown below, where X is a second period homologue of carbon, *i. e.* silicon, or any other second period element.

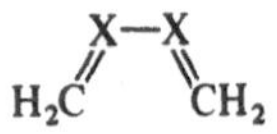

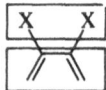

In comparing this system to 1,3-butadiene, one hopes that $CH_2...CH_2$ steric repulsions will be deemphasized due to the longer X—X bond while sigma nonbonded attraction will not be affected greatly due to the loss of $CH_2...CH_2$ overlap.

We shall now examine how substituents can modify the conformational preference of 1,3-butadiene. A typical case is that of butadiene bearing two substituents carrying lone pair electrons. We shall first consider the case of 2,3-disubstitution. In this case, the dissection employed is the one shown below:

The pi MO's of this system can then be constructed from the pi group MO's spanning the two X groups in a *cis* or *trans* conformation and the *cis* or *trans* pi MO manifolds of the butadienic fragment. These constructions are illustrated by means of the interaction diagram of Fig. 33. Proceeding as before we now compare the stabilizing interactions for the *cis* and *trans* conformations.

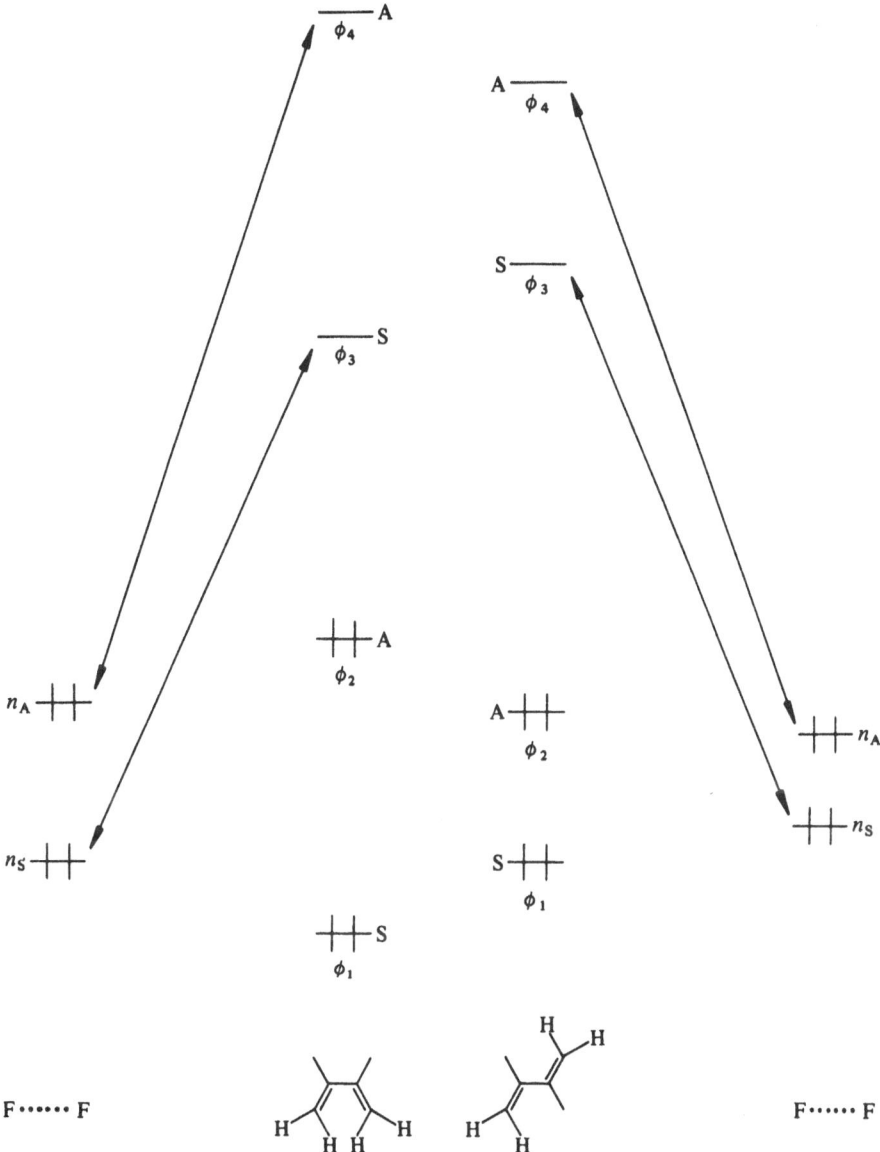

Fig. 33. Dominant stabilizing pi orbital interactions in *cis* and *trans* 2,3-difluorobutadiene. The symmetry labels are assigned with respect to a mirror plane (*cis* isomer) or a rotational axis (*trans* isomer)

Analysis of the two electron stabilizing interactions is complicated by the fact that in the transformation *cis* → *trans* the change in energy of the X----X group MO's and the butadienic pi MO's is not in the same direction. Consequently, the MO overlap integrals will play the key role in determining the relative stabilization of the two conformers. Specifically, the $n_S(cis)$ MO is lower in energy than the $n_S(trans)$ MO but in the transformation *cis* → *trans* the energy of ϕ_3 increases. Consequently, the energy gap between the n_S and ϕ_3 MO's will be comparable in both cases. However, the overlap integral $S_{n_S\phi_3}$ will be greater for the *cis* conformation than the *trans*. This arises because the antibonding interactions $Xp_z–C_4p_z$ and $Xp_z–C_1p_z$ are greater in the *trans* conformer as shown below:

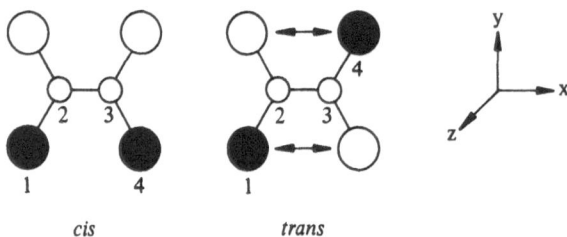

cis   trans

Hence, the *cis* conformer is stabilized more by the $n_S–\phi_3$ interaction than is the *trans* conformer.

Arguing as before, we see that the energy gap between the n_A and ϕ_4 will be comparable for both *cis* and *trans* conformers but $S_{n_A\phi_4}$ will be greater in the *trans* conformation due to a $X–C_4$ bonding interaction:

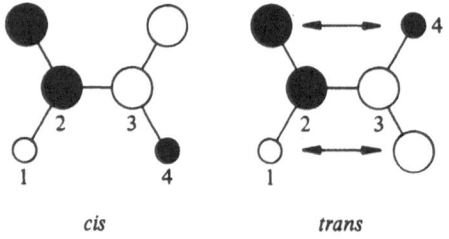

cis   trans

Consequently, the $n_A–\phi_4$ interaction stabilizes the *trans* conformation more than the *cis*.

The results of the above discussion are summarized below:

$$\Delta E^2_{n_S\phi_3}(cis) < \Delta E^2_{n_S\phi_3}(trans)$$

$$\Delta E^2_{n_A\phi_4}(cis) > \Delta E^2_{n_A\phi_4}(trans)$$

Once we realize that the $n_S–\phi_3$ interaction will tend to be stronger than the $n_A–\phi_4$ interaction, if we consider the proximity of the interacting levels, and weaker than the $n_A–\phi_4$ interaction, if we consider the AO coefficients at the points of substituent attachment, we conclude that these two interactions do not discriminate between the two isomers. In other words, the pi nonbonded inter-

106

actions introduced by the substituents will not play a fundamental role in determining the conformational preference of a 2,3-disubstituted diene.

Whether the $n_S-\phi_3$ or $n_A-\phi_4$ interaction dominates can be easily probed by means of calculations. We distinguish the following consequences:

a) If $n_S-\phi_3$ dominates, the pi F...F bond order should be negative and become increasingly negative in the transformation *trans →cis*.

b) If $n_A-\phi_4$ dominates, the pi F...F bond order should be positive and become increasingly positive in the transformation *cis → trans*.

Results of INDO calculations are shown below for *cis* and *trans* 2,3-difluoro-butadiene.

$p_{F,F}^{\pi}$	−.0012	.0011
$p_{F,F}^{\sigma}$	.0000	.0000
N_T^{π}	.6101	.6116

Computation: INDO, standard geometry

It is clear from the above data that neither the $n_S-\phi_3$ nor the $n_A-\phi_4$ interaction dominates overall in either geometry. It appears that the former dominates in the *cis* geometry and the latter in the *trans* geometry.

In summary, we expect that 2,3-substitution will not greatly alter the conformational preference of the diene system, *i. e.* conformational isomerism of 2,3-difluoro-butadiene will be subject to the same electronic factors as that of the unsubstituted diene and these factors operate in the same direction in both cases. It is expected that, by analogy to 1,3-butadiene, the order of conformer stability of 2,3-difluoro-butadiene will be dictated by steric effects, *i. e.* it will be *trans > gauche > cis*.

Experimental support for the above analysis of 2,3-disubstituted 1,3-dienes is available. For example, 2,3-dichlorobutadiene has been shown to exist preferentially in a planar *trans* geometry[205]. A more interesting case is the conformational isomerism exhibited by oxalyl bromide and chloride. It is found that in the gas phase, these molecules exist in an equilibrium mixture of *trans* and *gauche* isomers but no detectable concentration of the *cis* conformation was observed[206, 207]:

X = Cl, Br

The existence of a large concentration of the *gauche* conformation implies that pi nonbonded interactions of the type discussed previously play some role in the conformational isomerism of oxalyl bromide and chloride.

Ab initio calculations of 2,3-disubstituted dienes have predicted that the *trans* conformation is more stable than the *cis* conformation. Results of these calculations are shown in Table 21.

We will now consider an interesting case where sigma nonbonded interactions of the hydrogen bond type help to reinforce the conformational preference of sub-

Table 21. Computed relative energies of *cis* and *trans* diene systems

Molecule	E(cis)–E(trans) (kcal/mol)	Ref.
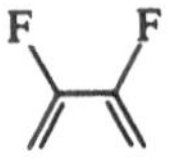 F F	2.88	199)
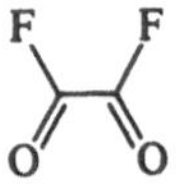 F F, O O	4.17	199)
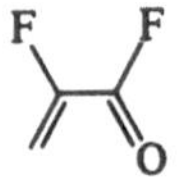 F F, O	0.14	199)

stituted 1,3-butadiene dictated by steric effects. For example, consider the systems shown below.

In a molecule of this type, *trans* conformational preference is not only due to effects inherent in any diene molecule but also to an attractive sigma nonbonded interaction between the Xp_x lone pair and the 1s AO of the hydrogen attached on C_3. Consequent-

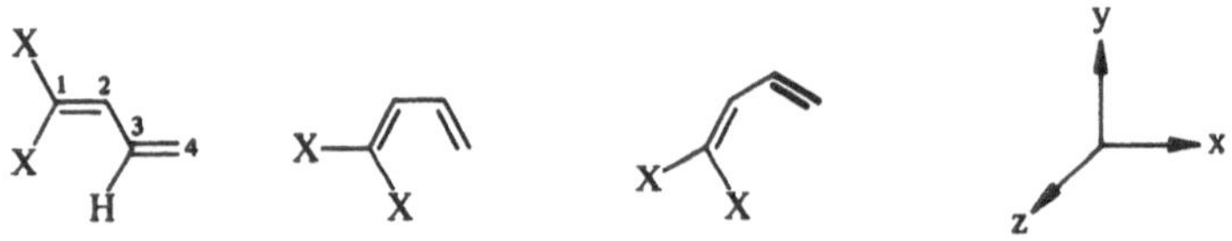

trans   *cis*   *gauche*

ly, it is expected that abolition of this attractive sigma nonbonded interaction by appropriate replacement of the hydrogen may force the molecule into a preferred conformation other than the *trans*. Of course, such a replacement will also bring into play additional steric effects destabilizing the *trans* conformation.

Typical experimental results for halo substituted 1,3-butadienes are shown in Table 22. It can be seen that when the crucial C_3 position is substituted the *gauche* conformer becomes the lowest energy conformation.

Finally, we wish to point out that a similar theoretical approach has been used by Hehre[199] in a discussion of 2,3-disubstituted butadienes and by Pople and Hehre in a work on C_4 molecules[214].

3.16. Nonbonded Interactions in Peptides and Polypeptides

The fundamental unit of proteins is the peptide linkage. An understanding of the electronic factors dictating conformational preference in the peptide linkage, therefore, can lead to further insight into polypeptide and protein torsional isomerism.

Table 22. Preferred conformations for substituted 1,3-butadienes

Molecule	Dihedral angle (θ)	Ref.
	180°	193, 208)
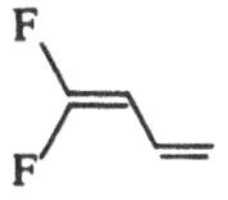	180°	209)
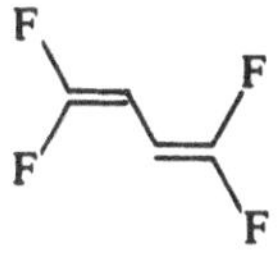	180°	210)
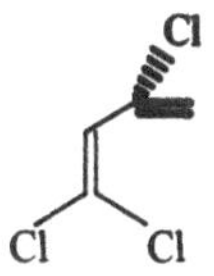	50°	211)
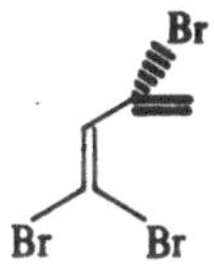	50°	211)
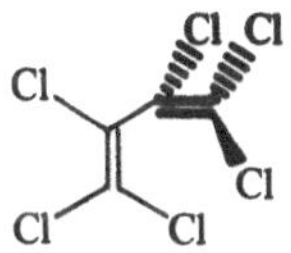	0° < θ < 90°	212)
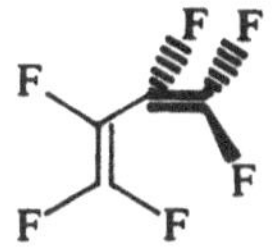	47°	213)

Our model for the peptide linkage is N-methyl formamide and the two conformations which we shall compare are the C_s and C_e conformations shown below.

As before, we can dissect the molecule into (CH_3————C=O) and (—NH—) fragments and construct the appropriate interaction diagrams which in this case are similar to that constructed for R'—X—R molecules (Fig. 30). Arguing as before, we conclude that the C_s conformation is the preferred conformation of N-methyl formamide.

109

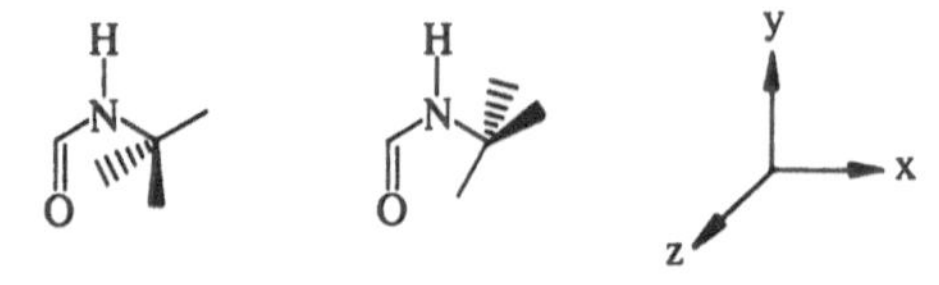

	C_s	C_e
N_T^π	.66301	.65982
$P_{(O--CH_3)}^\pi$	.06064	.00009

Computation: INDO, standard geometry

X-ray data suggest that the peptide linkage is planar in either small molecules or proteins themselves[215–220]. The planarity of the peptide linkage results into interesting structural patterns encountered in polypeptide molecules. Two important conformations of peptides are the α-helix and fully extended conformations[221]. Examination of the fully extended form of a polypeptide immediately reveals that this conformation enjoys considerable stabilization from the factors we have been discussing.

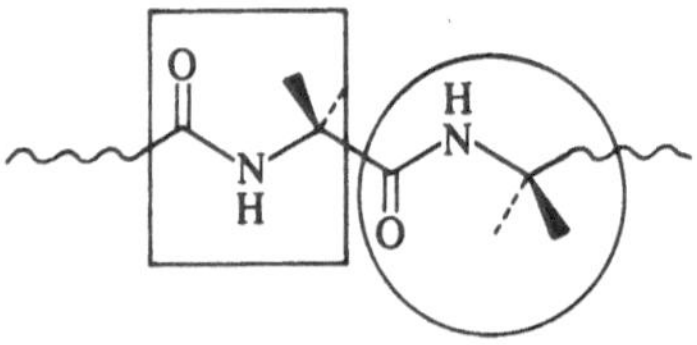

As can be seen above, the stabilization of the extended form arises from alternating Hückel pi aromatic structures as indicated by the square and circle.

In this connection, it is interesting to note that experimental evidence, including UV, IR and ORD spectroscopy, indicates that the fully extended form of oligoglycine and derivatives of polyglycine is more stable than the α-helix[222, 223]. In the crystalline state, polyglycine exists in two forms one of which is fully extended and the other nearly extended[224, 225].

Ab initio as well as semi-empirical calculations predict that the fully extended conformation of a simple polypeptide is the most stable conformation[226, 227]. INDO calculations of the C_s and C_e conformations of N-methyl formamide shown above clearly indicate that the greater stability of the C_s conformation is due to pi nonbonded attractive interactions.

3.17. Torsional Isomerism in Ring Systems

In the previous sections we have examined diverse acyclic molecular systems where sigma and/or pi nonbonded interactions obtain. The same considerations are applicable to cyclic and bicyclic systems. In the space below, we provide an overview of torsional isomerism in such systems.

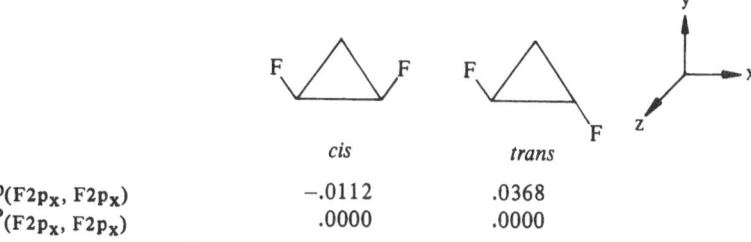

	cis	trans
$P(F2p_x, F2p_x)$	$-.0112$	$.0368$
$P_{(F2p_x, F2p_x)}$	$.0000$	$.0000$

Computation: CNDO/2, standard geometry

Sigma lone pair nonbonded interaction in 1,2-difluorocyclopropane presents a similar situation to that observed in 1,2-difluoroethane (see Pattern b Scheme 1). Specifically, sigma nonbonded attraction may favor the *trans* isomer. This can be seen in the results of CNDO/2 calculations shown above where it appears that the sigma nonbonded interaction is repulsive in the *cis* isomer.

A similar discussion can be offered for other 1,2-disubstituted cycloalkanes.

The boat form of molecules of the type shown below may involve sigma nonbonded attractive interactions between the X and Y groups.

Obviously, steric interactions disfavor the boat relative to the chair form. Nonetheless, systems can be designed where the "bow to bow" nonbonded attraction may become dominant. It is interesting to discuss in detail one example in order to illustrate our approach.

Let us consider the molecule shown below dissected in the manner indicated.

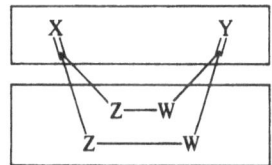

We first construct the group of MO's of X=C————C=Y as shown in Fig. 34. Obviously, when X and Y are nonidentical some stabilization resulting from the interaction is possible. Furthermore, coupling through the Z–W————Z–W system, as illustrated in Fig. 35, leads to sigma nonbonded attraction between the X=C and Y=C fragments favoring the boat form. This preference is maximized when the Z–W bond is characterized by a high energy HOMO.

Accordingly, we conclude that the best chance for obtaining boat preference in the system under consideration may materialize if X≠Y and Z and W are appropriately selected.

An interesting experimental result concerns the most stable conformation of 1,4-cyclohexadione. This molecule is found to exist preferentially in a skew boat conformation[228, 229]. This conformation may be a result of a compromise between nonbonded attraction, which is maximized in the boat conformation, and steric effects which favor the chair conformation. Obviously, this molecule does not

111

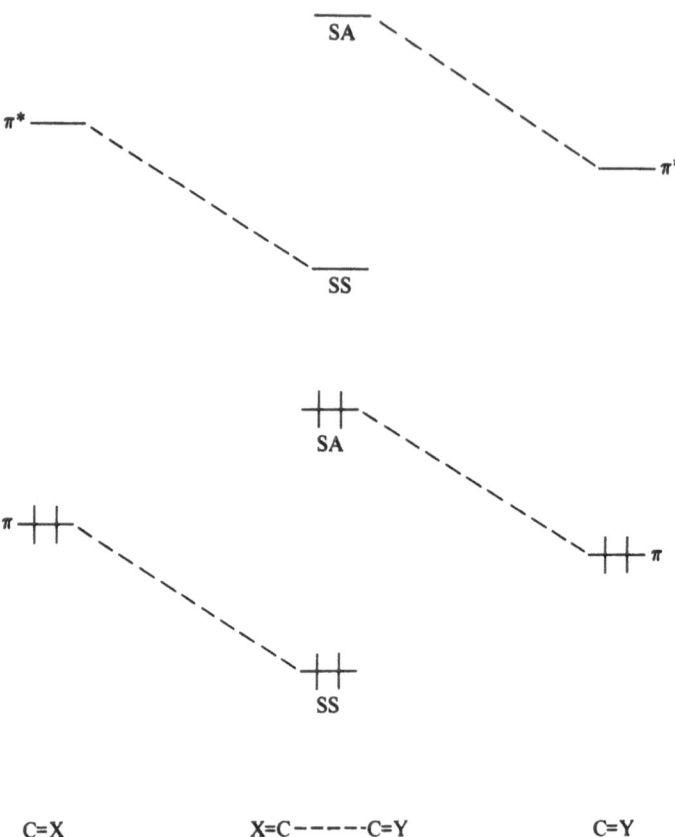

C=X   X=C-----C=Y   C=Y

Fig. 34. Pi type group MO's of the X=C---C=Y component system constructed from the pi MO's of CX and CY

represent the ultimate in nonbonded attraction, so the very fact that it exists in a "near" boat conformation provides us with hope that the strategy outlined above for the construction of a boat ring system may succeed.

Similarly, bicyclic systems of the type shown below enjoy sigma nonbonded attractive interactions in the *syn* form.

Experimental results indicate that steric effects dominate and the *anti* form is more stable by 2.1 kcal/mol[230]. However, by following the strategy illustrated for the cyclic system discussed above, we can design systems standing a good chance of being thermodynamically more stable in the *syn* form.

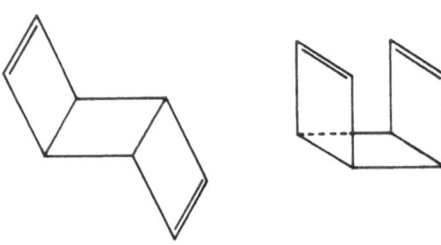

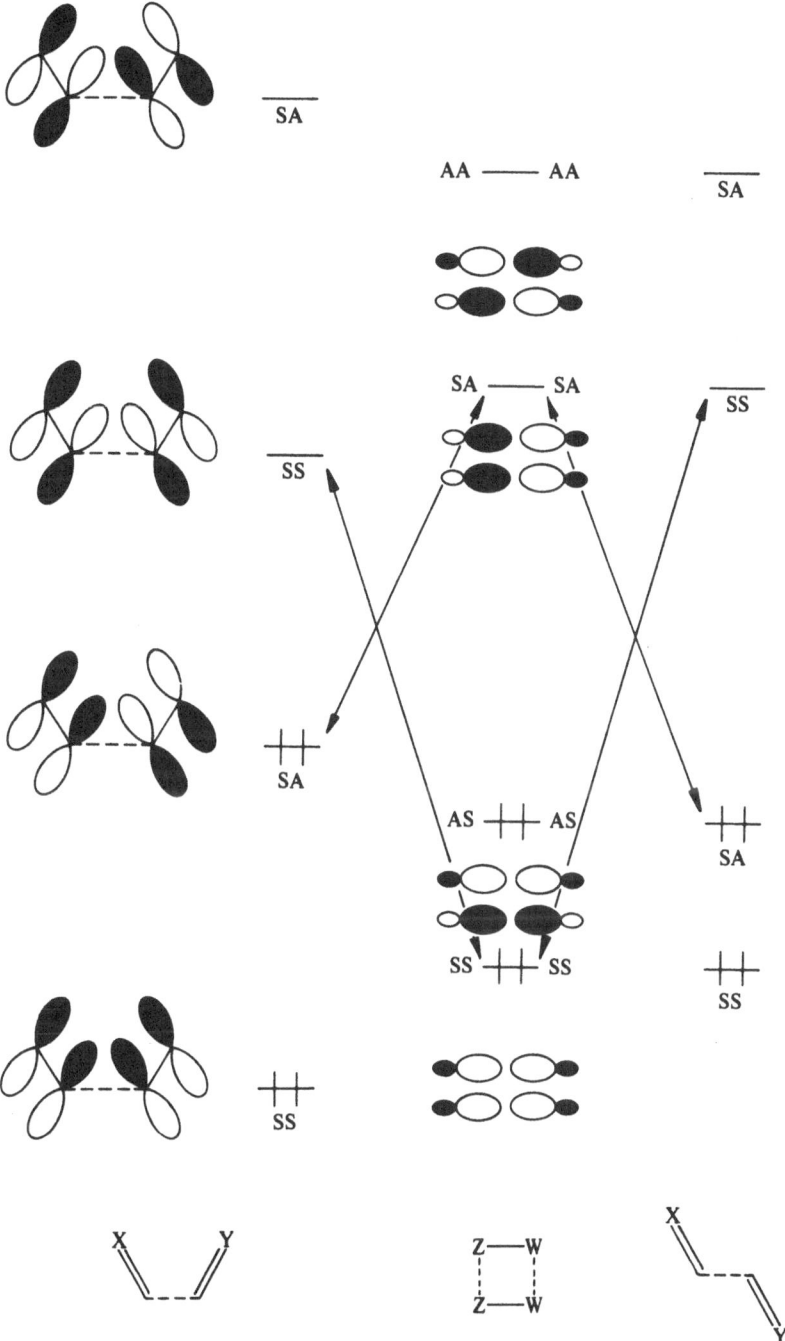

Fig. 35. Dominant stabilizing orbital interactions in the boat and chair conformers of

3.18. The Concept of the Isoconjugate Series

In our discussion of pi nonbonded attractive interactions we have sought to provide representative cases where these interactions obtain. We can further systematize the mass of experimental evidence regarding conformational preferences by seeking analogies among apparently unrelated molecules and geometries. One fruitful approach to this task involves the concept of the isoconjugate series exemplified below.

Consider the molecules *cis* 2-butene and *cis* 1, 3, 5-hexatriene. By realizing that a $-CH_3$ substituent contributes a π and a π^* type orbital, *i. e.*, that $-CH_3$ can be simulated by $H \diagdown C=X$, we can immediately anticipate that the preferred geometry of the two molecules will be the same, *i.e. trans.*

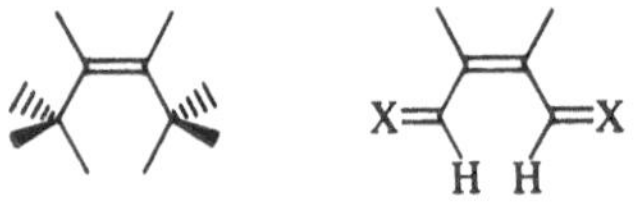

A further example is given by consideration of the preferred conformation of *anti* n-butane. This system is isoconjugate to *trans* 2,3-dimethyl-butadiene or *trans* 2,3-divinylbutadiene. Now, *trans* 2,3-dimethyl-butadiene is expected to have the conformation shown below.

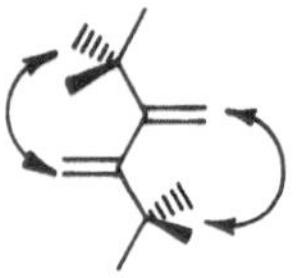

Accordingly, the predicted conformation of *anti* n-butane is the one shown below.

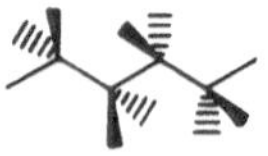

It would be interesting to test systematically the extent to which such analogical reasoning can be carried out. Some problems can arise in a predictable manner. For example, in the comparison of 2,3-dimethyl-butadiene and n-butane, the C_2-C_3 bond lengths are different. This has an impact upon the steric interaction of the attached groups which, in turn, may ultimately be responsible for the groups adopting a different conformation in the two isoconjugate systems.

3.19. Assorted Systems

We end the discussion of the structural effects of nonbonded interactions by noting that the ideas illustrated in the previous sections can also be applied to inorganic systems. For examples, octachlorodimolybdate(II), $Mo_2 Cl_8^{-4}$, is predicted to have an eclipsed geometry in order to maximize the nonbonded attractive interactions between the chlorine atoms:

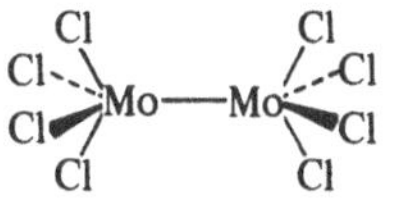

Recent SCF–XαSW calculations indicate that the above geometry is stabilized by attractive ligand-ligand nonbonded interactions[231, 232].

4. Tests of Nonbonded Interactions

In the following sections, further experimental tests of nonbonded interactions will be discussed. Specifically, we shall examine the following topics:
a) Physical Manifestations of Nonbonded Interactions.
b) Spectroscopic Probes of Nonbonded Interactions.
c) Reactivity Probes of Nonbonded Interactions.

4.1. Physical Manifestations of Nonbonded Interactions

In this section we shall determine the effect of nonbonded interactions on the physical properties of related molecules. As our model compounds we will utilize difluoroethylene, a molecule having substituents bearing only lone pairs and dicyano-ethylene, a molecule bearing substituents with both filled and unfilled MO's. In particular, we focus on the three possible isomers of a disubstituted ethylene *i.e.* the 1,1, *trans*-1,2- and *cis*-1,2-isomers. The three possible isomers are shown below.

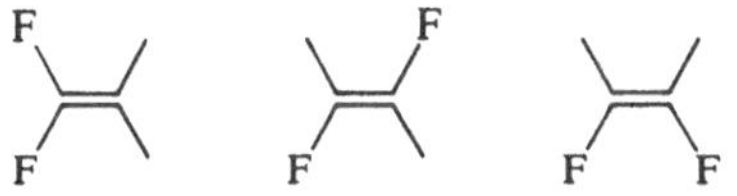

As was done previously, we can dissect the molecules into an F⋯F fragment and a HC=CH fragment. The appropriate interaction diagram for the three isomers of difluoroethylene is shown in Fig. 36. In constructing these diagrams, it is assumed that the splitting between the group MO's of the substituents is much larger in the 1,1 case than in the *cis* 1,2-case and almost zero in the *trans* 1,2 case. Typical AO overlap integrals are shown in Table 23 and support our assumption. A consideration of these diagrams leads to the following conclusions:
a) the interaction between the n_S MO and the ethylenic π MO increases in the order

trans 1,2 > *cis* 1,2 > 1,1

and, consequently, the HOMO energies of the composite molecules will follow the order:

$$\epsilon_{1,1}^{HO} < \epsilon_{cis}^{HO} < \epsilon_{trans}^{HO}$$

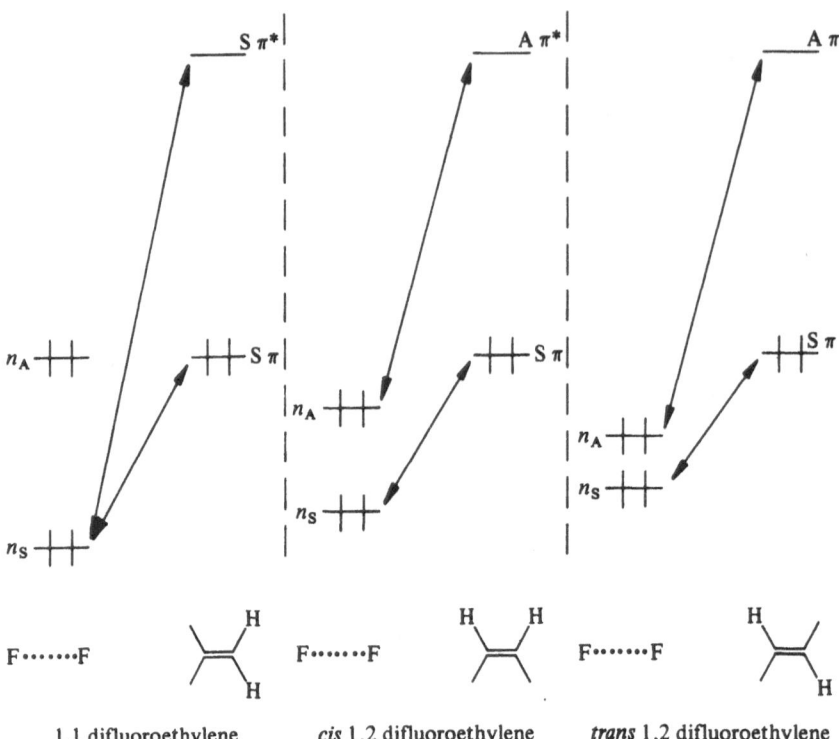

1,1 difluoroethylene   cis 1,2 difluoroethylene   trans 1,2 difluoroethylene

Fig. 36. Dominant pi orbital interactions in the three structural isomers of difluoroethylene. The symmetry labels are assigned with respect to a mirror plane (1,1 and *cis* isomers) or a rotational axis (*trans* isomer). The diagrams are schematic

Table 23. Pi overlap integrals between the substituents in the three possible isomers of disubstituted ethylenes[a]

X	1,2-*cis*	1,2-*trans*	1,1
F	0.0004	0.0001	0.0036
Cl	0.0048	0.0004	0.0123
NH_2	0.0041	0.0003	0.0117
OH	0.0016	0.0002	0.0048

[a] Overlap integrals calculated by a CNDO/2 program. Standard geometries.

b) the interaction between the n_A MO, in the *cis* and *trans* 1,2 isomers, and the n_S MO, in the case of the 1,1 isomer, with the ethylenic π^* MO follows the order

cis 1,2 > *trans* 1,2 > 1,1

and consequently, the energy of the LUMO in the composite molecules will follow the order:

$$\epsilon_{1,1}^{LU} < \epsilon_{trans}^{LU} < \epsilon_{cis}^{LU}$$

An analogous procedure can be applied to the dicyanoethylene. We can dissect the molecule into an NC···CN fragment and a HC=CH fragment. The appropriate interaction diagrams are shown in Fig. 37. A consideration of these diagrams leads to the following conclusions.

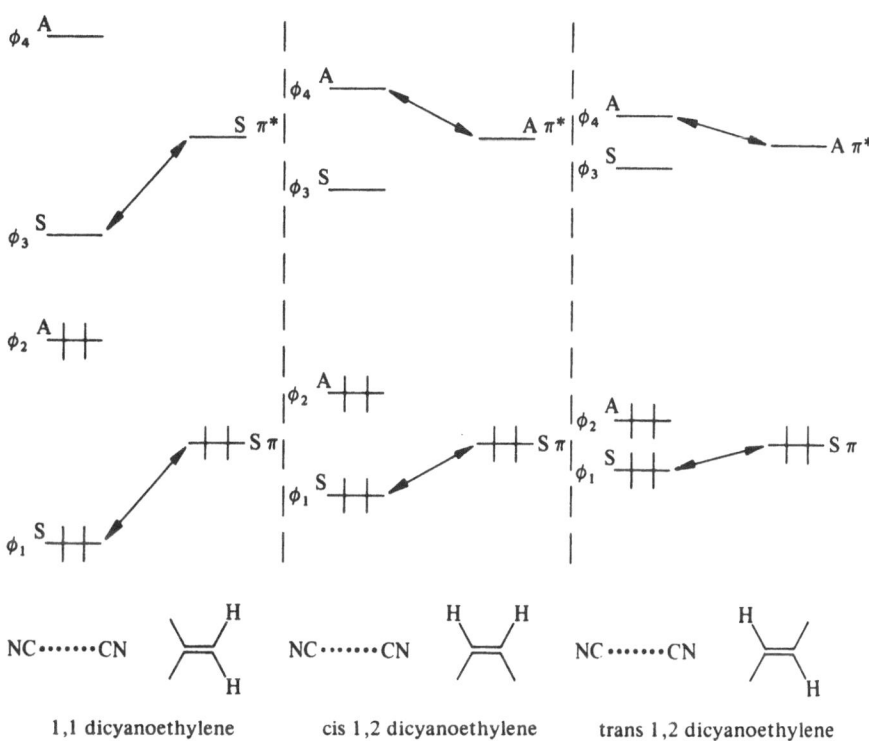

Fig. 37. Dominant pi orbital interaction in the three structural isomers of dicyanoethylene. The symmetry labels are assigned with respect to a mirror plane (1,1 and *cis* isomers) or a rotational axis (*trans* isomer). The diagrams are schematic

a) the $\phi_1 - \pi$ interaction strength varies in the order:

trans 1,2 > *cis* 1,2 > 1,1

Consequently, the energy of the HOMO in the composite systems follows the order:

$$\epsilon_{1,1}^{HO} < \epsilon_{cis}^{HO} < \epsilon_{trans}^{HO}$$

This is completely analogous to the situation in difluoroethylene.

b) the $\phi_4 - \pi^*$ interaction in both 1,2 isomers and the $\phi_3 - \pi^*$ interaction in the 1,1 isomer have the consequence that the 1,1 isomer will have a lower LUMO even

117

though the $\phi_3-\pi^*$ interaction will be smaller than the $\phi_4-\pi^*$ interaction present in the 1,2 isomers, *i.e.*, the inherently lower energy of ϕ_3 in the 1,1 isomer rather than the interaction effect produces the final result. Thus, the energy of the LUMO in the various isomers will follow the order

$$\epsilon_{1,1}^{LU} < \epsilon_{trans}^{LU} < \epsilon_{cis}^{LU}$$

An alternative approach to 1,2-disubstituted ethylenes is within the framework of open shell-open shell interactions. Specifically, the dissection used in this theoretical approach is shown below for the model system 1,2-difluoroethylene:

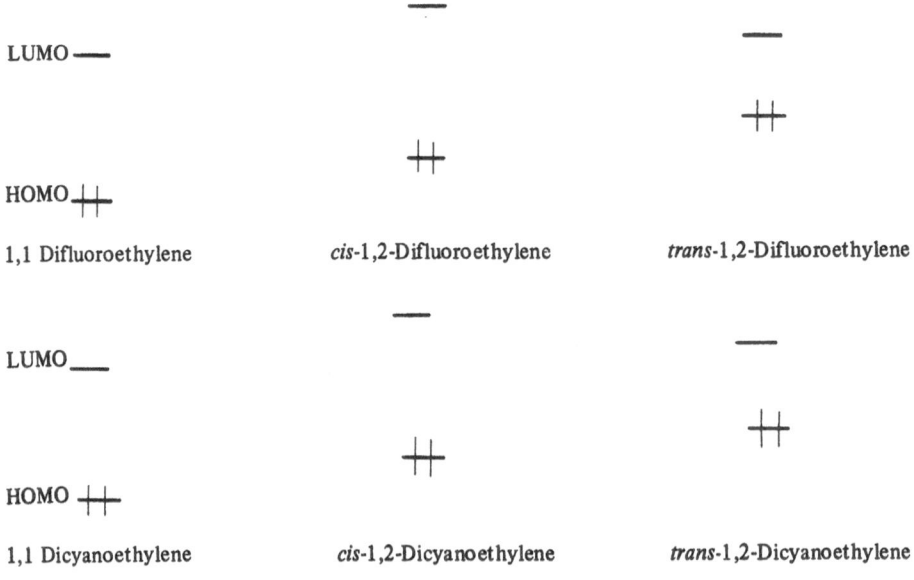

1,2−*trans*   1,2−*cis*

The interaction diagram for the *cis* and *trans* union of two FHC · radical fragments is shown in Fig. 38. From Fig. 38 it is obvious that the pi HOMO-LUMO splitting is larger for the case of *cis* union than for *trans* union. The final conclusions is that the HOMO and LUMO energies in *cis* and *trans* 1,2-difluoroethylene vary in the following way:

$$\epsilon_{cis}^{HO} < \epsilon_{trans}^{HO}$$
$$\epsilon_{trans}^{LU} < \epsilon_{cis}^{LU}$$

In summary, the relative HOMO and LUMO energies of the three isomers of difluoroethylene and dicyanoethylene are as shown below.

LUMO ——

HOMO ╫

1,1 Difluoroethylene   *cis*-1,2-Difluoroethylene   *trans*-1,2-Difluoroethylene

LUMO ——

HOMO ╫

1,1 Dicyanoethylene   *cis*-1,2-Dicyanoethylene   *trans*-1,2-Dicyanoethylene

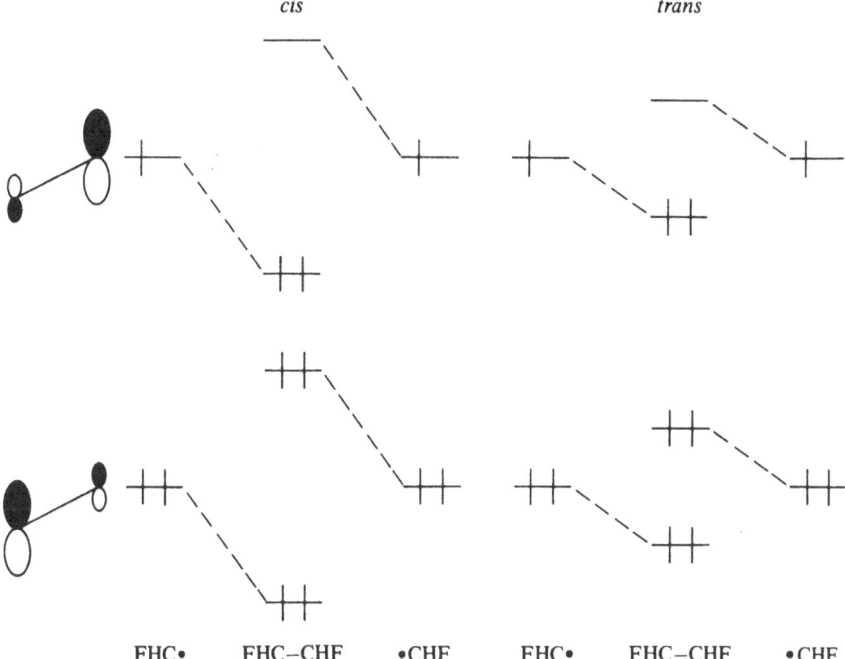

FHC•   FHC–CHF   •CHF   FHC•   FHC–CHF   •CHF

Fig. 38. The pi MO's of the *cis* and *trans* isomers of 1,2-difluoroethylene as constructed from the union of two CHF fragments in the appropriate geometry

We now examine the available experimental evidence in the light of the above discussion. Specifically, we shall consider physical properties such as (a) ionization potentials and (b) electron affinities.

A. Ionization Potentials

According to Koopmans' theorem[28], the first ionization potential of a disubstituted ethylene corresponds to the negative of the pi HOMO energy. From the previous discussion, the pi HOMO energy is predicted to vary in the following order:

$$\epsilon_{1,1}^{HO} < \epsilon_{cis}^{HO} < \epsilon_{trans}^{HO}$$

Hence, the predicted ionization potentials for the three molecules will vary in the following way:

$$IP(trans) < IP\ (cis) < IP(1,1)$$

It should be pointed out that the predicted higher ionization potential and lower electron affinity of a *cis* 1,2-disubstituted ethylene relative to the corresponding *trans* isomer is based upon the assumption that substantial pi nonbonded inter-

119

Table 24. Ionization potentials of disubstituted ethylenes

Molecule	I. P. (vertical) (eV)	I. P. (adiabatic) (eV)	Method	Ref.
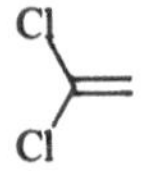 1,1-dichloroethylene (Cl, Cl geminal)	10.00	9.74 9.83	PES PES	233) 234)
 cis-1,2-dichloroethylene	9.93 9.83	9.68 9.65	PES PES	233) 234)
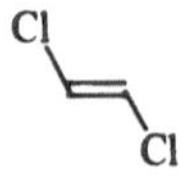 trans-1,2-dichloroethylene	9.93 9.81	9.69 9.64	PES PES	233) 234)
 cis-1,2-dibromoethylene (Br, Br)	9.69	9.45 9.46	EI, PI PI	235) 236)
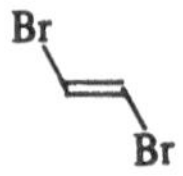 trans-1,2-dibromoethylene	9.54	9.46 9.46	EI, PI PI	235) 236)
 cis-1-chloro-2-fluoroethylene (Cl, F)	10.30	9.87	EI, PI	235)
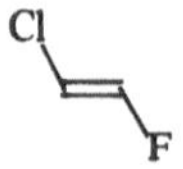 trans-1-chloro-2-fluoroethylene	10.30	9.87	EI, PI	235)
 H_3CO—CH=CH—OCH_3		7.97	PES	237)
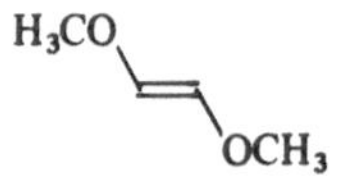 H_3CO—CH=CH—OCH_3 (trans)		8.04	PES	237)
 H_3C—CH=CH—CH_3 (cis)		9.13	PES	238)
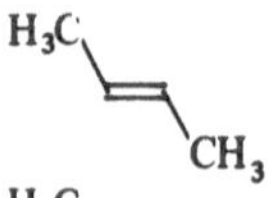 H_3C—CH=CH—CH_3 (trans)		9.13	PES	239)
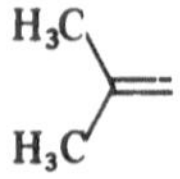 $(H_3C)_2C$=CH$_2$		9.23	PES	238)
 NC—CH=CH—CN (cis)		11.15	PES	240)
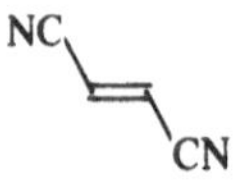 NC—CH=CH—CN (trans)		11.15	PES	240)

action between the two substituents obtains in the *cis* isomer. However, it is possible that conformational effects dictate otherwise as, for example, in the cases of 1,2-dialkylethylenes, 1-substituted propenes, 1,3,5-hexatrienes, etc. where the preferred conformation of the *cis* isomer features sigma nonbonded attraction and weak pi nonbonded interaction. In such molecules, the difference in the ionization potential of the *cis* and *trans* isomers will be expected to be very small, perhaps negligible.

Ionization Potentials (IP) for various disubstituted ethylenes are shown in Table 24. As can be seen, the IP's for 1,2 *cis* and *trans* isomers are usually very close but the IP's of 1,1-disubstituted ethylenes relative to their 1,2-disubstituted counterparts are consistently higher, in agreement with our predictions.

We now consider the effect of nonbonded interaction on the energies of pi lone pair MO's in *cis* and *trans* disubstituted ethylenes. Our model systems are again *cis* and *trans* 1,2-difluoroethylene. The consequence of the interaction of the pi lone pair electrons in the *cis* and *trans* geometries is shown in Fig. 36. We distinguish the following three hypothetical cases and work out their consequences.

a) Through space interaction is appreciable in the *cis* isomer but negligible in the *trans* isomer and through bond interaction is negligible in both isomers. This would lead to the orbital pattern shown below:

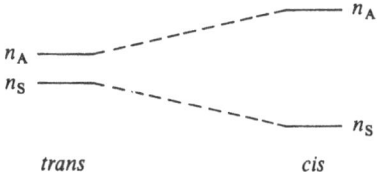

b) Through space interaction is negligible in both *cis* or *trans* isomers but through bond interaction is important in both isomers. This would lead to the orbital pattern shown below:

n_A — — — — n_A

n_S $\overline{trans}$ — — $\overline{cis}$ n_S

c) Through space interaction is appreciable in the *cis* isomer, but negligible in the *trans* isomer and through bond interaction is appreciable in both isomers. In this case n_S will attain a lower energy in the case of the *trans* isomer. This latter result would be compatible only with through space *and* through bond coupling and would constitute evidence in support of our proposal that overlap repulsion is greater in the *trans* isomer. On the other hand, n_A may or may not have lower energy in the *cis* isomer relative to the *trans* depending upon the extent to which the effect of through space interaction is counteracted by the effect of through bond interaction. The appropriate orbital pattern is shown below

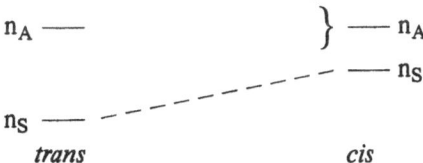

The energy ordering of the lone pair pi MO's of *cis* and *trans*, 1,2-dichloroethylene as determined by photoelectron spectroscopy is shown below:[234]

$$n_A \underline{\quad} \quad \overline{\quad} n_A$$
$$-11.93eV \qquad -11.85eV$$

$$n_S \underline{\quad} \quad \overline{\quad} n_S$$
$$-12.61eV \qquad -12.51eV$$

$$\text{\textit{trans}} \qquad\qquad \text{\textit{cis}}$$

As can be seen, the n_S MO of the *trans* isomer is lower in energy than the n_S MO of the *cis* isomer, a result consistent only with the presence of strong through space and through bond interaction of the two Cl atoms. This is an important result insofar as it indicates that four electron overlap repulsion is greater for *trans* than for cis 1,2-disubstituted ethylenes.

B. Electron Affinities

Linear correlations between the energy of the LUMO of a molecule and its electron affinity are known[39]. Since the energy of the LUMO of the disubstituted ethylenes varies, according to our model, in the order, $\epsilon_{1,1} < \epsilon_n < \epsilon_{trans} < \epsilon_{cis}$ we expect that the corresponding molecular electron affinities will follow the order:

$$A_{1,1} > A_{trans} > A_{cis}$$

Experimental electron affinity data of disubstituted ethylenes is not abundant but Table 25 lists the A's for some typical examples. As can be seen, the A of the 1,1-isomer is larger than the A's for the 1,2-isomers, in accord with our prediction.

Table 25. Electron affinities of disubstituted ethylenes

Substituent	Electron affinity (eV)			Ref.
	cis	trans	1,1	
CN, CN	0.78	0.78	1.54	34)
	0.78	0.78	–	241)
CO_2CH_3, CO_2CH_3	0.60	0.60	–	242)

An indirect probe of the A's for a series of compounds is obtained from a consideration of the half wave potentials. The reduction potential is related to the electron affinity by the following equation:[243]

$$E^{\circ} = A + \Delta E_{solv} - E^{\circ}_{ref} - X_{Hg}$$

where $E°$ is the cathodic one electron reduction potential, X_{Hg} is the electron work function of the mercury surface, $E°_{ref}$ is the potential of the reference electrode, and ΔE_{solv} is the solvation energy. Since $E°_{ref}$ and X_{Hg} can be considered constant for a series of molecules and ΔE_{solv} is nearly constant for closely related molecules, we expect that the half wave reduction potential will reflect the EA of the molecule. Consequently, we predict that $E°$ for the disubstituted ethylenes will vary in the following way:

$$E°_{1,1} > E°_{trans} > E°_{cis}$$

The above predictions are partially confirmed by the available electrochemical data. Specifically, the half wave reduction potentials for *cis* and *trans* 1,2-disubstituted ethylenes have been measured and it is found that it is easier to reduce the *trans* isomer, *i.e.* $E°_{trans} > E°_{cis}$ [243−248].

4.2. Spectroscopic Probes of Nonbonded Interactions

The spectroscopic probes to be examined in this section involve: (a) ultraviolet spectroscopy and (b) vibrational spectroscopy.

A. Ultraviolet Spectroscopy

We first consider the relative electronic transition energies in *cis* and *trans* 1,2-disubstituted ethylenes. From Fig. 28 we can clearly see that the pi HOMO-LUMO energy gap is larger for the case of the *cis* isomer relative to the *trans* isomer. Hence, the $\pi\pi^*$ transition is expected to occur at shorter wavelengths in *cis* 1,2-disubstituted ethylenes.

UV data shown in Table 26 clearly support our expectations. For example, the $\pi\pi^*$ transition energy for *cis* 1,2-dichloroethylene is 42300 cm^{-1} while the same quantity is 41700 cm^{-1} in the corresponding *trans* molecule. The same trend is observed in the 1,2-difluoroethylenes, as well as in the 1,2-dialkylethylenes, where a distinct hypsochromic shift is observed in the comparison of *trans* and *cis* isomers.

An interesting example is found in the $n\pi^*$ transition of *ortho* and *para* benzoquinone. The two molecules are dissected as shown below:

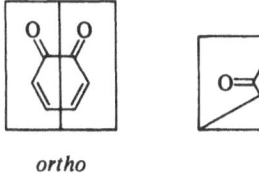

| *ortho* | *para* |

The relative energies of the π LUMO's in the two cases can be deduced by evaluating the overlap integral $S_{\psi_3 \psi_3'}$ as shown schematically below:

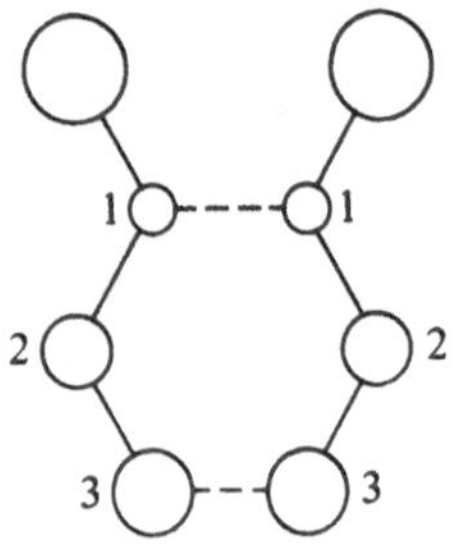

Part II. Nonbonded Interactions

ortho 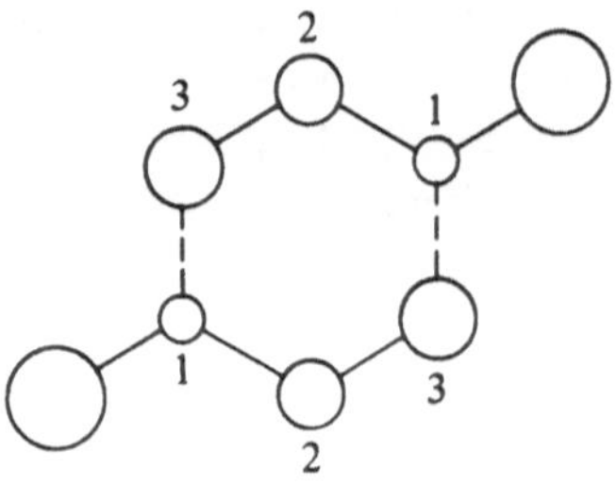   para

Table 26. Ultraviolet spectral data for *cis* and *trans* 1,2-disubstituted ethylenes

Molecule	$E_{\pi\pi}*(cm^{-1})$	Ref.
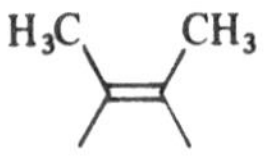 H$_3$C CH$_3$ (alkene)	57471	249)
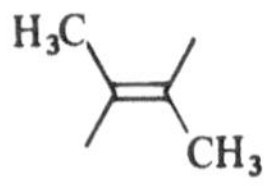 H$_3$C CH$_3$ (alkene)	56179	249)
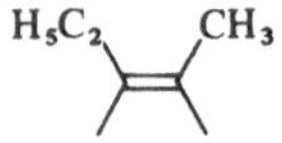 H$_5$C$_2$ CH$_3$ (alkene)	56401	249)
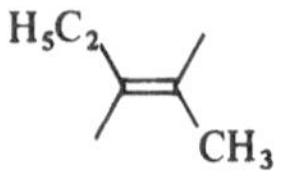 H$_5$C$_2$ CH$_3$ (alkene)	55188	249)
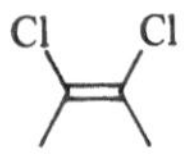 Cl Cl (alkene)	42300	251)
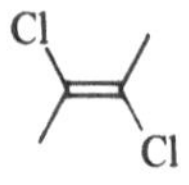 Cl Cl (alkene)	41700	251)
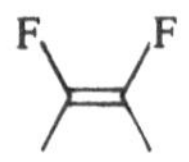 F F (alkene)	63003	250)
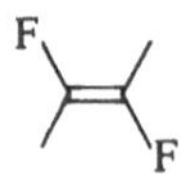 F F (alkene)	58728	250)

124

The overlap integral $S_{\psi_3 \psi_3'}$ for the two cases is given by the following expressions

$$S_{\psi_3 \psi_3'} \ (ortho) \ \alpha \ a^2 S_{11} + b^2 S_{33}$$
$$S_{\psi_3 \psi_3'} \ (para) \ \alpha \ abS_{13} + abS_{31}$$

where S_{ij} is the 2p AO overlap integral between carbon centers and a and b are the 2p AO coefficients of C_1 and C_3, respectively. Assuming that $S_{11} = S_{33} = S_{13} = S_{31} = K$, the above equations become:

$$S_{\psi_3 \psi_3'}(ortho) = (a^2 + b^2)(K)$$
$$S_{\psi_3 \psi_3'}(para) \ = (2ab)(K)$$

Since $a^2 + b^2 > 2ab$ for all positive values of a and b, except $a = b \geq 0$, we conclude that $S_{\psi_3 \psi_3'} \ (ortho) > S_{\psi_3 \psi_3'} \ (para)$. Consequently, the pi LUMO of *ortho* benzoquinone will be lower in energy than the pi LUMO of *para* benzoquinone. Furthermore, the degeneracy of the oxygen lone pair AO is lifted more in the case of *ortho* benzoquinone. The final conclusion is that the $n\pi^*$ transition is predicted to occur at longer wavelengths in the *ortho* isomer relative to the *para* isomer.

The experimental λ_{max}'s for the $n\pi^*$ transition of various quinones are shown below and are in accord with our expectations:[25]

| 610 nm | 450 nm | 540 nm | 420 nm |

B. Vibrational Spectroscopy

Raman spectroscopy and infrared spectroscopy are important tools to test qualitative theories since they can give us an idea about the relative strengths of bonds[252]. An interesting vibrational mode of 1,2-disubstituted ethylenic molecules involves torsion about the C=C bond. In the *trans* isomer, the torsional frequency is dependent only on the "stiffness" of the double bond. In the *cis* isomer, however, there is the additional factor of a nonbonded attractive interaction between the substituent groups which will decrease as the substituents move away from each other. Clearly, this is an unfavorable process for the molecule and the energy for "twisting" should be greater for the *cis* isomer relative to the *trans* isomer where a decrease in stabilizing nonbonded interactions do not obtain. Hence, *cis*-1,2-disubstituted ethylenes should have a higher C=C torsional frequency than the corresponding *trans* isomers.

Recent Raman studies of *cis* and *trans* 2-butenes show that the torsional frequency of the *cis* isomer is greater than that of the *trans* ($cis = 394$ cm^{-1}, $trans = 294$ cm^{-1}) which is in agreement with our expectations[253, 254]. Furthermore, IR and Raman studies of the *cis* and *trans* isomers of 1,2-difluoroethylene and 1-fluro-2-chloro-ethylene revealed larger force constants for the torsion around the double bond in the *cis* isomer[255, 108].

4.3. Reactivity Probes of Nonbonded Interactions

Relative reactivities of molecules can also shed light upon the validity of our general theoretical approach. In this section we shall examine the following reactivity probes of nonbonded interaction:

A. Diels Alder Reactivity

The Diels Alder reaction has been an important testing ground for qualitative theories for many years[256, 257]. In this section we consider the importance of nonbonded attractive interactions on Diels Alder reactivity. Our model reaction system is butadiene and the three isomers of dicyanoethylene:

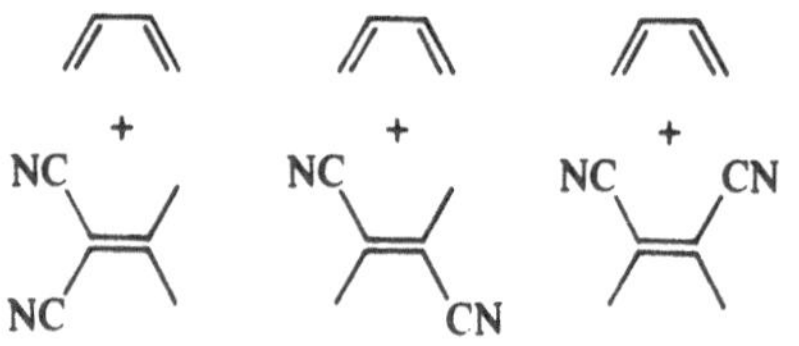

The key stabilizing interaction in the above reactions is between the pi HOMO of butadiene and the pi LUMO of the isomeric dicyanoethylenes. The pi LUMO's of the isomeric dicyanothylenes vary in the order $1,1 < 1,2$-*trans* $< 1,2$-*cis*. Accordingly, on the basis of Eq. (1), we conclude that the stabilization of the reaction complex and, consequently, the rate of the reaction will vary in the order $1,1 < 1,2$-*trans* $< 1,2$-*cis*.

Experimental evidence supporting the above prediction can be found in Table 27. In both examples, the 1,1 isomer reacts much faster than either the 1,2-*cis* or *trans* isomers which exhibit comparable reactivity.

B. The Stereochemistry of the S_N2' Reaction

The importance of nonbonded interactions in chemical systems is illustrated by the stereochemistry of the abnormal bimolecular substitution reaction (S_N2'). It has been suggested that the nucleophile attacks the same side from which the leaving group departs[259, 260]. However, the mechanism of the reaction is still unclear[261–263] and an investigation of the reaction utilizing a theoretical approach can be useful in shedding light on the factors which determine the preferred stereochemistry.

We illustrate our theoretical approach by reference to the two model transition states shown below:

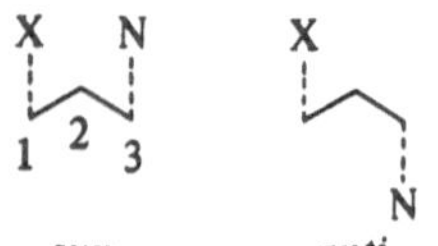

Table 27. Relative rates of Diels Alder addition for dicyanoethylenes[a]

		CH_3
	⬠	⬡⬡⬡ (with CH₃ above and below)
		CH_3

		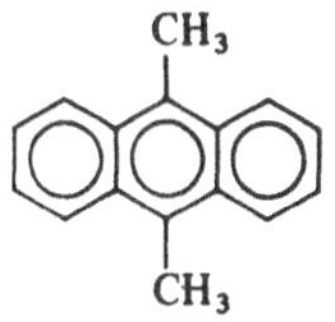
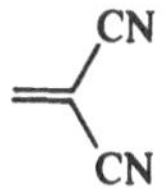 CN ⟋⟍ CN	45,000	127,000
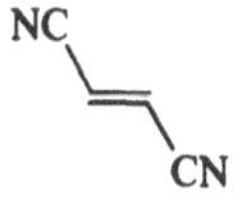 NC ⟍⟋ CN	81	139
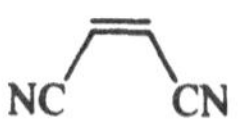 NC⟍⟋CN	91	131

[a] Ref.[258].

Here X is the leaving group, N is the nucleophile and the C_3---N and C_1---X bond distances are taken to be equal.

The pi MO's of the reaction complex can be constructed from the group MO's spanning the two p_z lone pair AO's of X and N and the pi MO's of the allyl cation. First the X---N group orbitals are constructed for the stereochemical modes of reaction in the usual manner. In the case of *syn* attack, the X and N $2p_z$ AO's overlap and their through space interaction becomes appreciable. In anti attack, overlap is nearly zero and the lone pair MO's remain unaffected. The next step in the analysis is the interaction of the lone pair MO's with the pi MO's of the allyl cation. This can be understood by means of the interaction diagram of Fig. 39. Proceeding as before, the stabilizing interactions can be discussed as follows:

a) A two electron stabilizing interaction between n_A and ϕ_2. In this case, the overlap integral $S_{n_A \phi_2}$ is larger in the case of *syn* compared to *anti* attack. Furthermore, $\epsilon_{n_A} - \epsilon_{\phi_2}$ is smaller and $(k - \epsilon_{n_A})^2$ is larger for the syn transition state and, hence, on the basis of Eq. (1'), the two electron stabilization will be greater for the *syn* than the *anti* case.

b) A two electron stabilizing interaction between n_S and ϕ_3. By going through the same arguments as before, we conclude that this two electron stabilizations will favor *anti* attack. However, due to the large energy gap between the interacting orbitals we expect that the contribution of the $n_S - \phi_3$ interaction to the total two electron stabilization energy will be small. The results of the above discussion are summarized below:

$$\Delta E^2_{n_A \phi_2} (syn) < \Delta E^2_{n_A \phi_2} (anti)$$

$$\Delta E^2_{n_S \phi_3} (anti) < \Delta E^2_{n_S \phi_3} (syn)$$

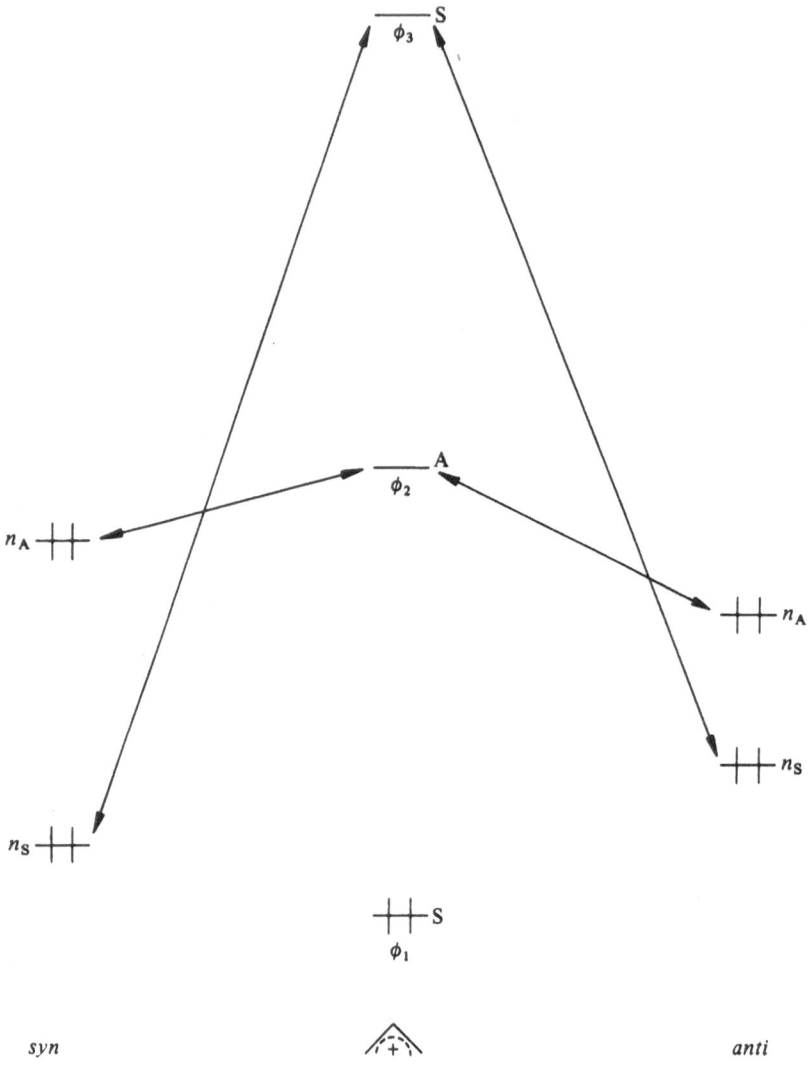

Fig. 39. Stabilizing orbital interactions involved in the *syn* and *anti* transition states of the S_N2' reaction

The quantity $(\Delta E^2_{n_A\phi_2} + \Delta E^2_{n_S\phi_3})$ will be more negative (*i.e.*, more stabilizing) for *syn* than *anti* attack since the dominant contributor to the stabilization energy is the $n_A - \phi_2$ interaction. It is clear, therefore, that nucleophilic attack *syn* to the leaving group will be favored over *anti* attack.

An additional factor in determining the stereochemistry of the S_N2' reaction is whether the attacking nucleophile is charged or uncharged. That is, the charge distributions in the transition state complex will differ depending on the nature of the nucleophile as illustrated below for the case of *syn* attack, assuming X is more electronegative than N.

Neutral nucleophile   Charged nucleophile

For neutral nucleophiles the electrostatic interaction between N and X will stabilize the *syn* more than the *anti* geometry due to the proximity of the two oppositely charged groups while the opposite will obtain when N is charged.

In summary, there are two principal factors which determine the stereochemistry of an $S_N 2'$ reaction:

a) an attractive nonbonded interaction factor and

b) an electrostatic interaction factor.

For convenience, we tabulate the direction of these factors for the two classes of nucleophiles below:

	Neutral	Charged
Nonbonded interaction	*syn > anti*	*syn > anti*
Electrostatic interaction	*syn > anti*	*anti > syn*

Ab initio calculations (STO-4G basis set) as well as semiempirical computations (CNDO/2) confirm the above analysis[7]. Specifically, it is found that a *syn* geometry is favored when the attacking nucleophile is neutral and this was shown to result from an increase in nonbonded attraction in the transformation *anti → syn* augmented by electrostatic interactions. For charged nucleophiles it was concluded that nonbonded attraction, *although present,* is dominated by electrostatic effects and, consequently, the anti geometry is lower in energy than the model *syn* transition state.

Part III. Geminal Interactions

5. Definitions

In the previous sections, we focused our attention upon nonbonded interactions and their effect upon the relative stability of conformational and geometric isomers. We shall now use OEMO theory in order to examine how geminal lone pair-lone pair, bond pair-lone pair and bond pair-bond pair interactions can affect the shape of a molecule[264].

Some clarifying statement as to the meaning of the term "geminal interaction" is in order. As an illustrative case, we consider the linear water molecule and inquire as to whether it will tend to be linear or bent. We can approach this latter problem by inquiring about the effect of the geminal interaction between the $2p_y$ lone pair and the two O—H bonds. In MO theoretical terms, we shall inquire about the effect of the interaction between the $2p_y$ lone pair AO and the sigma MO's which obtains upon bending. If this interaction is stabilizing the molecule will tend to bend and vice versa.

5.1. Theory of Lone Pair-Sigma Bond Geminal Interactions

We shall first illustrate how the OEMO approach can be used to elucidate the effect of geminal interactions on the stability and shape of molecules. We shall utilize the molecule H_2O as the model system and compare the relative stabilization of the linear and bent geometries. The analysis involves the construction of the sigma MO's of H_2O from the AO's of O and the group orbitals spanning the two hydrogens. The interaction diagrams of Fig. 40 show the various orbital interactions which obtain in the two geometries. However, before we make a one-to-one comparison of these interactions, a few remarks regarding the effect of bending upon the hydrogen group orbitals is necessary.

Upon bending, the two hydrogens of H_2O come into greater proximity and according to Eq. (3'), this is expected to lower the energy of the σ orbital and raise the energy of the σ^* orbital. Simultaneously, the normalization constant of σ decreases and that of σ^* increases. This latter effect will have definite consequences with respect to the degree of spatial overlap between the hydrogen group orbitals and the AO's of the central atom as a function of geometry.

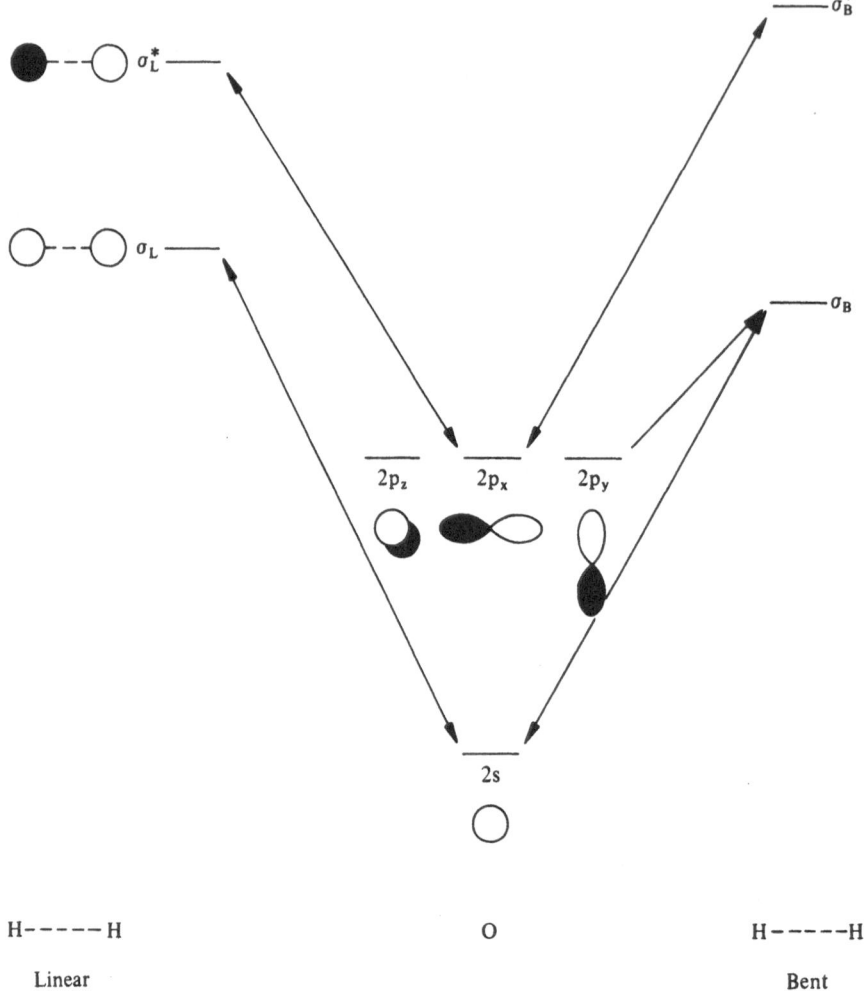

Fig. 40. The interactions between the H---H group MO's and the oxygen AO's in linear and bent H_2O. Notice that the $2p_y-\sigma_B$ interaction is "turned on" as bending occurs

We shall now compare the various MO interactions in the two geometries. The following trends become immediately apparent:

a) The $2s-\sigma_B$ interaction is favored relative to the $2s-\sigma_L$ interaction by the energy gap.

b) The $2p_x - \sigma_L^*$ interaction is favored relative to the $2p_x - \sigma_B^*$ interaction by the energy gap term as well as the overlap integral term. We conclude that the $2s-\sigma$ and $2p_x - \sigma^*$ interactions vary in opposite directions in the two geometries and, hence, are not expected to give rise to a preference for the linear or bent form[g].

_____

[g] Actually, the $2p_x-\sigma^*$ interaction changes significantly and, consequently, AH_2 molecules with four valence electrons are linear.

c) The $2p_y-\sigma$ interaction can only occur in the bent form and the resulting stabilization will give rise to a preference for the bent geometry.

By reference to Fig. 40, we can conclude that the incremental change of the stabilization energy upon bending, SE, is given by the following equations, for two molecules A and B.

$$SE_A = -\Delta E_A^2 \tag{22}$$

$$SE_B = -\Delta E_B^2 \tag{23}$$

where the quantities ΔE_A^2 and ΔE_B^2 are calculated at a fixed angle θ.

On the basis of the above discussion, we can define the following quantity:

$$F_\theta = \frac{-k\,H_{mn}^2}{\epsilon_m - \epsilon_n} \tag{24}$$

In the equation above, F_θ is termed the bending tendency, m and n are the two orbitals which interact only in the bent form and k is the occupation number. For a series of molecules, H_{mn} and $\epsilon_m - \epsilon_n$ are evaluated at a constant angle θ. Note that in the above equation the orbitals m and n are group MO's.

The quantities SE and F_θ have the meaning of a slope. Thus, by saying that F_θ for a certain molecule A is larger than F_θ for another molecule B we mean that the rate of change of the stabilization energy upon bending is larger for A and, thus, the equilibrium geometry is reached at a smaller angle. These basic assumptions of our model are illustrated by the diagram of Fig. 41. The derivative $\dfrac{d\,E}{d\,\theta}$ becomes zero at an increasingly smaller θ as F_θ increases.

We shall now examine what happens when the two hydrogens become artificially more electronegative. Fig. 42 shows that, as a result of the increased electronegativity, the σ and $\sigma*$ MO's decrease in energy and the interaction matrix element $<2p_y\,|\,H\,|\,\sigma>$ becomes more negative. This will tend to increase the stabilization energy which obtains upon bending due to the $2p_y-\sigma$ interaction up to the point that σ has become degenerate with $2p_y$. A further increase of hydrogen electronegativity will have a two fold effect, i.e. the $2p_y-\sigma$ energy gap will now begin to increase while the interaction matrix element will continue to become more negative. Accordingly, we may be inclined to think that these antagonistic effects will level the stabilization due to $2p_y-\sigma$ interaction which occurs upon bending. However, this is not an appropriate conclusion for the very simple reason that the evaluation of the stabilization energy has been carried out with respect to two different zero order electronic configurations. Thus, as long as σ remains above $2p_y$ the stabilization energy is evaluated with respect to an electronic configuration of the type $2s^2\,2p_x^2\,2p_y^2$ while as long as σ stays below $2p_y$ it is evaluated with respect to, e.g. $2s^2\,\sigma^2\,2p_x^2$, i.e. the zero point for our comparisons is not the same in the two different cases. This is a very crucial point because we are primarily interested in developing a theory of substituent effects on molecular shape. Thus, we shall return to this problem in a later section when a different formulation will allow us to escape the ambiguity of defining the magnitude of the stabilization energy upon bending as a function of substituent electronegativity.

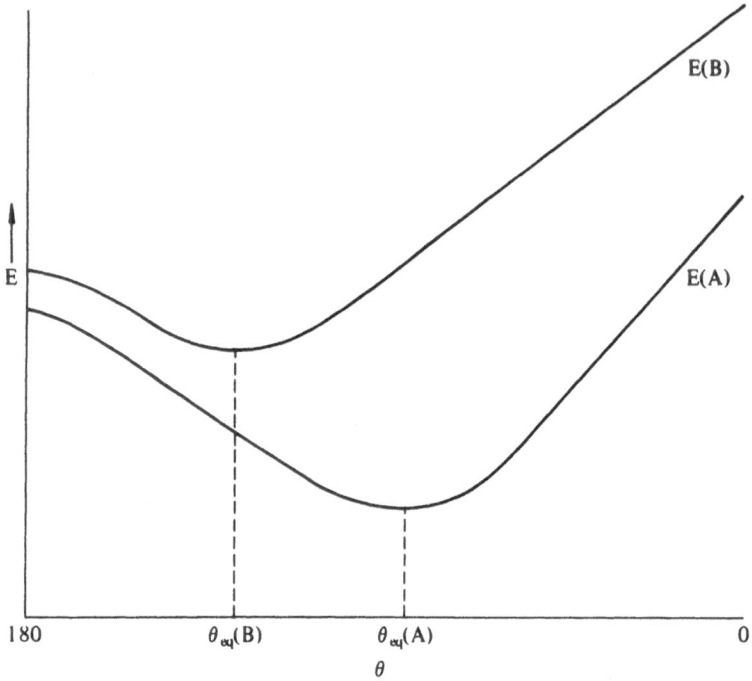

Fig. 41. The variation of the total energy E $F_\theta(A) > F_\theta(B)$ due to a dominant orbital interaction turned on by bending as a function of the angle θ

The electron occupancy of the lone pair AO which can interact with the σ hydrogen MO upon bending in AX_2 molecules is crucial for molecular geometry. Thus, F_θ is maximal when the lone pair AO is occupied by two electrons, *i.e.* $k = 2$, and becomes zero when no electrons are contained in this AO, *i.e.* $k = 0$. Accordingly, we formulate the following rule: as the electron occupancy of a lone pair orbital which can interact with an unoccupied orbital upon bending is depleted, the tendency for angle shrinkage decreases. Typical applications of this rule are shown in Table 28.

We can use an alternative scheme in order to predict the effect of the nature of the atoms A and X on the preferred geometry of AX_2 molecules. Thus, for example, consider the MO's of linear H_2O which are shown in Fig. 43. Upon bending, the key stabilizing interaction introduced is the interaction between the original lone pair HOMO and the original sigma LUMO. This will increase as the HOMO-LUMO gap in the linear molecule decreases and the corresponding interaction matrix element increases or remains constant.

As in the previous case, we can define a bending tendency, F'_θ, given by the equation below:

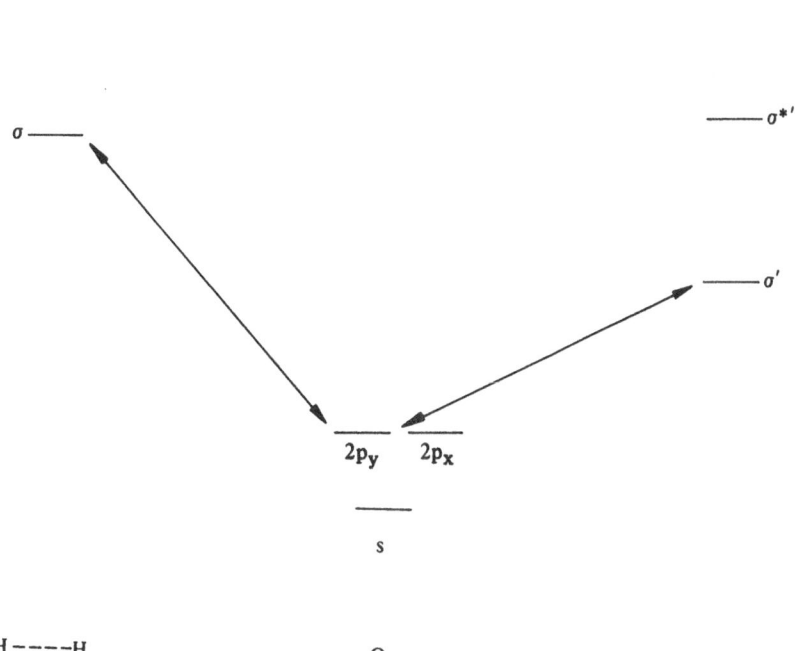

Fig. 42. The effect of electronegativity on the $2p_y - \sigma$ interaction. H' is assumed to have an artificially greater electronegativity than H. The $2p_y - H'$ interaction is stronger than the $2p_y - H$ interaction

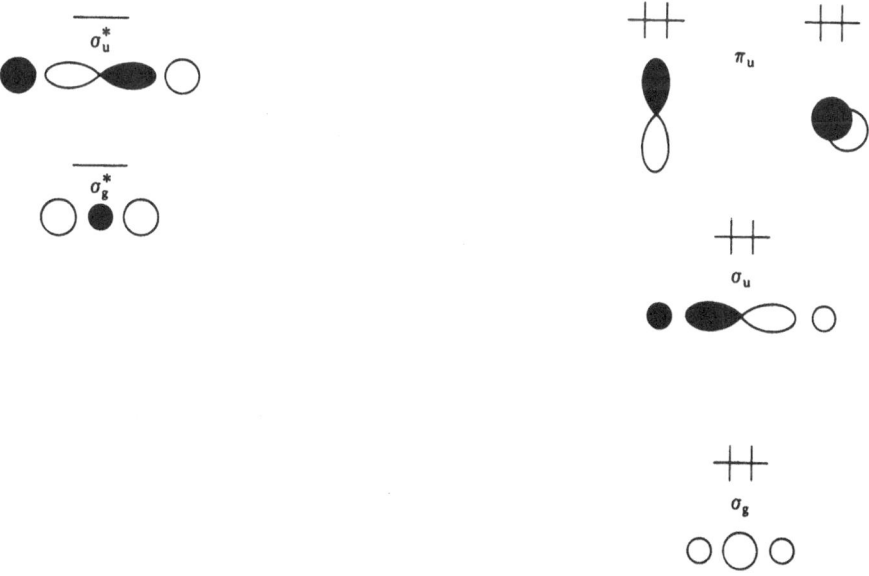

Fig. 43. The valence MO's for linear H_2O

Table 28. Variation of central angle with change of electron occupancy (k) of stabilized p orbital in AB_2 molecules

Molecule	k	Angle[a]
BeH_2	0	180
BeH_2^-	1	115–125
BH_2^+	0	180
BH_2	1	131
BH_2^-	2	100
CH_2^+	1	140
CH_2	2	105
NO_2^+ (crys.)	0	180
NO_2	1	134
BeH_2		
$^2A_1$	1	115–125
$^2B_1$	0	180
BH_2		
$^2A_1$	1	131
$^2B_1$	0	180
CH_2^+		
$^2A_1$	1	140
$^2B_1$	0	180
BH_2^-		
$^1A_1$	2	100
$^3B_1$	1	125–130
CH_2		
$^1A_1$	2	105
$^3B_1$	1	134
NH_2^+		
$^1A_1$	2	115–120
$^3B_1$	1	145–155
SiH_2		
$^1A_1$	2	90–95
$^3B_1$	1	120–125

a) Ref.[265].

$$F'_\theta = \frac{-k\, H'^2_{pq}}{E_p - E_q} \tag{25}$$

In this case, the ϕ_p and ϕ_q are no longer group MO's at a certain fixed angle. Rather, they represent two MO's of the linear system the interaction of which is "turned on" by bending. Of course, this is an approximate equation which is very valuable in terms of relating the results of the calculation of a linear molecule to its tendency to bend. Clearly, as the nature of A and X varies in a way which makes the $E_p - E_q$ smaller and H'_{pq} more negative, the bending force F'_θ will increase.

It should be noted that F'_θ has the meaning of a slope of the energy of the MO resulting from the interaction "turned on" by bending plotted as a function of angle θ. For reasons which will become apparent, we can rewrite Eq. (25) as follows:

$$F'_\theta \simeq -\frac{C \, kH_{pq}^2}{E_p - E_q} = -\frac{C \, kH_{pq}^2}{\Delta E_{pq}} \tag{26}$$

In the above equation C is the coefficient of the central atom orbital which becomes stabilized upon bending. Thus, for example, C = 1 for linear H_2O.

We shall first examine the effect of the central atom and ligands on the linearity and stability of AX_2 systems. In comparing two different types of central atoms or ligands, we shall be referring to the Eq. (26) for the tendency of bending or pyramidalization. This equation can be rewritten as follows:

$$F'_\theta \simeq -C \left(\frac{kH_{pq}^2}{\Delta E_{pq}} \right) = CL \tag{27}$$

A given substituent may affect C, H_{pq}^2, ΔE_{pq}, or any combination thereof. If the variation of each and every one of the three terms favors an increase of F'_θ, an unambiguous prediction can be made. The same is true if the variations uniformly favor a decrease of F'_θ. On the other hand, if the C, H_{pq} and ΔE_{pq} terms vary in opposite directions, the variation in F'_θ will depend upon the rate of change of the opposing terms.

We now proceed to give simple rules for predicting how a change in A or X will affect the various terms of Eq. (26).

a) The C term. The magnitute of C equals one when X = H. However, substitution of H by a group X may alter it drastically. In this case, the doubly occupied $2p_y$ AO of A of AH_2 should be replaced by a doubly occupied pi MO of HAX.

b) The ΔE_{pq} term. The magnitude of ΔE_{pq} decrease as the "lone pair" MO increases in energy and the crucial vacant sigma MO decreases in energy. The energy of the lone pair MO increases as the electronegativity of A and/or X decreases. On the other hand, the energy of the vacant sigma MO decreases as the electronegativity of X increases along a row or decreases down a column of the Periodic Table (constant A). A similar pattern is expected in the case of the variation of A for constant X.

c) The H_{pq} term. The size of the matrix element remains relatively constant as X becomes more electronegative along a row and decreases as X varies down a column of the Periodic Table (constant A). A similar pattern is expected in the case of variation of A for constant X.

The above rules are derived on the basis of our discussion of the effect of heteroatomiç substitution (Section 1.1). Further evidence of their validity will be provided in Part IV in connection with our discussion of the intrinsic donor-acceptor properties of lone pairs arìd pi or sigma bonds.

We are now prepared to embark upon a consideration of the various possible cases which may arise on the basis of Eq. (26). These are:

a) Case I: C, ΔE_{pq} und H_{pq} all favor larger F'_θ. This condition is met by the typical series shown in Scheme 4. As the electronegativity of X increases, F'_θ increases and, hence, the XAX or XAH angles are expected to shrink.

In order to illustrate these points, we shall consider as an example OF_2 vs. $O(NH_2)_2$, the latter taken in the all planar geometry. The values of C and ΔE_{pq}

Scheme 4.

Increasing angle (deg.)

$\longrightarrow$

HOF	HOOH	HONH$_2$	HOCH$_3$
97[266)	100[267)	103[267)	107.3[267)
OF$_2$			O(CH$_3$)$_2$
103[268)			111[267)

are given below and H$_{pq}$ is expected not to vary appreciably. We predict a smaller angle in OF$_2$ than in O(NH$_2$)$_2$ subject to the geometrical constraints stated above.

	OF$_2$[269)	O(NH$_2$)$_2$[269)
C	0.9565	0.5810
$-\Delta E_{pq}$ (eV)	0.89	13.04

b) Case II: C favors small F$_\theta'$ and ΔE_{pq} and H$_{pq}$ have the opposite effect, *i.e.* C and L vary in opposite directions. This condition is met by the series shown in Scheme 5. Let us consider as an example the case of H$_2$O vs F$_2$O and Cl$_2$O. Here, C favor a smaller angle for H$_2$O while H$_{pq}$ and ΔE_{pq} favor a smaller angle for F$_2$O or Cl$_2$O. The difference in C is smallest for the comparison H$_2$O vs F$_2$O and the difference in ΔE_{pq} and H$_{pq}$ is largest also for the same comparison. Typical data is given below. Accordingly, we predict that F$_2$O will have a smaller angle

	C[269)	$-\Delta E_{pq}$ (eVs)[269)
H–O–H	1	24.99
F–O–F	0.9565	0.89
Cl–O–Cl	0.4767	5.17

than H$_2$O but the latter may very well have a smaller angle than Cl$_2$O. Pertinent data is given in Scheme 5.

Scheme 5.

Increasing angle (deg.)

$\longrightarrow$

HOF	HOH
97[266)	105[270)
HOCl	HOH
104[271)	105
FOF	HOH
103[268)	105
ClOCl	HOH
111[272, 281)	105

c) Case III: C and H$_{pq}$ favor small F$_\theta'$ and ΔE_{pq} has an opposite effect so that C varies while L remains roughly constant. This condition is met by any comparison of FAH vs ClAH or F$_2$A vs Cl$_2$A. In this case, the larger C for, *e.g.*, FAH relative to ClAH will cause a smaller angle in the fluoro derivative. Typical series are shown in Scheme 6.

Scheme 6.

Increasing angle (deg.)

———————————————————————————→

FOH
$97^{266)}$

ClOH
$104^{271)}$

F_2O
$103^{268)}$

Cl_2O
$111^{272, 281)}$

F_2S
$96^{273)}$

Cl_2S
$101^{274)}$

d) Case IV: C is constand and H_{pq} and ΔE_{pq} vary in opposite directions so that C is constant while L varies. This condition is met by any comparison of XA^1H vs XA^2H or X_2A^1 vs X_2A^2, where A^1 is a first row and A^2 a second row heteroatom. In this case, the variation of ΔE_{pq} outweighs the variation of H_{pq} and the bending tendency increases as a first row atom is replaced by a second row atom. Typical series are shown in Scheme 7.

Scheme 7.

Increasing angle (deg.)

———————————————————————————→

H_2S
$92.2^{275)}$

H_2O
$105^{270)}$

F_2S
$96^{282)}$

F_2O
$103^{268)}$

Cl_2S
$101^{274)}$

Cl_2O
$111^{272)}$

At this point, the reader may wonder whether our prediction of the greater variation of ΔE_{pq} is somewhat arbitrary. However, it must be realized that the decrease in ΔE_{pq} is the result of two combined effects, *i.e.* a lowering of the vacant sigma MO and the simultaneous raising of the lone pair AO as A^1 is replaced by A^2. Typical calculated ΔE_{pq} values are shown in Table 29. It can be seen that the variation is large.

Table 29. Variation of the HOMO-LUMO energy gap (ΔE_{pq}) as the central atom changes from a first to a second row element

	$-\Delta E_{pq}$ (eV)[a]
H_2O	24.99
H_2S	10.86
F_2O	0.89
F_2S	0.08
Cl_2O	5.17
Cl_2S	2.96

[a] From extended Hückel calculations for linear molecules at standard geometries.

Another situation where the aforementioned condition is met is along a series of H_2A isoelectronic molecules where A varies along a row of the Periodic Table. Here, H_{pq} remains roughly constant and, thus, the bending force increases as ΔE_{pq} decreases, a trend which obtains as the lone pair ionization potential of the central atom decreases. Typical data are shown in Scheme 8.

Scheme 8.

Increasing angle (deg.)

$$\xrightarrow{\hspace{10cm}}$$

H_2N^-	H_2O	H_2F^+
110[276)	115[276)	
100–105[265)	104[265)	105[265)
86.4[277)	88.4[277)	92.2[277)

5.2. The Pyramidality of AX_3 Molecules

The treatment of eight valence electron AX_3 molecules and the effect of the nature of A and X on the degree of pyramidality of these molecules, *i.e.* the effect on the XAX bond angle, proceeds along the same lines as the treatment of the angle problem in eight valence electron AX_2 systems.

The MO's of flat NH_3 are shown in Fig. 44. Upon pyramidalization, a stabilizing interaction is introduced due to the interaction between the original HOMO and original LUMO of the planar molecule which will tend to render NH_3 pyramidal. Various predictions along with pertinent experimental and computational results for Cases I–IV are given in Schemes 9–13 and Table 30.

Case I: C, ΔE_{pq} and H_{pq} all favor larger F'_θ. Pertinent examples are shown in Scheme 9.

Case II: C favors small F'_θ and ΔE_{pq} and H_{pq} have the opposite effect. Pertinent examples are shown in Scheme 10.

Case III: C and H_{pq} favor small F'_θ and ΔE_{pq} has opposite effect *i.e.* C varies while L remains relatively constant. Pertinent data are shown in Scheme 11.

Case IV: C constant while H_{pq} and ΔE_{pq} vary in opposite directions. Pertinent data are shown in Scheme 12 for the case where the central atom is changed from a first row to a second row element. Examples of varying central atom along a row of the Periodic Table are shown in Scheme 13. Additional data are given in Table 30.

Finally, the change in the HXH angle as the electron occupancy of the HOMO is decreased is illustrated by the examples in Table 31. As before, our predictions agree with the available experimental data.

Certain features of geminal interactions in AX_3 molecules need emphasis:

a) The effect of geminal interactions upon the XAX angle parallels the effect upon inversion barriers, *i.e.* as the strength of the geminal interaction increases, the pyramidal form of AX_3 becomes more stabilized leading to a *smaller* XAX angle and a *higher* inversion barrier. Typical results concerning inversion barriers are given in Tables 30 and 32.

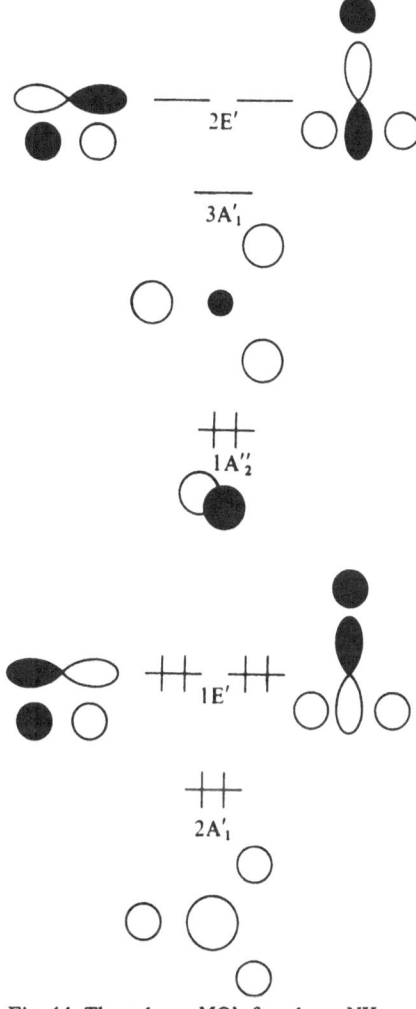

Fig. 44. The valence MO's for planar NH₃

Scheme 9.

Increasing XNH angle (deg.)

H₂NF	H₂NOH	H₂NNH₂	H₂NCH₃
99.8[279)	103[267)	112[267)	112[267)
NF₃			N(CH₃)₃
102.5[267)			108.7[267)

Scheme 10.

Increasing angle (deg.)

NF₃	NH₃
102.5[267)	106.6[278)

Part III. Geminal Interactions

Scheme 11.

Increasing angle (deg.)

$$\longrightarrow$$

PF$_3$
96[280)

PCl$_3$
100[267)

Scheme 12.

Increasing angle (deg.)

$$\longrightarrow$$

H$_3$P
93.2[267)

H$_3$N
106.8[278)

F$_3$P
96[280)

F$_3$N
102.5[267)

Scheme 13.

Increasing angle (deg.)

$$\longrightarrow$$

H$_3$C$^-$

H$_3$N
106.6[278)

H$_3$O$^+$
115–117[267)

105[265)
87.5[277)

107[265)
87.6[277)

118[265)
90.2[277)

Table 30. Theoretical inversion
barriers for AH$_3$ molecules

AH$_3$	Inversion Barrier[a]
	Ab initio
CH$_3^-$	5.46
SiH$_3^-$	39.60
NH$_3$	5.08
PH$_3$	37.20
OH$_3^+$	
SH$_3^+$	30.00

a) Ref.[281].

b) Geminal interactions can be very strong and able to compete with strong pi conjugative interactions. A manifestation of this is the pyramidal nature of phosphole as opposed to the planar nature of pyrrole. In both cases, Hückel aromaticity tends to flatten the molecule while the strong geminal interaction tends to enforce pyramidality about the phosphorous center. The nonplanarity of phosphole is testimony to the importance of geminal interactions.

The analysis of the effect of A and X upon the pyramidality of six valence electron AX$_3$ molecules is similar, in principle, to that offered for the previous systems.

Table 31. Angle variation with change of electron occupancy of stabilized orbital in AB_3 molecules

Molecule	k[a]	BAB angle[b]
BH_3	0	120
BH_3^-	1	110–115
CH_3^+	0	120
CH_3	1	115–120
CH_3^-	2	105
NH_3^+	1	120
NH_3	2	107
PH_3^+	1	110–120
PH_3	2	93
BH_3^-		
$^2A_1$	1	110–115
^2E	2	85–90
CH_3		
$^2A_1$	1	115–120
^2E	2	90–95
NH_3^+		
$^2A_1$	1	120
^2E	2	95–100
PH_3^+		
$^2A_1$	1	110–120
^2E	2	85–90

[a] Electron occupancy of stabilized orbital.
[b] Ref. [265].

5.3. Miscellaneous Problems

The ideas discussed above have very general applicability. For example, they can be utilized in order to predict the shape of AYX_3 molecules as a function of A, Y and X, the shape of planar AXYZ systems as a function of X,Y,Z, *etc.* In the space below we restrict our attention to only one further example of the applicability of the concept of geminal interactions to problems of molecular structure.

A fascinating situation arises when one inquires as to what happens when a sigma lone pair is replaced by a sigma bond, *i.e.* when the sp^2 lone pair of H_2O is protonated giving rise to H_3O^+. In order to answer this question, one has to compare the bending tendencies in linear H_2O and in a T shaped H_3O^+.

The interaction diagram of Fig. 45 provides the basis for our discussion. Here, the AO's of the central atom interact with the three group MO's which span the three hydrogen atoms in the T geometry. The following trends can be noted:

a) As bending occurs, the energy of σ_2 increases, that of σ_3 decreases and that of σ_1 remains roughly constant.

b) As a result of the variations of σ_1, σ_2 and σ_3 accompanying bending as well as the change in the appropriate overlap integrals, the energy of ϕ_1 remains relatively

143

Table 32. Experimental Inversion Barriers of
Selected Molecules

	$\Delta G^{\ddagger}$ (T °C) kcal/mol	Ref.
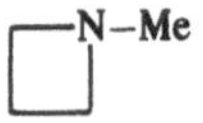 N–Me (azetidine)	10.2 (−69)	282)
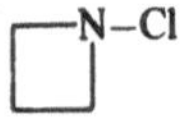 N–Cl (azetidine)	13.4 (−20)	282)
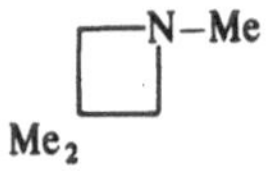 N–Me, Me$_2$ (azetidine)	8.85 (−98)	283)
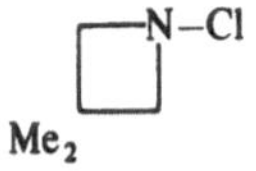 N–Cl, Me$_2$ (azetidine)	11.5 (−54)	283)
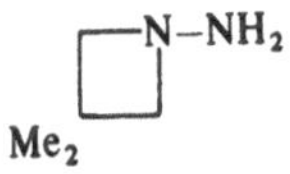 N–NH$_2$, Me$_2$ (azetidine)	10.2 (−65)	284)
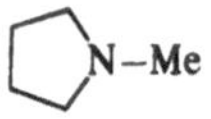 N–Me (pyrrolidine)	8.4 (−98)	282)
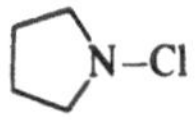 N–Cl (pyrrolidine)	10.3 (−68)	282)
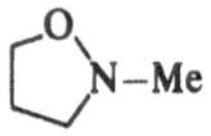 O, N–Me (isoxazolidine)	16.9 (62)	285)
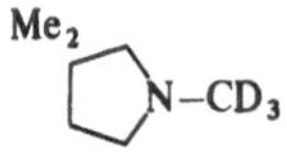 Me$_2$, N–CD$_3$	7.4 (−117)	286)
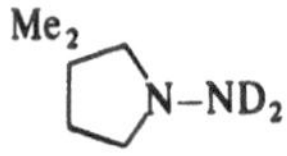 Me$_2$, N–ND$_2$	3.5 (−98)	286)
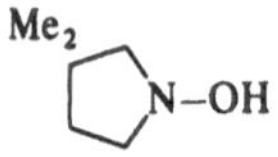 Me$_2$, N–OH	13.0 (−11)	286)
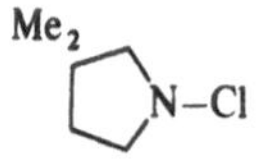 Me$_2$, N–Cl	9.0 (−87)	286)
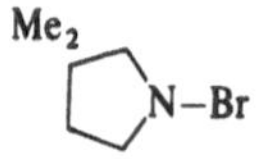 Me$_2$, N–Br	8.5 (−98)	286)

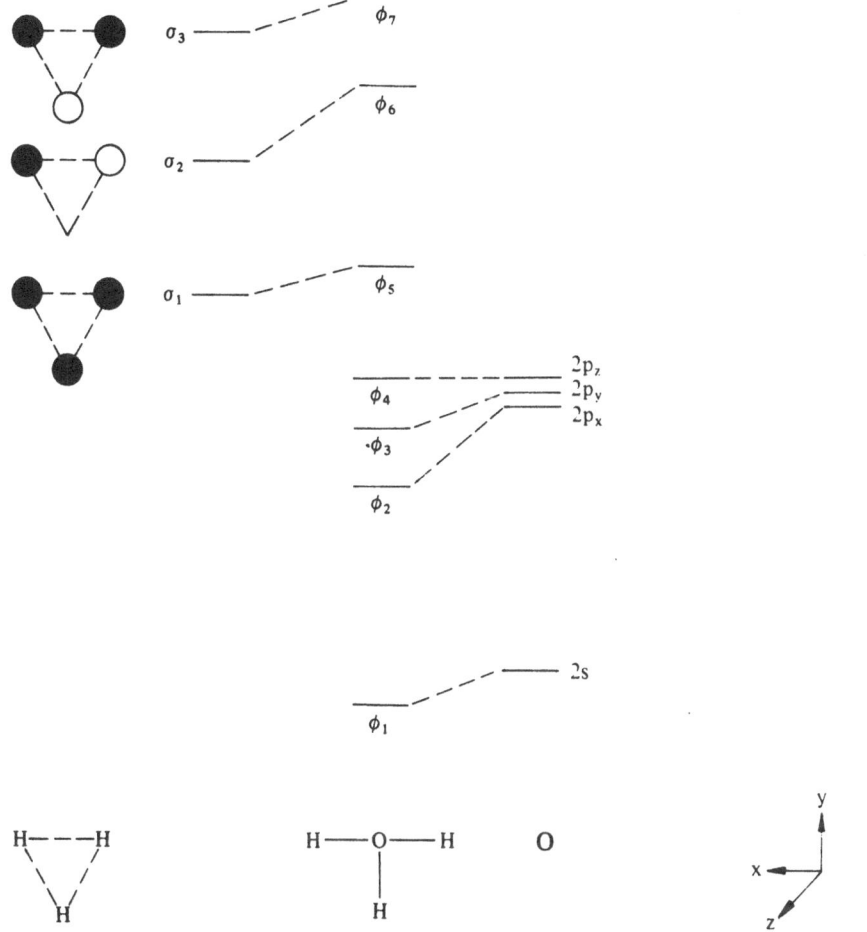

Fig. 45. The interactions of the H_3 group MO's and the oxygen AO's in the T geometry of H_3O^+

constant, that of ϕ_2 increases and that of ϕ_3 decreases. The energy lowering of ϕ_3 is interesting because it arises from two conflicting effects, an increasingly weaker $Op_y-\sigma_1$ interaction and an increasingly stronger $Op_y-\sigma_3$ interaction, the latter varying faster than the former. On the basis of the above considerations, the following conclusions become apparent:

a) The $O2s-\sigma_1$ interaction in H_3O^+ favors bending to a smaller extent than the comparable interaction in H_2O.

b) The $O2p_x-\sigma_2$ interaction in H_3O^+ disfavors bending to the same extent as the comparable interaction in H_2O.

c) The $O2p_y-\sigma_1$ and $O2p_y-\sigma_3$ interactions in H_3O^+ have a cumulative effect which favors bending to a less extent than the single comparable interaction of $O2p_y$ in H_2O.

On the basis of the above discussion, we can formulate the following general rule: replacement of a sigma lone pair by a sigma bond in any AX_2 system to yield an AX_2Y system will be accompanied by an opening of the XAX angle contrary to any expectations based upon consideration of steric effects.

Naturally, a similar analysis can be given concerning the influence of replacing a lone pair by a sigma bond in AX_3 to yield AX_3Y upon the XAX angle. Typical results which verify these rules are given in Table 33.

Table 33. Variation of angle in B_2A: vs B_2AX molecules

Fragment		Molecule	
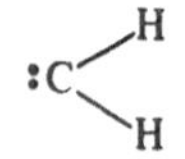 $:C(H)(H)$	102.5 [288)]	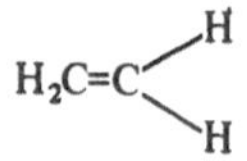 $H_2C=C(H)(H)$	116.2 [289)]
 $:C(F)(F)$	104.9 [289)]	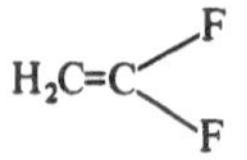 $H_2C=C(F)(F)$	109.3 [289)]
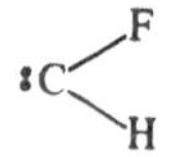 $:C(F)(H)$	101.8 [265)]	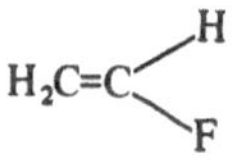 $H_2C=C(H)(F)$	117.5 [289)]
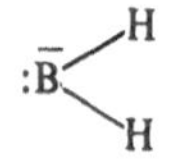 $:B^-(H)(H)$	100 [265)]	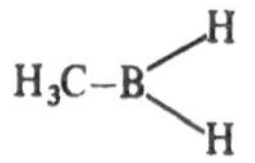 $H_3C-B(H)(H)$	118.5 [290)]
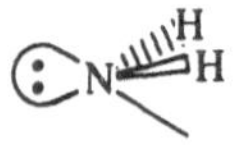 $:N(H)(H)(H)$	106.6 [278)]	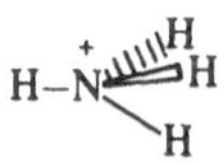 $H-N^+(H)(H)(H)$	109.47 [267)]
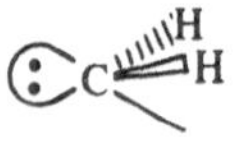 $:C^-(H)(H)(H)$	105 [265)]	 $H_3C-C(H)(H)(H)$	109.7 [291)]

Before departing temporarily this subject, we should point out that the models which form the basis for the equations of F_θ and F'_θ are not completely equivalent but rather constitute approximations to the solution of a very complicated problem. Thus, for example, the coefficient effect which comes in F'_θ is not seen in F_θ and the role of the energy gap involved in the equation for F'_θ is exaggerated relative to that in the equation for F_θ. Nonetheless, both models lead to similar predictions, at lest in most cases we have studied and, thus, both merit attention.

6. Donor and Acceptor Molecular Fragments and the Question of *Syn* vs. *Anti* Overlap

In this part we employ the OEMO method in connection with bond MO's rather than delocalized MO's. There are three types of bond MO's: n, pi and sigma. The interactions between bond MO's can be classified as follows:

a) $\pi - \pi$ interactions
b) $\pi - n$ interactions
c) $\pi - \sigma$ interactions
d) $n - n$ interactions
e) $n - \sigma$ interactions
f) $\sigma - \sigma$ interactions

where π stands for a pi type AO or MO, n for a lone pair AO or MO, and σ for a sigma type AO or MO. These interactions can be stabilizing or destabilizing depending upon orbital occupancy. In general, as we have stated before, the interaction between an occupied and an unoccupied orbital is stabilizing. Such interactions can be of the $\pi-\pi^*$, $n-\pi^*$, $\pi-\sigma^*$, $\sigma-\pi^*$, $n-\sigma^*$, and $\sigma-\sigma^*$ variety, where the asterisk denotes an antibonding vacant orbital and the absence of an asterisk implies a bonding doubly occupied orbital. The interaction between two occupied orbitals, however, is destabilizing. Such interactions can be of the $\pi-\pi$, $n-\pi$, $\sigma-\pi$, $n-n$, $n-\sigma$, and $\sigma-\sigma$ variety. Now, these destabilizing interactions play a definite role in determining geometrical preferences, but two electron stabilizing interactions are relatively more important. Accordingly, we shall focus our attention on two electron stabilizing interactions and how they influence geometrical preferences.

In our subsequent discussions, we shall be using consistently hybrid lone pair AO's and hybrid bond MO's. The explicit forms of these orbitals are given below, for the case of any lone pair, located on atom Y, and the case of an X–H bond:

$$n_Y = h_Y \, (sp^2)$$
$$\text{or} \quad n_Y = h_Y \, (sp^3)$$
$$\sigma_{X-H} = h_X \, (sp^2) + \lambda \, ls_H$$
$$\text{or} \quad \sigma_{X-H} = h_X \, (sp^3) + \lambda \, ls_H$$
$$\sigma^*_{X-H} = h_X \, (sp^2) - \lambda \, ls_H$$
$$\text{or} \quad \sigma^*_{X-H} = h_X \, (sp^3) - \lambda \, ls_H$$

The hybrid sp^2 and sp^3 AO's can be simply written in terms of s and p AO's as follows[292]:

$$h \, (sp^2) = .577 \, s + .408 \, p_x + .707 \, p_y$$
$$h \, (sp^3) = .500 \, s + .288 \, p_x + .816 \, p_y$$

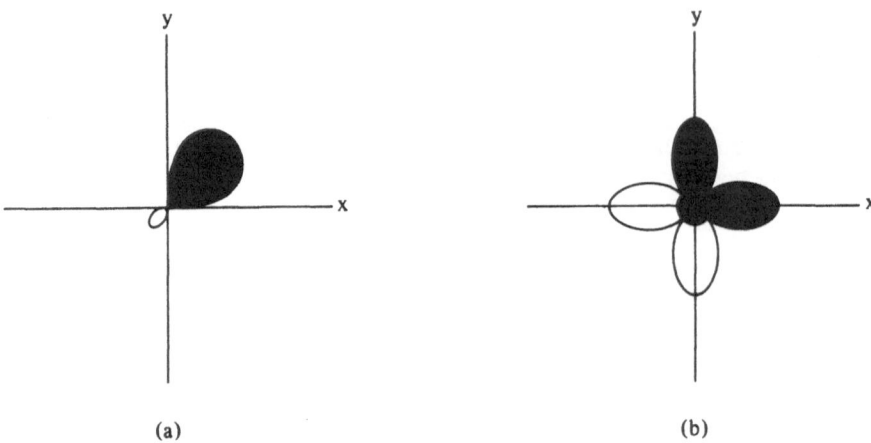

(a)   (b)

Fig. 46. Hybrid orbital and its expanded form

A hybrid orbital is drawn in Fig. 46a and is shown in its expanded form in Fig. 46b.

An important point to be stressed before we proceed further with our analysis concerns the formal correspondence, of hybrid orbitals and delocalized group orbitals. For example, consider the model system HN=NH which can exist in a cis or a *trans* geometry. We can understand whether sigma interactions favor one or the other geometry be means of one of the following two approaches:

a) The group MO's of HN- are constructed and their stabilizing interactions are assessed for the cis and the trans geometry. This is shown in Fig. 47.

b) The sp^2 hybrid lone pair AO of one nitrogen and the corresponding N–H bond MO's, made up of an sp^2 hybrid nitrogen AO and a hydrogen ls AO, are assumed to interact with the sp^2 hybrid lone pair AO and the corresponding N–H bond MO's of the second nitrogen. The attendant stabilization of the *cis* and *trans* geometries can then be assessed. This is shown in Fig. 48.

The reader can easily see that ϕ_2 is essentially an sp^2 lone pair AO, ϕ_1 essentially a σ_{N-H} MO, and ϕ_3 essentially a σ^*_{N-H} MO. In short, the two approaches are equivalent and we have chosen to use the bond orbital description for conceptual simplicity.

We now return to the discussion of the preferred geometry of HN=NH. In the *cis* geometry there are two syn $\sigma-\sigma^*$ interactions and two *anti* n–σ^* interactions, while in the *trans* geometry there are two *anti* $\sigma-\sigma^*$ interactions and two *syn* n–σ^* interactions. Now, a lone pair has higher energy than a bonding sigma MO, or, in other words, Ṅ is a better donor fragment than N–H. Hence, we focus attention on the dominant interaction between the best donor orbital n_N and the best acceptor orbital σ^*_{NH}. The stabilization energy difference for the *cis* and *trans* geometries due to the $n_N-\sigma^*_{NH}$ interaction is approximated by the expression

$$SE_{cis} - SE_{trans} \, \alpha \, (S^2_{anti} - S^2_{syn}) \tag{28}$$

In other words, we need to evaluate the S_{anti} and S_{syn} overlap integrals in order to determine which of the two geometries is favored. Clearly, interactions of the n–π^*,

148

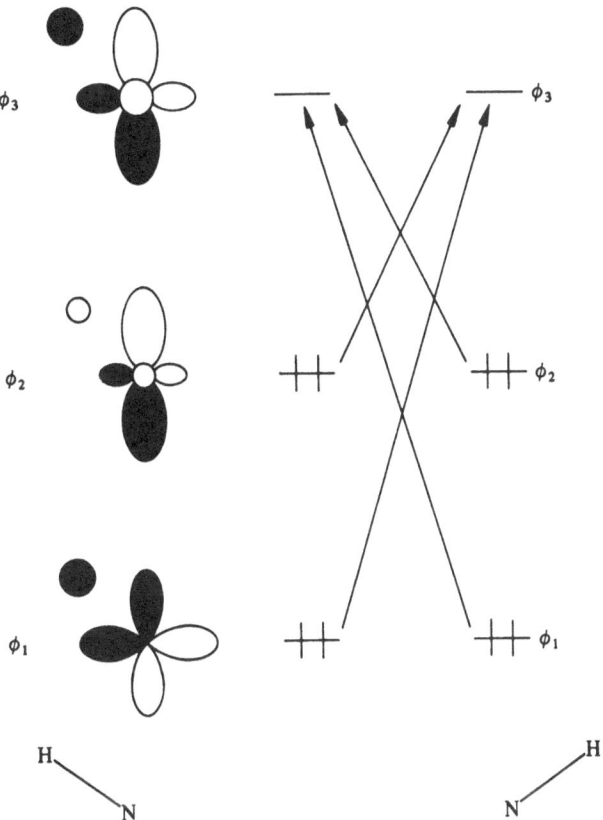

Fig. 47. The interaction of delocalized N–H MO's in HN=NH

σ–π* and π–σ* type do not exhibit *syn-anti* directional preference. The situation is completely different in the case of n–σ* and σ–σ* interactions where *syn* overlap is expected to be different from *anti* overlap. This is the fundamental problem which we shall encounter from now on in geometric or conformational isomerism dominated by n–σ or σ–σ interactions. The question arises: are there any general rules which can allow the qualitative prediction of the relative sizes of the *syn* and *anti* overlap integrals between n and σ* or σ and σ*. The answer is affirmative and we proceed to give a detailed discussion.

The n_N–σ^*_{NH} overlap integrals for the *syn* and *anti* arrangements in HN=NH are given below. In these equations the overlap integrals are all taken as positive and the sign of each term is determined from consideration of the phases of the overlapping AO's as shown in Figure 49.

$S_{syn} \propto Q + (.707)^2 (p_y \, | \, p_y) - \lambda (.707) (1s \, | \, p_y)$

$S_{anti} \propto Q - (.707)^2 (p_y \, | \, p_y) + \lambda (.707) (1s \, | \, p_y)$

where Q is constant for both cases and is given by the expression shown below:

$Q \propto (.577)^2 (s \, | \, s) - (.577) (.408) (s \, | \, p_x) - (.408) (.577) (p_x \, | \, s)$

149

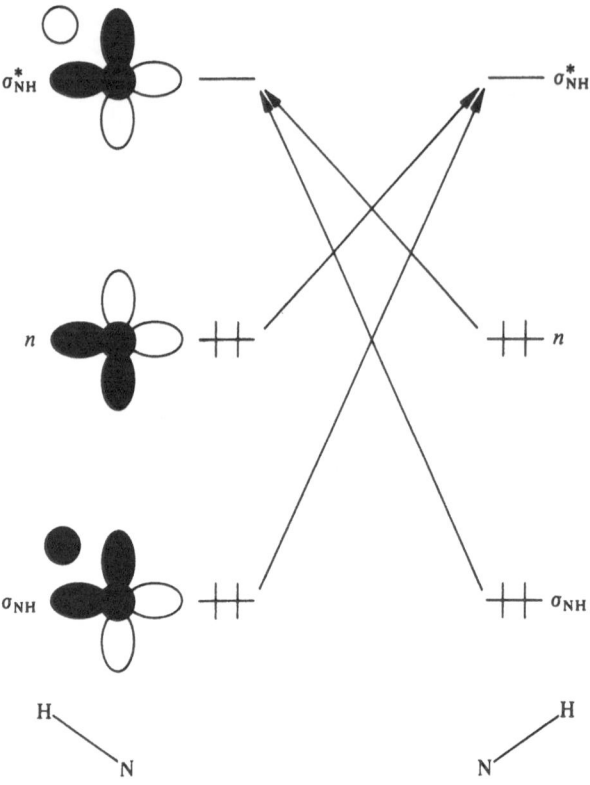

Fig. 48. The interaction of hybrid N–H bond and lone pair orbitals in HN=NH

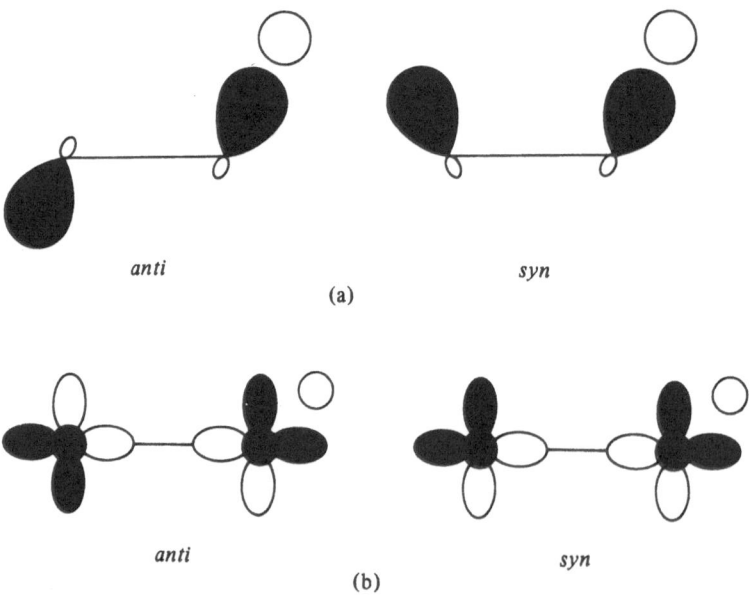

anti

syn

(a)

anti

syn

(b)

Fig. 49. Hybrid bond and lone pair orbitals (a) and their expanded forms (b)

$+ (.408)^2 \, (p_x \mid p_x) - \lambda \, (.577) \, (1s \mid s) + \lambda \, (.408) \, (1 \, s \mid p_x)$

In general, Q is a negative number. The quantities
$[(.707)^2 \, (p_y \mid p_y) - \lambda \, (.707) \, (1s \mid p_y)]$ and $[-(.707)^2 \, (p_y \mid p_y) + \lambda \, (.707) \, (1s \mid p_y)]$
are equal in absolute magnitude, but have opposite signs. Since the $Np_y - Np_y$ overlap is greater than the $Np_y \cdots H1s$ overlap, the quantity
$[(.707)^2 \, (p_y \mid p_y) - \lambda \, (.707) \, (1s \mid p_y)]$ has a positive sign for the *syn* case while the quantity $[-(.707)^2 \, (p_y \mid p_y) + \lambda \, (.707) \, (s' \mid p_y)]$ has a negative sign for the *anti* case, assuming reasonable values for λ.

We now define:
$\pm [(.707)^2 \, (p_y \mid p_y) - \lambda \, (.707) \, (1s \mid p_y)] = \pm R$

Accordingly, the previous expressions are reduced to the form

$S_{syn} \, \alpha \; - \mid Q \mid + \mid R \mid$

$S_{anti} \, \alpha \; - \mid Q \mid -- \mid R \mid$

Clearly then, the absolute magnitude of S_{anti} is greater than S_{syn}.
    The important points to be emphasized here are that:
    a) the invariant quantity Q is negative, and
    b) the variable quantity R will always be negative for the *anti* geometry and positive for the *syn* geometry due to the dominance of the $p_y - p_y$ overlap.
    Table 34 shows calculated overlap integrals for various combinations of hybridized atomic centers. In all cases, the absolute magnitude of the overlap for the *anti* alignment is greater.
    We can work in a similar manner to examine the relative absolute magnitudes of *syn* and *anti* $\sigma - \sigma^*$ overlaps.
    We obtain the following equations:

$S_{syn} \, \alpha \; Q + R - M$

$S_{anti} \, \alpha \; Q - R - N$

where

$Q = (.577)^2 \, (s \mid s) - (.577)(.408)(s \mid p_x) - (.408)(.577)(p_x \mid s)$
$\quad + (.408)^2 \, (p_x \mid p_x) - \lambda \, (.577)(1s \mid s) + \lambda \, (.408)(1s \mid p_x)$
$\quad + \lambda \, (.577)(1s \mid s) - \lambda \, (.408)(1s \mid p_x)$
$R = [(.707)^2 \, (p_y \mid p_y) - (.707)(1s \mid p_y) + \lambda \, (.707)(1s \mid p_y)]$
$M = \lambda^2 \, (1s \mid 1s) \, (syn)$
$N = \lambda^2 \, (1s \mid 1s) \, (anti)$

Table 34. $< n \mid \sigma^* >$ Overlap integrals

Interacting orbitals	Overlap integrals[a]		Difference
	Syn overlap	*Anti* overlap	
$<n_{Nsp^2} \mid \sigma^*_{HNsp^2} >$	.0357	.0867	.0510
$<n_{Nsp^3} \mid \sigma^*_{HNsp^3} >$	.0251	.0623	.0372
$<n_{Nsp^2} \mid \sigma^*_{HCsp^2} >$	.0567	.0977	.0410
$<n_{Nsp^3} \mid \sigma^*_{HCsp^3} >$	.0554	.0840	.0286
$<n_{Csp^2} \mid \sigma^*_{HNsp^2} >$	.0321	.0817	.0496
$<n_{Csp^3} \mid \sigma^*_{HNsp^3} >$	.0243	.0677	.0434
$<n_{Csp^2} \mid \sigma^*_{HCsp^2} >$	.0689	.1173	.0484
$<n_{Csp^3} \mid \sigma^*_{HCsp^3} >$	.0570	.0874	.0304

[a] Computed with a CNDO/2 program. Values given are absolute.

In this case the invariant term Q can be either positive or negative simply because a strong bonding X–H overlap at one site is counteracted by a strong antibonding X–H overlap at the other site and the remaining terms can sum to either a positive or negative number depending on interatomic distances. Either way, the absolute magnitude of Q is much smaller than that of R, M, or N. R and M are close in absolute magnitude with R usually, but not always, being the larger. In the case of *syn* overlap R and M tend to cancel each other, so that a small value for the overlap integral is obtained. R and N reinforce each other in the case of *anti* overlap, the result being that S_{anti} is greater in absolute magnitude than S_{syn}. Overlap integrals for various combinations of hybridized atomic centers are collected in Table 35.

Table 35. $<\sigma \mid \sigma^* >$ Overlap integrals

Interaction orbitals	Overlap integrals[a]	
	Syn	*Anti*
$<\sigma_{NHsp^2} \mid \sigma^*_{NHsp^2} >$	.0076	.1420
$<\sigma_{NHsp^3} \mid \sigma^*_{NHsp^3} >$	.0166	.1176
$<\sigma_{CHsp^2} \mid \sigma^*_{NHsp^2} >$	.0066	.1271
$<\sigma_{CHsp^2} \mid \sigma^*_{NHsp^3} >$	.0161	.1176
$<\sigma_{NHsp^2} \mid \sigma^*_{CHsp^2} >$	.0555	.1592
$<\sigma_{NHsp^3} \mid \sigma^*_{CHsp^3} >$	.0461	.1500
$<\sigma_{CHsp^2} \mid \sigma^*_{CHsp^2} >$	.0667	.1799
$<\sigma_{CHsp^3} \mid \sigma^*_{CHsp^3} >$	.0472	.1437

[a] Computed with a CNDO/2 program. Values given are absolute.

In the previous section we have identified the orientation of two orbitals which maximizes their interaction. In the model system HN=NH, the only difference between the *cis* and *trans* geometric isomers is the orientation of the inter-acting orbitals. In this case, a prediction of the relative stabilization of the *cis* and *trans* geometries due to donor-acceptor interactions hinges upon a determination of the dominant orbital interaction. Thus, depending on whether the $n-\sigma_{NH}^*$ or $\sigma_{NH}-\sigma_{NH}^*$ interaction dominates, the *cis* isomer, which maximizes the former, or, the *trans* isomer, which maximizes the latter interaction, will be favored.

A similar problem arises when ine considers torsional isomerism where a prediction hinges upon the determination of the dominant interaction. For example, consider the molecule NH_2-CXYZ. Here, it is apparent that the most important orbital interactions will involve the nitrogen lone pair. These are the $n_N-\sigma_{CX}^*$, $n_N-\sigma_{CY}^*$ and $n_N-\sigma_{CZ}^*$ interactions. A determination of the relative strength of these interactions should be based upon consideration of the energies of the orbitals involved and the size of the corresponding matrix elements. Accordingly, we distinguish the following possibilities:

a) The energies of the σ^* orbital vary in the order $\sigma_{CX}^* > \sigma_{CY}^* > \sigma_{CZ}^*$ and the absolute magnitude of the matrix elements varies in the reverse order. In such a case, the dominant interaction is unequivocally the $n_N-\sigma_{CZ}^*$ interaction and the conformation which maximizes it is the one shown below.

b) The energies of the σ^* orbitals vary as in (a) and the absolute magnitudes of the matrix elements also vary in the same order. In such a case we deal with an interaction which may be energy gap or matrix element controlled.

As we have already discussed in Section 1.2, there are several cases which may obtain in comparing two orbital interactions. Since all of them cannot be incorporated in any single simple framework, we have chosen to develop a model which leads to correct predictions whenever $\Delta\delta \epsilon_{ij} (A,B) < 0$ and $\Delta H_{ij} (A,B) > 0$ or $\Delta\delta \epsilon_{ij} (A, B) < 0$ and $\Delta H_{ij} (A, B) \simeq 0$ (see Section 1.2). Accordingly, interactions, which are matrix element controlled should be anticipated and treated separately.

The basic features of the model are the following:

a) Lone pair and bonds are classified according to their intrinsic donor and acceptor abilities. Tables 36 and 37 summarize the relative intrinsic donor and acceptor strengths of various lone pairs and bonds.

b) By reference to Tables 36 and 37, one can identify the fragments which are capable of interacting in a dominant fashion, *i.e.* select the best intrinsic donor lone pair or bond and the best intrinsic acceptor bond.

Table 36. Intrinsic donor ability of lone pairs and bonds[a]

---

a) Lone pairs

$\bar{C} \gg \ddot{N} > \ddot{O}$ (p) $> \ddot{O}$ (sp^2) $> \ddot{F}$
$\ddot{I} > \ddot{B}r > \ddot{C}l > \ddot{F}$
$\ddot{T}e > \ddot{S}e > \ddot{S} > \ddot{O}$
$\ddot{A}s > \ddot{P} > \ddot{N}$

b) Bonds

C–H > N–H > O–H > F–H
H–I > H–Br > H–Cl > H–F
H–S > H–O
H–P > H–N
N–Si > H–C
C–I > C–Br > C–Cl > C–F
C–Cl > C–C > C–H > C–F

---

[a] The intrinsic donor ability of lone pairs and bonds is evaluated with reference to Tables 3 and 4.

Table 37. Intrinsic acceptor ability of C–X and Y–H sigma bonds[a]

---

a) C–X bonds

C–F > C–O > C–N > C–C
C–I > C–Br > C–Cl > C–F
C–S > C–O
C–P > C–N
C–Si > C–C

b) Y–H bonds

H–F > H–O > H–N > H–C
H–S > H–O
H–Si > H–C

---

[a] The intrinsic acceptor ability of the Y–H and C–X bonds is evaluated with reference to Table 4.

In general, the strength of an orbital interaction varies in the order
$\pi-\pi^* > n-\pi^* > n-\sigma^*$, $\pi-\sigma^* > \sigma-\sigma^*$ reflecting the fact that the energy separation between the interacting orbitals is, in most cases, smallest for $\pi-\pi^*$ or $n-\pi^*$ and largest for $\sigma-\sigma^*$. If the dominant interaction is of the $n-\sigma^*$ or $\sigma-\sigma^*$ type, the geometry which allows for interaction is either *syn* or *anti*. Intermediate geometries

will be assumed to be unfavorable for appreciable interaction, *i.e.* overlap in such cases is poor, and will be considered only if coplanar arrangements are not possible.

c) If the dominant interaction is of the n–σ* type, where n is a hybridized lone pair AO, the geometry which places the two orbitals in an *anti* geometry will be preferred. The preference will increase as one of the fragments becomes an increasingly better donor and the other an increasingly better acceptor. The same will be true in the case of the dominant interaction being of the σ–σ* type.

d) Other factors, such as nonbonded attraction or repulsion and steric effects should be considered in conjunction with conjugative effects.

The Prediction of Gross Atomic Charges and Bond Overlap Populations

In discussing gross atomic charges it must be realized that a given pattern of atomic charge densities in a molecule can be imposed by several factors. We identify three effects responsible for a molecular charge distribution:
1. Sigma conjugative effect.
2. Nonbonded interaction effect.
3. Electrostatic effect.

The mechanism of charge reorganization attending the interconversion of geometric isomers due to nonbonded interaction effects has already been discussed before[1]. The electrostatic effect can be thought of as the effect which forces the distribution of charge in such a way that electrostatic repulsions are minimized.

As an example, let us consider the cases of N_2H_2 and N_2F_2. The *cis* and *trans* isomers of N_2H_2 are shown below.

If sigma conjugative interactions were the dominant factor then we would predict that the *cis* hydrogens should be less *positive* than the *trans* hydrogens since charge transfer from n_N to σ^*_{N-H} is greater in the *cis* isomer. Based solely on electrostatic considerations one would also predict less positive charge on the *cis* rather than the *trans* hydrogens in order to minimize the repulsive H––H interaction which is more severe in the *cis* than in the *trans* isomer. In the case of N_2F_2, the nonbonded interaction effect predicts that the fluorines in the *cis* isomer will be less *negative* than in the *trans* isomer because of greater charge transfer from an antisymmetric F–––––F pi group MO into the antibonding pi MO of the N=N moiety. A similar prediction is made on the basis of the electrostatic effect. However, if a sigma conjugative interaction is the dominant factor, then the cis fluorines should be more negative than the fluorines in the trans isomer.

In the light of the above two examples, it is imperative that caution be exercised in the interpretation of charge densities in terms of one effect only. A comparison of calculated charge densities with those predicted by the three approaches will be useful in some cases in pinpointing the factor primarily responsible for the relative stability of a certain geometric isomer.

The factors which determine bond overlap population are as follows:
1. Sigma conjugative effect.
2. Nonbonded interaction effect.
3. Electrostatic or steric effect.

Once more, let us consider the case of N_2F_2 where all the above effects obtain. If the sigma conjugative effect is dominant, we expect the N–F overlap population to be smaller in the *cis* isomer where the $n_N-\sigma^*_{NF}$ interaction is maximized. On the other hand, if the nonbonded interaction effect is dominant, we expect exactly the opposite trend, *i.e.* larger N–F overlap population in the *cis* isomer, due to a stronger mixing of the antisymmetric F–––F pi group MO with the antibonding pi MO of the N=N moiety. Following similar reasoning, we predict that the N=N bond overlap population will be smaller in the *cis* isomer if steric effects or nonbonded interaction effects obtain and larger if sigma conjugative effects dominate.

7. Structural Effects of n–π, σ–π and π–π Interactions

7.1. n–π Interactions

We first turn our attention to the structural effects of n–π orbital interactions. Our approach can be illustrated by reference to the model system hydroxyethylene. The two conformations we shall compare are shown below:

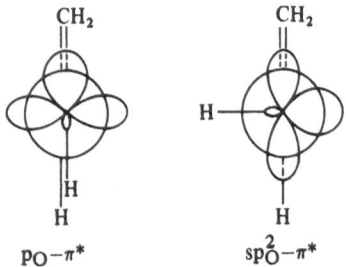

$$p_O-\pi^* \qquad\qquad sp^2_O-\pi^*$$

The n–π stabilizing interaction which obtains in each conformation is listed above. Since the oxygen 2p lone pair AO is a better intrinsic donor orbital than the oxygen hybrid sp^2 lone pair AO, we conclude that n–π^* interactions favor the conformation in which all atoms are contained in the same plane.

7.2. σ–π Interactions

In this section the structural consequences of σ–π interactions are examined. Our model system is 1-fluoro-2-propene.

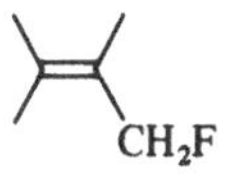

$$\text{CH}_2\text{F}$$

The two conformations of 1-fluoro-2-propene which we shall compare are the *cis* and *gauche* conformations shown below along with the dominant stabilizing donor-acceptor interactions which obtain in each case.

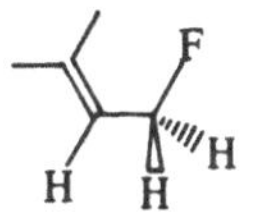

cis	*gauche*
σ_{CH}–π^*	π–σ^*_{CF}

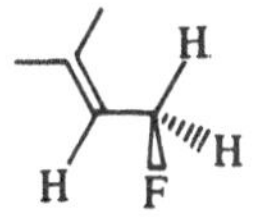

Approximate estimates of the relative strengths of the σ_{CH}–π^* and π–σ^*_{CF} interactions show that they are not widely different. Hence, nonbonded attraction present in the *cis* form is expected to give rise to preference for the *cis* conformation.

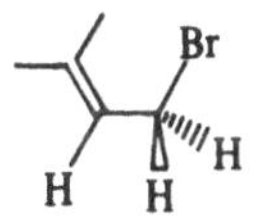

cis	*gauche*
σ_{CH}–π^*	σ_{CBr}–π^* and π–σ^*_{CBr}

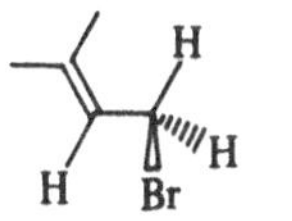

Replacement of fluorine by a halogen atom of a higher period will result into greater σ–π stabilization of the *gauche* relative to the *cis* conformer.

Since the C–Br bond is a much better donor and a much better acceptor than the C–H bond, the prediction is that the *gauche* conformation will be stabilized more than the *cis* conformation by σ–π interactions. These interactions may then dominate nonbonded attractive effects present in the *cis* isomer.

Microwave spectroscopic studies of 1-fluoro-2-propene show that the *cis* conformation is more stable than the *gauche* conformation by approximately 306 cal/mol[293]. However, when fluorine is replaced by chlorine, bromine, or iodine, the gauche conformation becomes more stable[294]. These results confirm our expectations that the conformational preferences of allyl halides may depend on a balance of nonbonded attractive effects and σ–π interaction effects.

7.3. σ–p^+ Interactions

Conjugative interactions can be important in influencing the geometry as well as rotational barriers of carbocations. We will first consider examples where a sigma bond conjugates with an empty carbon $2p$ AO.

We begin by considering the relative energy of the conformations of a 1-substituted ethyl cation shown below:

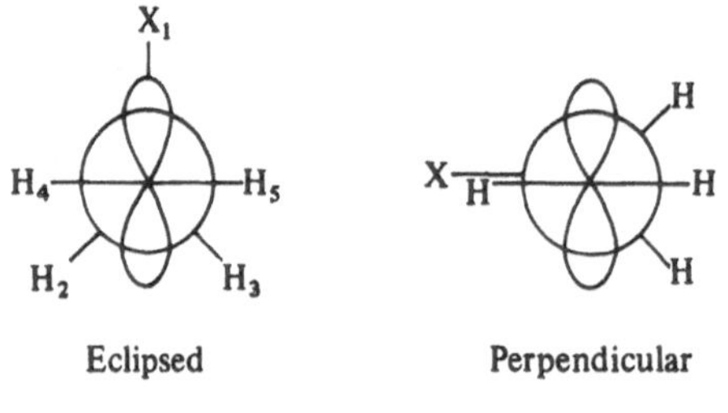

Eclipsed   Perpendicular

In the eclipsed conformation the stabilizing interaction is σ_{CX} –p^+ and in the perpendicular conformation the principal stabilizing interaction is the σ_{CH} – p^+ interaction. Arguing as before we conclude the following:

a) When the C–X bond is a better donor than the C–H bond, e. g. X=CH$_3$, the eclipsed conformation will be preferred to the perpendicular conformation.

b) When X is a highly electronegative atom or group, the C–H bond becomes the best donor bond and the perpendicular conformation will be more stable than the eclipsed conformation. The energy difference between the perpendicular and eclipsed conformation should increase as X becomes increasing electronegative along a MO of the Periodic Table, i. e., as the C–X bond becomes an increasingly poor donor relative to the C–H bond.

Ab initio calculations on substituted ethyl cations confirm the above predictions. Typical results are shown in Table 38. As can be seen, the eclipsed conformation is

Table 38. Rotational barriers for substituted ethyl cations

X	$\Delta E^{a)}$ (kcal/mol)
H	0
CH$_3$	2.52
CN	−1.95
OH	−7.67
F	−9.31

$^{a)}$ $\Delta E = E_{Perpendicular} - E_{Eclipsed}$, see Ref.[295].

favored when X=CH$_3$ while the perpendicular conformation is favored when X is a highly electronegative first ion atom or group. Also, the energy difference between the two conformations increases as the electronegativity of X increases, *i.e.*, F > OH > CN[296)].

The above analysis also applies to substituted 1-propyl cations which can exist in either of the geometries shown below.

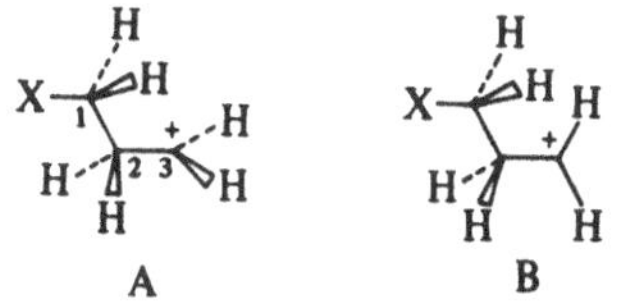

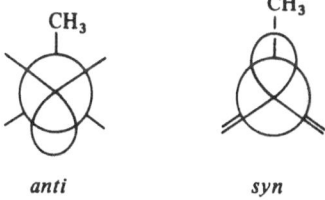

The preferred conformer will be (A) since the C–C bond is a better donor. However, as the electronegativity of the substituent increases, the donating ability of the C–C sigma bond should decrease. Consequently we expect the energy difference between (A) and (B) will decrease as the substituent X is made more electronegative.

The results of recent *ab initio* computations are shown in Table 39. As the substituent electronegativity increases the energy difference between (A) and (B) does indeed decrease. However, this is a limited trend since OH leads to a smaller difference than F. Additional factors must be involved.

Table 39. Rotational barriers for substituted 1-propyl cations[a)]

X	ΔE[b)](kcal/mol)
CH$_3$	3.73
H	2.52
F	2.11
OH	.91

a) See Ref.[297)].
b) $\Delta E = E_{(B)} - E_{(A)}$.

We now consider an example of a non-planar carbocation, *i. e.*, the tetrahedral 1-propyl cation. The two conformers to be considered are the *syn* and *anti* conformers shown below:

CH$_3$

CH$_3$

anti

syn

159

From the previous discussions, we expect the *anti* geometry to be stabilized due to the larger overlap integral between the C–C sigma bond and the empty sp³ orbital.

Ab initio calculations on these conformers indicate that the *anti* geometry is 6.04 kcal/mol more stable than the *syn*[297]. Some of this energy difference is un-doubtedly due to the eclipsing interaction of the hydrogens but this can not account for the total rotational barrier. Thus, the σ–p$^+$ interaction effect must be responsible for some major component of the rotational barrier.

At this point we pause to examine the controversy surrounding the so called Baker-Nathan order in light of our previous discussion concerning the donor and/or acceptor capabilities of bonds and lone pairs. The Baker-Nathan effect refers to certain experimental observations which suggest that in highly electron demanding reactions in solution the effect of alkyl substituents is not that expected on the basis of inductive effects, *i. e. t*–Bu $>$ *i*–Pr $>$ Et $>$ Me[298]. In the 1950's the Baker-Nathan order was, at first, interpreted in terms of hyperconjugation[299]. Specifically, C–H hyperconjugation was hypothesized to be more important than C–C hyperconjugation, *i. e.* the C–H bond was assumed to be a better donor than the C–C bond. This view is in opposition to the arguments presented in this work as well as to *ab initio* calcu-lations which support the contention that the C–C bond is a better electron donor than the C–H bond[295]. It is apparent, therefore, that the Baker-Nathan effect is due primarily to some other factor. A likely candidate is differential inhibition of solvation by the alkyl substituents. This interpretation of the Baker-Nathan order was proposed twenty years ago by Schubert and co-workers on the basis of chemical and physical measurements designed to test their hypothesis[300–304]. Their emphasis on the importance of steric hindrance to solvation in accounting for the Baker-Nathan order has found recent support from ion cyclotron resonance experiments[305].

7.4. π–π Interactions

We have already discussed the structural effects of π–π interactions in Part. II. The reader is referred to Section 3.15 for a discussion of pi interactions in butadiene which represents a model system for these interactions.

7.5. Competitive n–π, σ–π, and π–π Effects

In Part II we commented that pi and sigma nonbonded interactions can reinforce or oppose each other. The same situation obtains in the case of pi conjugative inter-actions. As an example consider the three conformations of diazabutadiene:

cis	gauche	trans
π–π* (*syn*)	π–π* (*gauche*)	π–π* (*anti*)
n_N–σ^*_{NC} (*anti*)	n_N–π*	n_N–σ^*_{NC} (*syn*)

The stabilizing bond orbital interactions which obtain in each conformation are listed above. From our discussion of butadiene (Section 3.15), we conclude that the two electron stabilization energy arising from the π–π^* interaction will vary in the following way *gauche > trans > cis*. Since a pi bond is a better acceptor than a sigma bond we conclude that the n_N–π^* stabilizing interaction can dominate the n_N–σ^*_{NC} interactions. It is clear, therefore, that the *gauche* conformation will be more stable than the *cis* or *trans* conformers when only electronic effects are considered.

8. Tests of n–π, σ–π and π–π Interactions

8.1. Physical Manifestation of n–π, σ–π, and π–π Interactions

In this section we shall examine the effects of n–π and π–π interactions on the ionization potentials of substituted ethylenes and benzenes. A theoretical analysis has already been given in section 1.1. In the space below we survey some pertinent data.

A. First Period Pi Electron Releasing Substituents (R)

From the ionization potential listed in Table 40 it can be seen that substitution of ethylene or benzene by an electron releasing group of the first period lowers the ionization potential. Also, as the energy gap between the ethylene π MO and the substituent p_z AO decreases the change in ionization potential increases. Thus, the energy change for F substitution is 0.22 eV while that for OMe substitution is 1.59 eV.

B. Second Period Pi Electron Releasing Substituents (R′)

The expected lowering of the ionization potential is revealed by the data in Table 40.

C. Pi Electron Withdrawing Substituents (W)

From the ionization potential data for acrolein and acrylonitrile, it can be deduced that in both cases the major interaction of the ethylenic π MO is with the π* of the CH=O and C≡N groups, respectively.

D. Unsaturated Substituents (U)

The data shown in Table 40 for butadiene and biphenyl support our analysis of Section 1.1.

Table 40. Ionization potentials of substituted ethylenes and benzenes (eV)

X	 $\diagdown$ X	 X ⬡	Ref.
H	10.52	9.40	239, 306)
F	10.3	9.50	234, 306)
Cl	9.9	9.31	234, 306)
Br	9.80	9.25	37, 306)
OMe	8.93	8.54	237, 306)
CH_3	9.73	8.82	30a, 306)
CHO	10.93	9.80	30a, 306)
CN	10.92	10.02	34, 306)
C=C	9.07	8.48	30a, 306)
 ⬡	8.48	8.20	30a, 37)

8.2. Spectroscopic Probes of n–π, σ–π, and π–π Interactions

A. UV Spectroscopy

Using the simple analysis outlined in Section 8.1, it is possible to predict the effects a substituent will have on the observed ultraviolet absorption bands of benzene. As was done before, substituents as well as their effects can be classified into four major groups.

   a) First Period Pi Electron Releasing Substituents. Since the effect of this substituent is more pronounced on the π than the $\overset{*}{\pi}$ MO, the UV absorption band should be red shifted. As can be seen from Table 41, both –OH and –NH$_2$ group cause a bathochromic shift and, as expected, the magnitude of the shift is greater in aniline than in phenol.

   b) Second Period Pi Electron Releasing Substituents. If d orbital participation is important, the absorptions should be red shifted by an amount substantially more than in case (a). From the data in Table 41 for chloro and bromo benzene, it is concluded that d orbital participation in these molecules is small.

   c) Pi Electron Withdrawing and Unsaturated Substituents. In both cases, the UV absorption should be red shifted substantially. Again, this is borne out by the data in Table 41.

Table 41. UV absorption of substituted benzenes[a]

X	$^1A_{1g} \rightarrow {}^1B_{1u}$(nm)	$^1A_{1g} \rightarrow {}^1B_{2u}$(nm)
H	203.5	254
Cl	209	263.5
Br	210	261
CH_3	206.5	261
OH	210.5	270
OMe	217	269
NH_2	234	284
C=C	248	282, 291
CHO	249.5	

[a] Ref.[307].

8.3. Reactivity

It is well known that the rate of electrophilic addition to olefins generally increases as the HOMO of the alkene is raised energetically, *i. e.*, its ionization potential is lowered. Pertinent experimental data have recently been reviewed[308].

9. Structural Effects of n–σ Interactions

We will illustrate the importance of n–σ orbital interactions by discussing the geometrical and conformational preferences of the following types of molecules:
a) $R_2-A-A-R_1$
b) $R_3R_2-A-A-R_1$
c) $R_4R_3R_2-A-A-R_1$
d) $R_4R_3-A-A-R_1R_2$
e) $R_5R_4R_3-A-A-R_1R_2$
f) Saturated heterocycles.

A. R_2A-A-R_1 Molecules

The first R_2AAR_1 molecule which we shall examine is diimide, N_2H_2. This molecule can exist in *cis* and *trans* geometries and the stabilizing interactions which obtain in the two geometries are specified below:

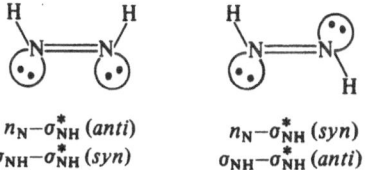

$n_N-\sigma^*_{NH}$ (*anti*)
$\sigma_{NH}-\sigma^*_{NH}$ (*syn*)

$n_N-\sigma^*_{NH}$ (*syn*)
$\sigma_{NH}-\sigma^*_{NH}$ (*anti*)

163

The dominant interaction is the n–σ* interaction and since it is maximized in an *anti* arrangement, it is concluded that HN=NH will be expected to exist in the sterically crowded *cis* geometry. Of course, if the n–σ* interactions are weak they will not be able to reverse the *trans* over *cis* preference dictated by steric effects.

The relative energies, atomic charges (in parenthesis) and bond overlap populations of *cis* and *trans* N_2H_2 are shown in Scheme 14.

Scheme 14. 4–31G Calculation at STO–3G optimized geometry

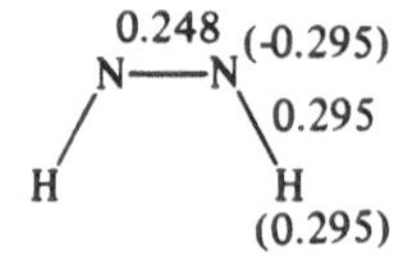

$$0.248 \quad (-0.295)$$
$$N—N$$
$$0.295$$
$$H \quad H$$
$$(0.295)$$

E_{total} = -109.792 a.u.

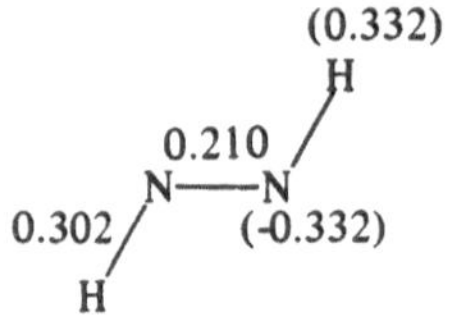

$$(0.332)$$
$$H$$
$$0.210$$
$$N—N$$
$$0.302 \quad (-0.332)$$
$$H$$

E_{total} = -109.805 a.u.

The relative magnitude of charges, overlap populations and total energies of the geometric isomers of N_2H_2 as predicted on the basis of each of the three important effects thought to determine molecular structure (see Section 6.0) are shown in Table 42. Comparison of the *ab initio* data with the various predictions reveals that

Table 42. Predicted charges, overlap populations (O. P.), and relative stability of the geometric isomers of N_2H_2

Property	Sigma conjugative effect	Nonbonded interaction effect	Electrostatic or steric effect	*Ab initio*
H Positive charge	*trans > cis*	–	*trans > cis*	*trans > cis*
N–H O.P.	*trans > cis*	–	*trans > cis*	*trans > cis*
N=N O.P.	*cis > trans*	–	*trans > cis*	*cis > trans*
Relative Stability	*cis > trans*	–	*trans > cis*	*trans > cis*

sigma conjugative effects are larger in the *cis* isomer, *e. g.* the N=N overlap population varies in the order *cis > trans*. However, the *trans* isomer is found to be more stable than the *cis* isomer[309]. Thus, we conclude that geometric isomerism in N_2H_2 is dominated by steric effects.

164

We can increase the magnitude of the n–σ* interactions by replacing one or both H's by a more electronegative atom or group. For example, consider the molecule difluorodiazene. The stabilizing interactions which obtain in the *cis* and *trans* geometries are specified below:

$n_N-\sigma^*_{NF}$ (anti) $n_N-\sigma^*_{NF}$ (syn)
$\sigma_{NF}-\sigma^*_{NF}$ (syn) $\sigma_{NF}-\sigma^*_{NF}$ (anti)

Again, the *cis* isomer is stabilized to a greater extent by sigma conjugative interactions than is the *trans* isomer.

The *ab initio* data for N_2F_2 are displayed in Scheme 15 and predictions based on considerations of each of the three important effects previously discussed are collected in Table 43. Comparison of the *ab initio* data with the

Table 43. Predicted charges, overlap populations (O. P.), and relative stability of the geometric isomers of N_2F_2

Property	Sigma conjugation effect	Nonbonded interaction effect	Electrostatis or steric effect	Ab initio
F negative charge	cis > trans	trans > cis	trans > cis	cis > trans
N–F O.P.	trans > cis	cis > trans	trans > cis	trans > cis
N=N O.P.	cis > trans	trans > cis	trans > cis	cis > trans
Rel. Stability	cis > trans	cis > trans	trans > cis	cis > trans

various predictions reveals that sigma conjugative effects are consistent with all *ab initio* data. Two slight anomalies are noted. First, the relative N=N overlap populations in the two isomers varies depending upon the basis set. However, the "superior" basis set in describing bonding, i. e. the 4–31G basis set, is the one which yields results consistent with expectations based on consideration of sigma conjugative effects. Second, the STO–3G basis set predicts greater stability of the *cis* isomer, in agreement with experimental results, while the 4–31G predicts the opposite. However, this latter basis set is known to exaggerate dipolar effects[310] and this is probably responsible for the underestimation of the stability of *cis* N_2F_2. An indication of this deficiency of the 4–31G basis set in describing interactions between highly polar bonds is provided by the fact that a 3 x 3 configuration interaction (CI) calculation of the relative stability of the geometric isomers of N_2F_2 including a ground, lowest singly excited and lowest doubly excited configurations reverses the relative stability order, i.e. *cis* becomes favored over *trans* by

Part IV. Conjugative Interactions

16.48 kcal/mol. This type of CI improves the ionicity of the ground state wavefunction and possibly removes some of the inadequacy of this basis set in treating polar bonds.

Scheme 15.
A. STO–3G Calculation at STO–3G geometry

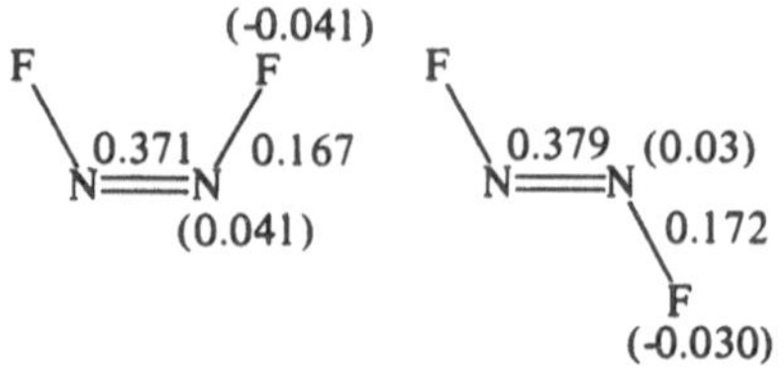

$E_{rel.}$(kcal/mol)	0.000	.634
N_T^π	.17057	.17520
P_{FF}^π	.00015	.00000
P_{FF}^σ	−.00016	.00000

B. 4–31G Calculation at STO–3G geometry

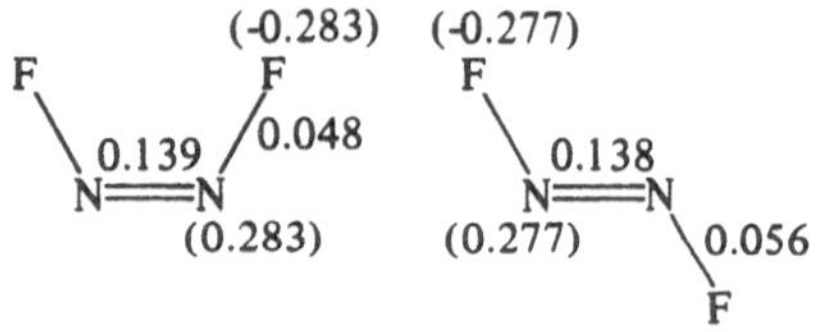

E_{rel}(kcal/mol)	2.535	0.000
N_T^π	.18317	.17959
P_{FF}^π	.00115	.00002
P_{FF}^σ	.00003	.00009

Cis N_2F_2 is found experimentally to be more stable than *trans* N_2F_2. The energy difference between the two isomers is found to be 3.0 kcal/mol[87]. On the basis of our discussion, we suspect that this substantial energy difference between the geometric isomers of N_2F_2 originates from both sigma conjugative and pi nonbonded interaction effects.

The demonstration of strong n–σ* conjugative effects in XN=NX allows one to make certain qualitative predictions of substituent control upon the relative stability of the geometric isomers. Thus, it is expected that as X becomes increasingly electronegative, thus improving the n–σ* interaction maximized in the *cis* geometry, the *cis-trans* energy difference will tend to favor the *cis* isomer. This prediction can be tested by studying the graded series $CH_3N=NCH_3$, $CF_3N=NCH_3$ and $CF_3N=NCF_3$. Available experimental results, though not conclusive yet, indicate that the first

166

two members of the series have a preferred *trans* geometry but the last is most likely more stable in the *cis* geometry[311, 312].

Another typical $R_1 AAR_2$ system is the molecule HO–OH, one of the simplest tetraatomic molecules where conformational preference can be observed. The dominant interaction is n_O–σ^*_{OH} and the preferred conformation of the molecule should be the one which allows an oxygen lone pair AO to interact maximally with an O–H antibonding MO. The preferred geometry can be predicted assuming an sp^2 hybridized oxygen.

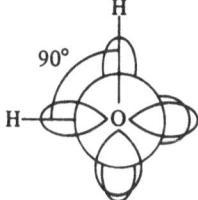

In this case, the most stable geometry is the one which allows the 2p lone pair of oxygen to interact with the O–H bond simply because the 2p lone pair is a better donor than the sp^2 lone pair [37]. We can then predict that the angle of the most stable conformer of HOOH will be near 90°.

We can now attempt something more ambitious. Specifically, we shall try to predict the relative stability of the *cis, trans* and *gauche* conformers of HOOH by drawing Newman projections and specifying the key sigma conjugative interactions in each case.

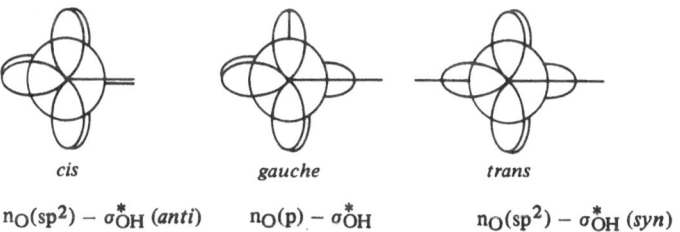

cis	gauche	trans
$n_O(sp^2) - \sigma^*_{OH}$ (*anti*)	$n_O(p) - \sigma^*_{OH}$	$n_O(sp^2) - \sigma^*_{OH}$ (*syn*)

Our qualitative analysis predicts an order of relative stability which is *gauche* > *cis* > *trans* and a *gauche* angle of 90°. The prediction of the relative stability of the conformers is based on the realization that a 2p lone pair is a better donor than an sp^2 lone pair and a n–σ* interaction is stronger for an *anti* than a *syn* orientation.

Ab initio calculations show that the relative energies of the *cis, trans* and *gauche* conformers of HOOH are as follows[313]:

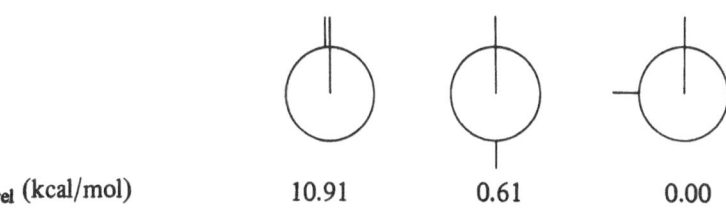

| E_{rel} (kcal/mol) | 10.91 | 0.61 | 0.00 |

The dihedral angle of the *gauche* structure has been found to be 123 degrees. Thus, the predicted order of relative stability of the three conformers differs from the one revealed by *ab initio* calculations by an inversion in the relative stability of the *cis* and *trans* forms. This arises because in HOOH the n–σ* interactions are not particularly strong and, thus, cannot overrule the *trans* over *cis* preference dictated by steric effects. We can increase the strength of the n–σ* interactions by replacing one or both H's by more electronegative atoms or groups, *e.g.* fluorines. Typical data are shown in Table 44. The order of stability in FOOH is *gauche* > *cis* > *trans*[314] indicating a

Table 44. Calculated relative energies of conformers of HO_2F and F_2O_2

FO–OH[a)	*gauche*	0.00
	cis	3.33
	trans	7.39
FOOF[b)	*gauche*	0.00
	cis	16.61
	trans	20.81

a) See Ref.[314].
b) See Ref.[315].

substantial n_O–σ^*_{OF} interaction which overcomes the unfavorable steric repulsions in the *cis* conformer. The order of stability in FOOF is also found to be *gauche* > *cis* > *trans*. In this latter case, an important n_O–σ^*_{OF} interaction as well as a non-bonded attractive interaction between the fluorines are at work in determining conformational preference.

How can we prove that the relative stabilities of the *gauche, cis* and *trans* conformers of HOOF are dictated by n–σ* interactions? A simple way is to follow the gross charges of oxygen, hydrogen and fluorine in a representative system such as HOOF as a function of conformation. Recalling that an increasingly strong n–σ* interaction will tend to deplete charge from the hydroxyl oxygen and accumulate charge on the fluorine and the adjacent oxygen atom, we expect the hydroxyl oxygen to be most electron rich in the *trans* conformation and least electron rich in the *gauche* conformation. The converse holds true for the fluorine and the oxygen attached to fluorine.

Typical CNDO/2 results are shown below confirming most predictions.

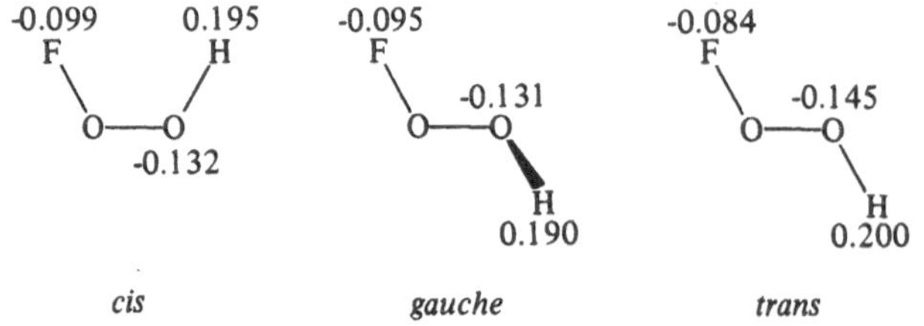

cis   gauche   trans

The only anomaly occurs in the case of the fluorine charge variation. However, recalling that electrostatic effects can be important in determining the charge distributions in molecules, this apparent anomaly can be explained. Specifically, the negative charge on fluorine becomes largest in the *cis* isomer in order to maximize an attractive F––H electrostatic interaction. This would explain why the calculated order for increasing negative charge on fluorine which we expected to be *gauche* > *cis* > *trans* on the basis of n–σ* interactions above, is *cis* ≈ *gauche* > *trans*. The gross charges on fluorine and oxygen in O_2F_2 are shown below[315]:

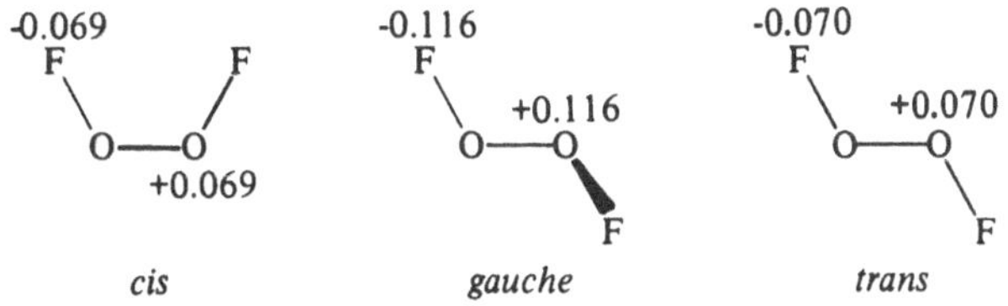

-0.069
F   F
O—O
+0.069

cis

-0.116
F
+0.116
O—O
F

gauche

-0.070
F
+0.070
O—O
F

trans

A prediction based only on the importance of n–σ* interactions would give an order of increasing positive charge on oxygen as *gauche* > *cis* > *trans* and an order of increasing negative charge on fluorine as *gauche* > *cis* > *trans*. We find, however, that the calculated order of increasing positive and negative charge of oxygen and fluorine respectively is *gauche* > *cis* ≈ *trans*. Once again, electrostatic effects seem to be involved.

In closing, we should mention that the importance of a hyperconjugative mechanism in enforcing *gauche* preference in molecules containing adjacent lone pairs, such as H_2O_2, N_2H_4, etc., has been discussed by Pople *at al*[310, 314]. However, these workers did not differentiate between *syn* and *anti* geometries on the basis of such a mechanism.

B. $R_3R_2A–A–R_1$ Molecules

Particularly intriguing molecules of the $R_3R_2A–A–R_1$ type are those containing adjacent nitrogen and oxygen atoms. Thus, NH_2OH is predicted to exist in a crowded W conformation which places the N lone Pair *anti* to the OH bond.

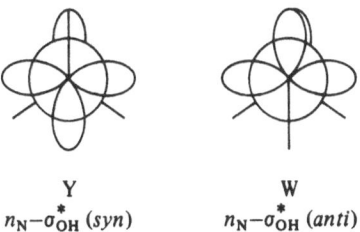

Y
$n_N–\sigma^*_{OH}$ (*syn*)

W
$n_N–\sigma^*_{OH}$ (*anti*)

However, *ab initio* calculations show that the Y conformation is more stable than the W conformation[314]. As in the case of HOOH, n–σ* interactions are present but weak and cannot overide the preference for the Y conformation dictated by conventional steric effects. Nonetheless, the increase in n–σ* conjugative interactions

169

in going from the Y to the W conformation should be manifested in a decrease of the gross charge in nitrogen and an increase, *i.e.* less positive, at the charge of the hydroxyl hydrogen atom. Results of CNDO/2 calculations for NH_2OH confirm our expectations:

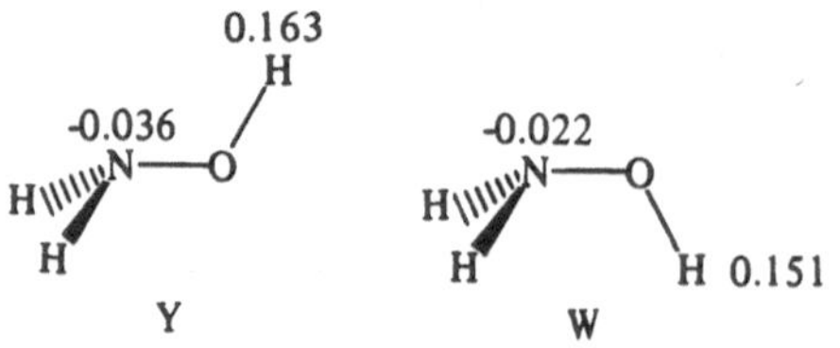

Once more, replacement of the hydroxyl hydrogen by a group or an atom more electronegative than hydrogen, replacement of the hydroxyl oxygen by its third period counterpart, or, a combination thereof, will enhance $n-\sigma^*$ interaction and should lead to a preference for the "crowded" W conformation over the Y conformation. This expectation is confirmed by *ab initio* calculations in which the preferred conformation for NH_2OF is found to be the more "crowded" W conformer[314].

$R_1R_2C=NR_3$ molecules represent another important class of $R_1R_2AAR_3$ systems. In the parent system shown below the strength of the various sigma interactions varies in the order:

$$n_N-\sigma^*_{CH} \; (anti) > n_N-\sigma^*_{CH} \; (syn) > \sigma_{CH}-\sigma^*_{NH} \; (anti)$$

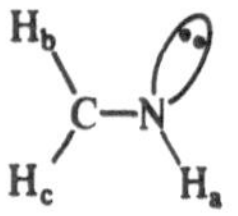

Accordingly, if sigma conjugative effects dominate, the hydrogen atoms are predicted to become increasingly positive in the order $H_a > H_b > H_c$ and the C–H overlap populations to vary in the order $C-H_c < C-H_b$.

The various atomic charges and bond overlap populations of $CH_2 = NH$ are shown below and confirm our expectations based on consideration of sigma conjugative effects. However, an anomaly is noted in the case of the 4–31G calculation of the C–H_b and C–H_c overlap populations. Since the STO–3G optimization leads to a longer C–H_c bond, as predicted, the anomaly represents most likely a computational artifact.

Scheme 16. 4–31G Calculation at STO-3G geometry

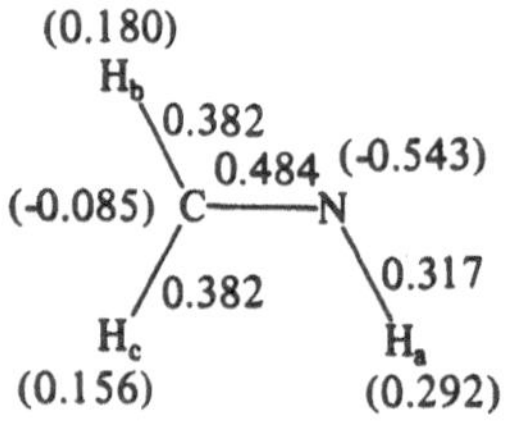

Accordingly, these intramolecular comparisons where sigma conjugative effects are not pitted against other important effects provide good evidence of the superiority of *anti* orbital overlap.

Turning now to the case of geometrical isomerism in CHF=NF we proceed as before by identifying the major stabilizing sigma conjugative interactions present in the *cis* and *trans* isomers.

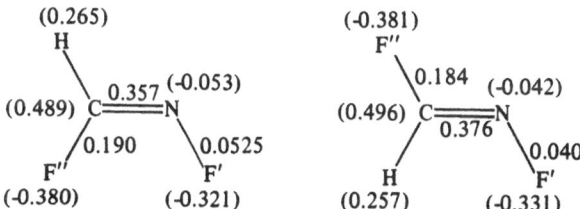

Arguing as before, we predict that the *cis* isomer of CHF=NF will be more stable than the *trans* isomer.

The *ab initio* data for CHF=NF are displayed in Scheme 17 and predictions based on consideration of each of the three important effects previously discussed are collected in Table 45. Comparison of the *ab initio* data with the various predictions

Table 45. Predicted charges, overlap populations (O. P.), and relative stability of the geometric isomers of CHF"= NF'

Property	Sigma effect	Nonbonded effect	Electrostatic or steric effect	*Ab initio*
H positive charge	*cis > trans*	–	*trans > cis*	*cis > trans*
F' negative charge	*cis > trans*	*trans > cis*	*trans > cis*	*trans > cis*
F" negative charge	*cis > trans*	*trans > cis*	*trans > cis*	*trans > cis*
C–F" O.P.	*trans > cis*	*cis > trans*	*trans > cis*	*cis > trans*
N–F' O.P.	*trans > cis*	*cis > trans*	*trans > cis*	*cis > trans*
C=N O.P.	*cis > trans*	*trans > cis*	*trans > cis*	*trans > cis*
Rel. stability	*cis > trans*	*cis > trans*	*trans > cis*	*cis > trans*

Scheme 17.
A. 4–31G Calculation at STO–3G geometry

(0.265)
H
|
(0.489) C ===N (-0.053)
   0.357
/   \
0.190   0.0525
F"   F'
(-0.380)   (-0.321)

(-0.381)
F"
|
(0.496) C ===N (-0.042)
   0.184   0.376
/   \
H   0.040
(0.257)   F'
            (-0.331)

Scheme 17. (continued)

E_{rel}(kcal/mol)	0.000	.856
N_T^π	.19489	.19104
P_{FF}^π	.00063	.00001
P_{FF}^σ	.00004	.00008

B. 4−31G Calculation at 4−31G geometry

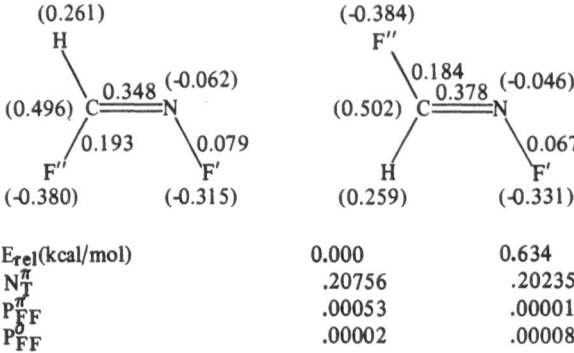

E_{rel}(kcal/mol)	0.000	0.634
N_T^π	.20756	.20235
P_{FF}^π	.00053	.00001
P_{FF}^σ	.00002	.00008

reveals that F————F nonbonded interaction dominates geometric isomerism in CHF=NF. The large difference in the total pi overlap populations and long range overlap populations between the *cis* and *trans* isomers (Scheme 17) provides yet another demonstration of the importance of nonbonded interaction in determining structural preferences.

The importance of sigma conjugative interactions can be realized by restricting ourselves to intramolecular comparisons. The order of strength of sigma conjugative interactions is $n_N-\sigma_{CF''}^* > n_N-\sigma_{CH}^* > \sigma_{CH}-\sigma_{NF'}^*$ and thus we predict that within a given isomer F' should be less negative than F'', as exactly revealed by the *ab initio* calculations. Once again, these data provide unmistakable support for the idea that *syn* and *anti* orbital overlap lead to differing stabilization[h].

---

[h] In a recent publication, Howell discussed *cis-trans* isomers of diazenes and substituted methyleneimides[411]. This author noted the charge effects which we have discussed in Part. IV. He was hesitant to attribute them to sigma conjugative effects although he did suggest this possibility. For example, this author stated "... we are hesitant to accept hyperconjugation as the sole, decisive factor in determining the energies of different isomers. If hyperconjugation were dominant, we would expect the C=N overlap population would be higher (and the bond length shorter) in H\C=N than in H\C=N–H

and in *cis*−FHC=NF than in *trans*−FHC==NF." However, we have seen that nonbonded attraction dominates sigma conjugative effects in the latter two molecules. In short, our demonstration of some key algebraic relationships which renders *anti* superior to *syn* n−σ* overlap, coupled with the realization of the importance of nonbonded attractive effects, may well remove any hesitancy to accept the importance of sigma conjugation.

172

One of the prime examples of the extreme importance of conjugative effects is the case of XYC=Z molecules. As Z becomes more electronegative, the following trends are expected:

a) The C–Z, C–Y and C–X bond lengths should become stronger on account of the effect of electronegativity on bond strengths which is discussed in Section 12.3.

b) The XĈY bond angle is expected to open up on account of geminal interactions discussed in Section 5.1.

c) If X and Y carry lone pairs, the change in the attractive X–––Y interaction will probably be small compared to the aforementioned effects.

The variation of the geometric parameters in XYC=Z which accompanies a variation of Z is instriguing to the extent that it cannot be understood solely in terms of the above effects. Thus, the following experimental trends have been recently projected to the attention of chemists:

a) $RC\overset{\overset{\displaystyle O}{\|}}{}$–H and $RC\overset{\overset{\displaystyle O}{\|}}{}$–Alkyl bonds have lower dissociation energies than the corre-

sponding bonds in $RC\overset{\overset{\displaystyle CH_2}{\|}}{}$–H and $RC\overset{\overset{\displaystyle CH_2}{\|}}{}$–Alkyl

b) C–X bond distances in $RC\overset{\overset{\displaystyle O}{\|}}{}$–X are longer than in $RC\overset{\underset{\displaystyle CH_2}{\|}}{}$–X

c) In comparing C–X distances in $RC\overset{\overset{\displaystyle O}{\|}}{}$–X relative to CH_3–X, the quantity Δr which equals r (X–COR) – r (X–CH$_3$) assumes values which range from positive to negative although one might have anticipated shorter C–X distances in RCO–X than in CH$_3$–X, i. e. negative Δr due to the different states of hybridization of the carbon atoms involved. Furthermore, the magnitude of the *shortening* in r($RC\overset{\overset{\displaystyle O}{\|}}{}$–X) varies in the order NH$_2$ > OH > F > CH$_3$[i)] while the magnitude of the *lengthening* in r($RC\overset{\overset{\displaystyle O}{\|}}{}$–X) varies in the order CF$_3$ > Br > Cl > CN[316)]. Typical data are collected in Table 46. It can be seen that the above sequences also correlate with the variation of the C=O distances.

The above trends can be easily explained if we focus upon the conjugative interaction of the in plane 2p lone pair of the carbonyl oxygen with the C–X bond in RCX=O. As X is varied along a row, increasing electronegativity along a row should lead to greater interaction and thus greater shortening of the C=O bond and greater loosening of the C–X bond relative to a standard. This is precisely what is observed (see Table 46). Furthermore, as X is varied along a column, decreasing electronegativity

---

[i)] *Ab initio* calculations of OHC–F and OHC–OH show that the σ*MO is mainly σ^*_{C-X}. See Ref. 20).

Part IV. Conjugative Interactions

$$\overset{O}{\underset{||}{}}$$

Table 46. Geometric parameters of $R-C-X$ molecules

R=H				R=CH$_3$			
X	Δr(x 10^{-3})[b]	r(C=O)[a]	$\widehat{OCX}$	Ref.	Δr(x 10^{-3})[b]	r(C=O)[a]	$\widehat{OCX}$	Ref.
NH$_2$	-122	1.219	124.7	317)	-87	1.220	122.0	320)
OH	- 66	1.217	123.4	318)	-62	1.215	122.8	321)
F	- 47	1.181	122.8	319)	-37	1.181	121.3	322)
Cl	--	--	--		+13	1.187	120.3	323)
Br	--	--	--		+37	1.184	120.8	323)

[a] In angstroms.
[b] $\Delta r = r(X-COR) - (X-CH_3)$.

should have the same effect. With the exception of the anomaly of the C=O bond length when X=F, these predictions are strikingly borne out by experiment as the data of Table 46 indicate. It should be noted that an extremely large and positive Δr value is obtained when X=CF$_3$, i. e. a group which strongly encourages "hyperconjugation". The effect of trifluoromethyl groups upon geometrical and conformational preferences has already been discussed.

Replacement of =O by =CH$_2$ will diminish the conjugative effect formerly expressed via the $n_O - \sigma^*_{CX}$ interaction. As a result, C–X bond shortening will occur.

It should be noted that the $\widehat{OCX}$ angle variation seems to be accountable in terms of geminal interactions for X varying along a row of the Periodic Table.

Conjugative interactions also play a role in determining bond angles. This can be best appreciated by comparison of the XCX angles in the two molecules shown below. As we have seen already, geminal interactions favor a smaller molecules shown below. As we have seen already, geminal interactions favor a smaller angle in X$_2$C=CH$_2$. On the other hand, conjugative interactions are stronger in X$_2$C=O and their maximization involves shrinkage of the XCX angle. These considerations are best understood in terms of the delocalized approach as illustrated in Fig. 15.

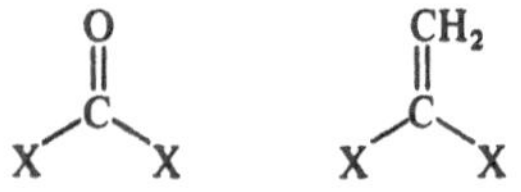

Table 47 shows typical results which illustrate that contrary to expectations based on consideration of geminal interactions alone, the XCX angle is smaller in CX$_2$=O than in CX$_2$=CH$_2$.

A fascinating corollary of the above analysis is the following: suppose that the HĈH angle in CH$_2$=O is progressively constrained to increasingly lower values. As a result, the conjugative interaction will increase and give rise to stronger sigma C=O bonding. This prediction can be tested by reference to the infrared absorption frequencies of cyclic carbonyl molecules constrained to have different RĈR angles. Typical data is shown in Table 48 and are in accord with our predictions. It appears

174

Table 47. XCX angles in $CX_2=O$ and $CX_2=CH_2$

X	$X\hat{C}X$ Angle in $CX_2=O$ (deg.)	$X\hat{C}X$ Angle in $CX_2=CH_2$ (deg.)	Ref.
H	116.5	116.2 + .8	326, 327)
F	108 ± .5	109.3 ± .6	328, 329)
Cl	111.3 ± .1	114.5 ± 1.0	330, 331)

Table 48. Infrared carbonyl stretching absorption frequencies of cyclic ketones[a]

Molecule	Angle (deg.)	ν, cm^{-1}
 (cyclopropanone)	60	1815
 (cyclobutanone)	90	1791
 (cyclopentanone)	108	1850–1730
 (cyclohexanone)	120	1717

[a] See Ref. 324 and 325.

that the Foote-Schleyer correlation[324, 325] may be based upon the electronic effect described here.

It should be noted that the proposed mechanism for the variation of the absorption frequency of the carbonyl group as a function of the angle, involves not only a progressive strengthening of the C=O bond but also the progressive weakening of the R–C bond as the RCR angle shrinks in $R_2C=O$, i. e. one should view the angle effect on carbonyl absorption frequencies as an overall geometry effect.

C. $R_4R_3R_2A-A-R_1$ Molecules

Turning our attention to a more complicated case, we examine rotational isomerism in $R_4R_3R_2A-A-R_1$ systems of which CH_3OH is the simplest example. Here, the

dominant interaction is $\sigma_{CH}-\sigma^*_{OH}$ despite the fact that an oxygen lone pair AO is a better donor than a C–H sigma bond (see Tables 3 and 4). That is, the $\sigma_{CH}-\sigma^*_{OH}$ interaction is more stabilizing than the $n_O-\sigma^*_{CH}$ interaction because the energy difference between the σ^*_{OH} and σ^*_{CH} MO's is expected to be greater than the energy difference between n_O and σ_{CH} (see Tables 3 and 4). Thus, the preferred conformation is predicted to be the one which places the C–H bond and the O–H bond in an *anti* arrangement, *i. e.* the staggered conformation. An identical prediction can be arrived at on the basis of steric effects.

Replacement of a methyl hydrogen by fluorine leads to a situation where the dominant interaction is $n_O-\sigma^*_{CF}$, since in this molecule the C–H bond acquires a donor ability comparable to the oxygen lone pair and the C–F bond is a better acceptor than an O–H bond. The molecule $CH_2F–OH$ can exist in the *syn, gauche* and *anti* conformations shown below.

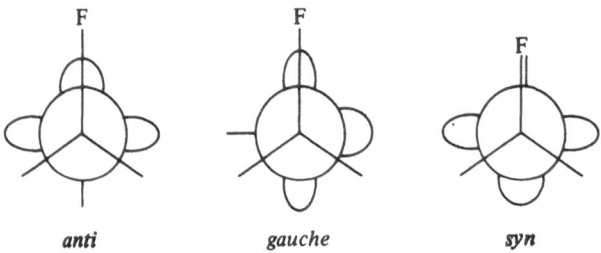

anti   *gauche*   *syn*

The simple enumeration of the type of dominant interactions which obtain in the three conformations leads to the prediction of the stability order *gauche > syn > anti*. This is indeed what *ab initio* calculations reveal[332]. This new order of stability is reflected in the fluorine gross charges as calculated at the CNDO/2 level. On the basis of the $n_O-\sigma^*_{CF}$ interaction, one expects the gross fluorine charge to increase in the order *gauche > syn > anti*. The computed order *gauche ≈ syn > anti* is understandable on the basis of an electrostatic effect which augments the fluorine charge in the *syn*-conformation in order to benefit from an H⋯F attractive interaction.

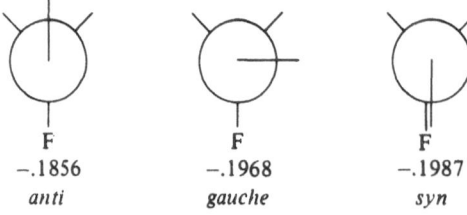

F   F   F
−.1856   −.1968   −.1987
anti   *gauche*   *syn*

Our discussion of the conformational preference of H_2CF-OH provides the simplest type of explanation for the so called anomeric effect[333] which refers to the tendency of an electronegative substituent at $C-1$ of a pyranose ring to exhibit a greater preference for the *axial* over the *equatorial* conformation than it does in cyclohexane.

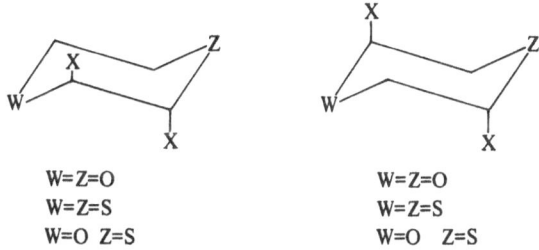

      axial   *equatorial*

According to our approach, it is predicted that, barring steric effects, the *axial* preference in pyranoses should increase in the order $X = F > OR > NR_2$ and also in the order $X = Cl > F$, $X = SR > OR$, etc., *i. e.* as the C_1-X bond becomes an increasingly better instrinsic acceptor.

Finally, mention should be made of the striking conformational preference of *trans* 2,3- and *trans* 2,5-dioxanes, dithianes, and thioxanes. Each of these compounds exists in a chair conformation with two *diaxial* halogens[334–341].

    W=Z=O   W=Z=O
    W=Z=S   W=Z=S
    W=O Z=S   W=O Z=S

Furthermore, the *axial* C–X bonds are unusually long in comparison with aliphatic C–X bonds[342]. An especially interesting observation is the difference between the *axial* and *equatorial* C–Cl bond lengths of the system shown below. This observation is in accord with greater $n_O-\sigma^*_{CCl}$ (*axial*) conjugation which cause elongation of the *axial* relative to the *equatorial* C–Cl bond.

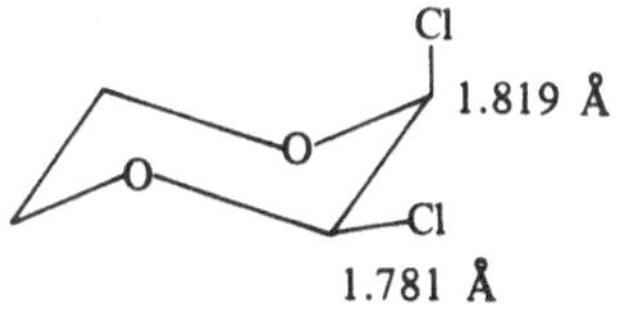

Cl
1.819 Å

Cl
1.781 Å

In conclusion, we would like to mention that a "hyperconjugative" interpretation of the anomeric effect has been offered by various workers such as Romers and Altona[333], Pople *et al.*[343] and Salem *et al.*[344].

D. $R_4R_3A-AR_1R_2$ Molecules

A typical $R_4R_3A–AR_1R_2$ molecule is hydrazine, a system which has attracted an enormous amount of theoretical interest. The *syn, anti* and *gauche* conformers of $NH_2–NH_2$ are depicted below along with an enumeration of the dominant interactions which obtain in each case:

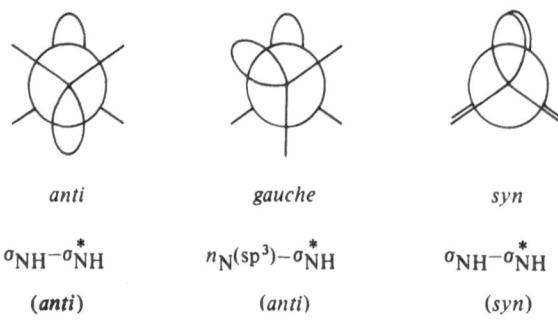

anti	gauche	syn
$\sigma_{NH}–\sigma^*_{NH}$	$n_N(sp^3)–\sigma^*_{NH}$	$\sigma_{NH}–\sigma^*_{NH}$
(anti)	*(anti)*	*(syn)*

Clearly, the preferred order of relative stability is *gauche > anti > syn*, since the n–σ* interaction is much stronger than the σ–σ* interaction. Furthermore, *anti* $\sigma_{NH}–\sigma^*_{NH}$ is more favorable than *syn* $\sigma_{NH}–\sigma^*_{NH}$ interaction. *Ab initio* calculations show that the preferred conformation of hydrazine is the *gauche* conformer with a dihedral angle of $100°$[314].

The analysis of the conformational preference of P_2H_4 proceeds along the same lines. It is again predicted that the relative stability of the various conformers will be *gauche > anti > syn*.

An interesting situation arises when one of the hydrogens of hydrazine is substituted by F. Here, we have a choice between two different *gauche* forms, one which places the lone pair of NH_2 *anti* to a N–F antibonding MO and one which places it *anti* to a N–H antibonding MO. Since the $n_N(sp^3)–\sigma^*_{NF}$ interaction is stronger than the $n_N(sp^3)–\sigma^*_{NH}$ interaction the *gauche* conformation which places the nitrogen lone pair anti to the NF bond will be the one preferred. This prediction is in agreement with the results of *ab initio* calculations in which the "internal" conformation is preferred[314].

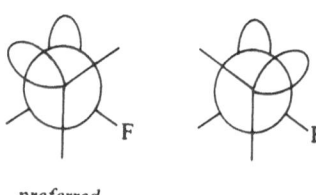

preferred

The molecule NHF–NHF can also exist in three *gauche* forms. Following the same line of reasoning, we predict a relative stability order A > B > C.

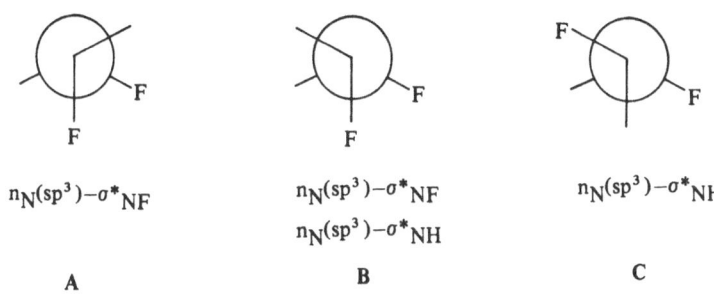

$$n_N(sp^3)-\sigma^*NF$$

A

$$n_N(sp^3)-\sigma^*NF$$
$$n_N(sp^3)-\sigma^*NH$$

B

$$n_N(sp^3)-\sigma^*NH$$

C

E. $R_5R_4R_3A-AR_1R_2$ Molecules

Our model system in this section is CH_3-NH_2. The dominant interaction is $n_N-\sigma^*_{CH}$ and the preferred conformation is predicted to be the one which places the lone pair *anti* to the C–H bond, i. e., the *anti* conformation. We note that the same conformation is predicted on steric grounds.

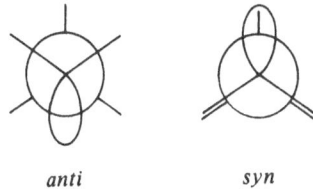

anti   *syn*

Now, consider replacement of one methyl hydrogen by an electronegative group or atom, i. e. F. The preferred conformation will then be the one which places the lone pair *anti* to the best acceptor bond, i. e. the C–F bond.

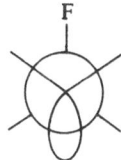

Again the dependence of the n–σ* conjugative interaction upon conformation can be seen in a comparison of the gross charge of F in the *anti*, *gauche*, and *syn* conformations of CH_2FNH_2. For CH_2FNH_2, the gross charge on fluorine should become more negative in the order *anti* > *syn* > *gauche*. Our expectations are confirmed by CNDO/2 calculations and the results are shown below:

–0.207   –0.195   –0.198

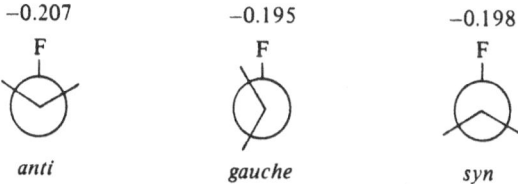

anti   *gauche*   *syn*

179

By following the strategy outlined previously, we predict that substitution of the three methyl hydrogens with groups of increasing electronegativity, $i.$ $e.$ F, OR, NR$_2$, should lead to the preferred conformation shown below:

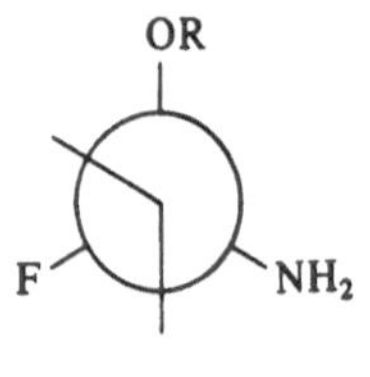

Similarly, substitution of the three methyl hydrogens by F, Cl, Br, should lead to the preferred conformation shown below.

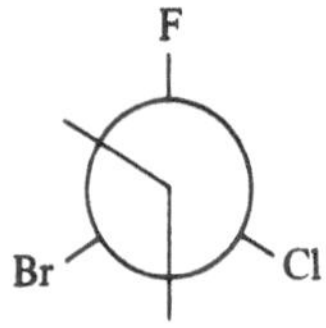

F. Saturated Heterocycles

In a previous section we have discussed the importance of n—σ* conjugative interactions in dictating the axial-equatorial preference of an electronegative substituent at C—1 of a pyranose ring, $i.$ $e.$ the "anomeric effect". Here, we shall extend the discussion of n—σ* conjugative interactions to other saturated heterocyclic molecules.

We first examine the case of piperidine. The two conformations, *axial* and *equatorial* are shown below along with an enumeration of the major stabilizing interactions, if we only consider vicinal bonds.

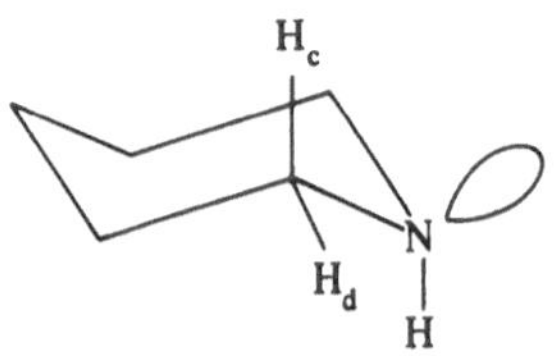

axial	*equatorial*

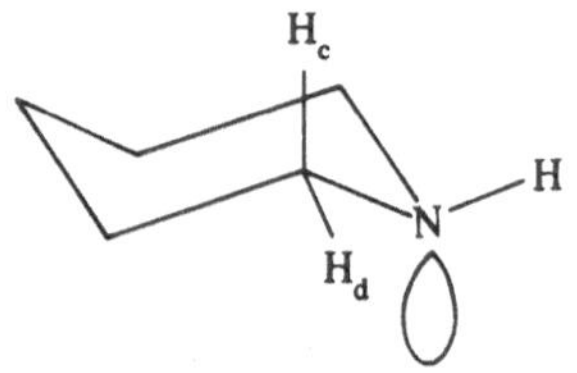

n_N—σ^*_{CC} (*anti*) n_N—σ^*_{CC} (*syn*)

n_N—$\sigma^*_{CH_d}$ (*gauche*) n_N—$\sigma^*_{CH_c}$ (*anti*)

n_N—$\sigma^*_{CH_c}$ (*gauche*) n_N—$\sigma^*_{CH_d}$ (*gauche*)

Since we expect the energy separation between σ^*_{CC} and σ^*_{CH} not to be very large it is obvious that, in this case, an unambiguous prediction cannot be made. In fact, the experimentally determined conformation of these types of six membered heterocyclic systems is found to be dependent on the heteroatom as illustrated below[345 – 347]:

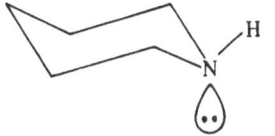

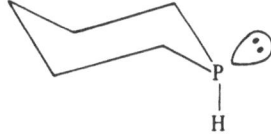

Most Stable Conformation   *Most Stable Conformation*

An *axial* or *equatorial* preference of the YH (Y=N, P) group can be dictated by a modification of the heterocyclic ring. This may be accomplished by substituting a more electronegative or electropositive atom for X=CH$_2$ in the system shown below:

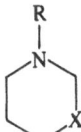

In the case of X=O , the energy separation between σ^*_{CH} and σ^*_{CO} is expected to be large and, consequently, the conformation is controlled by the $n_N-\sigma^*_{CO}$ interaction. Thus, we predict that the *axial* N–H conformation will be preferred since this is the geometry which maximizes the $n_N-\sigma^*_{CO}$ interaction.

However, in the cases shown below, the introduction of an electronegative atom, *i. e.* X=O for X=CH$_2$, is not expected to have a large effect:

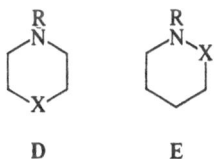

D   **E**

The –NH conformational preference in **D** and **E** is expected to be dictated by the same factors responsible for conformational isomerism in piperidine. The reasons for these expectations are as follows:

a) In **D**, the oxygen atom will lower the energy of the C$_2$–C$_3$ sigma antibonding orbital by an inductive effect. However, this effect will be small.

b) In **E**, the O–C$_3$ sigma antibonding orbital will be lower in energy than the C–H sigma antibonding orbital. However, the small oxygen AO coefficient in σ^*_{OC} will tend to diminish the acceptor ability of this bond.

Experimentally, it has been shown that the introduction of a heteroatom such as oxygen into the 4-position of piperidine has no appreciable influence on the conformational equilibrium[348]. Similarly, infrared overtone measurements suggest that in tetrahydro-1,2 oxazine the NH-*equatorial* conformer predominates[349]. On the other hand, many piperidines substituted by heteroatoms in the 3 position are found to exhibit a NR *axial* preference. For example, the preferred conformation for the molecules shown below is NR *axial*[350, 351]:

Part IV. Conjugative Interactions

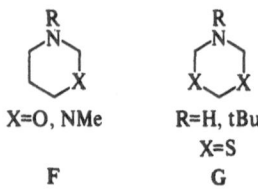

X=O, NMe   R=H, tBu
                      X=S
    F   G

The *axial* preference of R=*t*-Bu in G is dramatic evidence for the utility of recognizing donor-acceptor relationships in designing specific molecules.

Acidity of Hydrocarbons

The greater acidity of hydrogen in H than I has been interpreted by certain workers to imply predominance of inductive or field effects[352, 353]. While it is clear that in H lone pair-C–F interaction is not possible, it is also apparent that the interaction of the lone pair with an *anti* acceptor F_2C–CF_2 bond is optimized and can account for the relative rate of deuterium exchange.

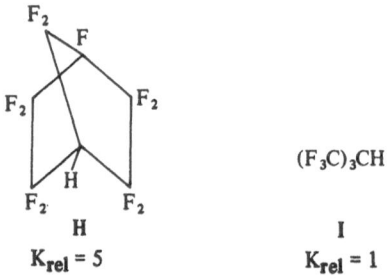

$(F_3C)_3CH$

H   I

$K_{rel} = 5$ $K_{rel} = 1$

The greater stabilization of K relative to J can also be attributed to the increasing numer of $-CF_2-CF_2$-bridges.

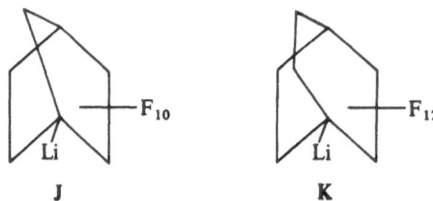

J   K

9.1. Possible Examples of Matrix Element Control of n–σ Interactions

Consider the two systems CH_2F–SH and CH_2F–OH. According to our approach both are predicted to exist in a preferred gauche conformation. However, the extent to which the $n_X-\sigma^*_{CF}$ interaction obtains in the two molecules may be subject to matrix element control simply because n_S is a better donor than n_O but yields a smaller interaction matrix element with σ^*_{CF}. The variation of these two effects may conceivably be comparable and subject to matrix element control due to the fact that the n–σ* orbital interaction involves well separated energy levels. Hence, one

182

may expect that delocalization in the case of the sulfur compound will be less than that in the case of the oxygen compound. This indeed seems to be the case judging from the structural data obtained by Altona et al.[339] for the molecule shown below.

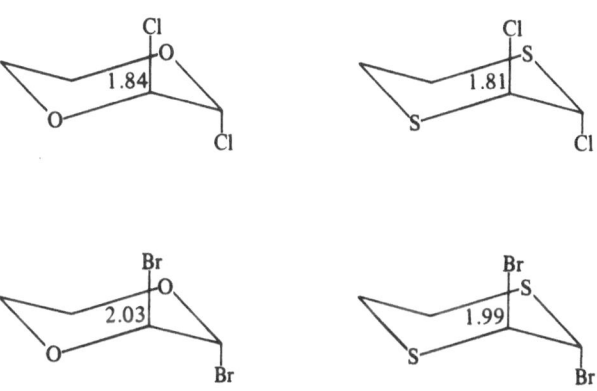

It was found that the C–Cl bond adjacent to O is longer than that adjacent to S as expected from greater charge transfer to a σ^*_{C-Cl} orbital in the case of oxygen. More examples can be found in the structural data shown below[335, 336, 339–341].

10. Tests of n–σ Interactions

In Part II we examined in detail various experimental tests of nonbonded interactions. Here, we again focus on specific physical and reactivity probes in order to test the importance of n–σ interactions in organic problems. Specifically, we shall examine the following two areas:
   a) Spectroscopic probes of n–σ interactions.
   b) Reactivity probes of n–σ interactions.

10.1. Spectroscopic Probes of n–σ Interactions

In this section, results of Photo Electron Spectroscopy (PES) as well as infrared spectroscopy shall be examined in terms of n–σ interactions.

The ionization potential of lone pair AO's is easily obtained by PES[35]. Consequently, this method can be a delicate probe of n–σ conjugative interactions. For example, consider the ionization energy of the nitrogen lone pair AO as a function of the substituent X in the model system shown below:

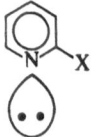

In the above molecule, we need only consider the effect of the $n_N - \sigma_{CX}$ and $n_N - \sigma^*_{CX}$ interactions on the energy of n_N. In most cases, the interaction which perturbs the n_N AO the most is the $n_N - \sigma_{CX}$ interaction due to the energy proximity of the orbitals. We can make the following predictions:

a) As X becomes increasingly more electronegative along a row of the Periodic Table, the ionization energy of the nitrogen lone pair AO will increase.

b) As X is varied down a column of the Periodic Table, the ionization energy of the nitrogen lone pair AO will decrease.

Experimental support for the above analysis is found in 2 substituted pyridines where X=CH$_3$ and X=Si(CH$_3$)$_3$. Specifically, the ionization potential of the nitrogen lone pair in the former case is 9–10eV and 8.5eV in the latter[355].

Another example can be found in the ionization energy of the in plane lone pair AO of X–CO–X molecules:

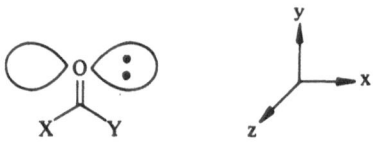

The analysis of these molecules is similar to the one presented above. Pertinent experimental PES data are displayed in Table 49. As can be seen, the ionization potential of the 2p$_x$ AO type lone pair is consistently lower when Y is a –SiR$_3$ group.

A series of interesting papers have appeared relating NH stretching frequencies to the geometrical relationship between the nitrogen lone pair and adjacent bonds.

Table 49. Ionization potential of the oxygen in plane lone pair in XCOY molecules[356]

X	Y	I.P.
CH$_3$	C(CH$_3$)$_3$	9.24
CH$_3$	Si(CH$_3$)$_3$	8.64
Ph	C(CH$_3$)$_3$	8.98
Ph	Si(CH$_3$)$_3$	8.41

It was found that the NH stretching frequency increases and the adjacent CH stretching frequency decreases more when the adjacent C–H bond is in an *anti* geometrical relationship to the nitrogen lone pair. The change in the NH and CH frequencies were ascribed to hyperconjugative interaction between the nitrogen lone pair and the adjacent C–H antibonding MO[357].

The infrared spectra of the following heterocyclic molecules were also studied and it was found that a higher NH stretching frequency was observed when the adjacent C–C bond is *cis* to the nitrogen lone pair[358]:

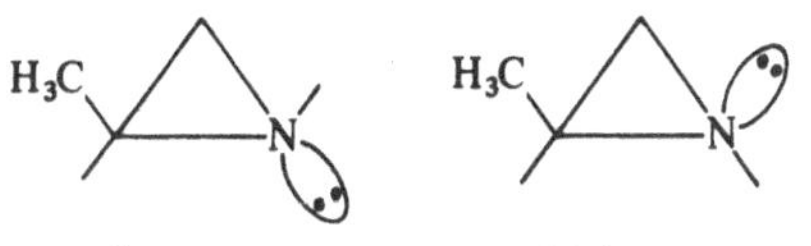

lower ν_{NH} higher ν_{NH}

This result is consistent with the idea that a C–C bond is a better instrinsic acceptor than a C–H bond.

10.2. Reactivity Probes of n–σ Interactions

In this section we shall examine in detail the role of n–σ interactions in carbanion chemistry. Carbanion stability as well as relative acidities of organic molecules are important probes of n–σ interactions.

It has been known for many years that hydrogens adjacent to a heteroatom are more acidic when the heteroatom is sulfur than when oxygen[359]. A striking example of this phenomenon is the behavior of dithiane, A, compared to its oxygen analog, B, when treated with strong base. The sulfur compound forms a carbanion while its oxygen counterpart fails to react[360]:

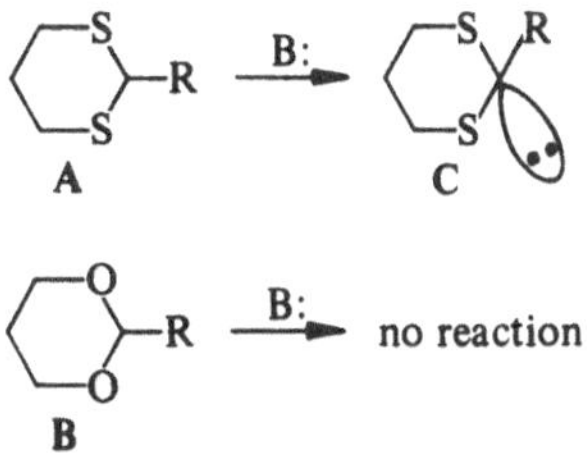

The stability of C can be attributed by our approach to delocalization of electron density from carbon into the adjacent S–C sigma antibonding orbitals. In the oxygen compound B the carbon-oxygen sigma antibonding orbital is too high in energy to participate in stabilizing a resultant carbanion.

The lone pair in the dithiane carbanion C can assume either an *equatorial* or *axial* position in the ring as shown below:

185

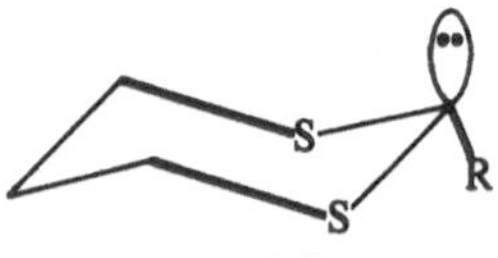

axial 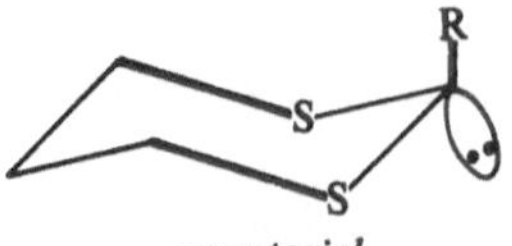   equatorial

It is readily recognized that the conformation in which the lone pair is *equatorial* corresponds to an *anti*-arrangement of the carbon lone pair in relation to the adjacent S–C bond (darkened for the sake of clarity) and the *axial* case a *syn* orientation. Based on our approach, we would predict that the carbanion with the lone pair *equatorial* will be the most stable conformation. This is indeed observed experimentally[361].

Another interesting case involves the base catalyzed H/D exchange of the systems **D** and **E** shown below[362]:

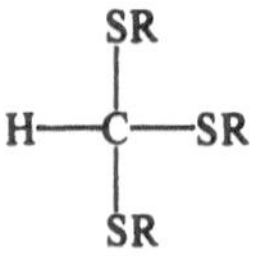

$k_{rel} = 1$ $k_{rel} = 10^3$
D   **E**

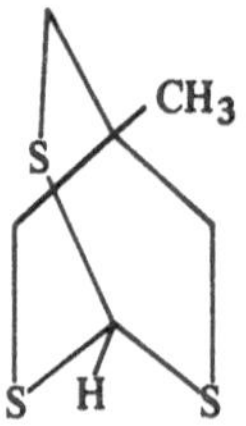

The fact that the H/D exchange of **E**, in which the lone pair formed in the transition state is forced to be in an *anti* conformational relationship to the S–C bonds, is so much larger than that of **D**, in which the *anti* conformational relationship is not imposed, clearly indicates that the *stabilization of a carbanion by adjacent –SR groups is conformationally dependent*.

We now turn our attention to the conformation of ⁻CH_2XH systems where X can be a first row or second row element, e. g. oxygen and sulfur. The two conformations to be considered are labeled Y and W and are shown below:

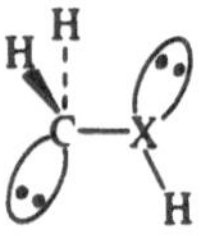

X=O, S

Y   W

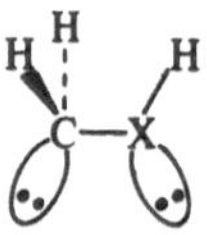

The dissection of the CH_2XH system into component fragments, A and B, is depicted below for the W conformation.

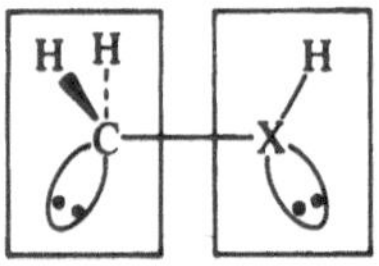

A   B

186

The appropriate interaction diagram is shown in Fig. 50. Arguing as before, we predict that the W conformation is preferred since there is an *anti* relationship between the carbon lone pair and the X–H bond in this conformation.

Ab initio calculations of the $^-CH_2XH$ carbanion show that the Y conformation is the most stable conformation for X=O[363]. However, in the case of X=S, the Y and W conformations are of comparable energy. Clearly, steric effects are important in determining the preferred conformations of these systems.

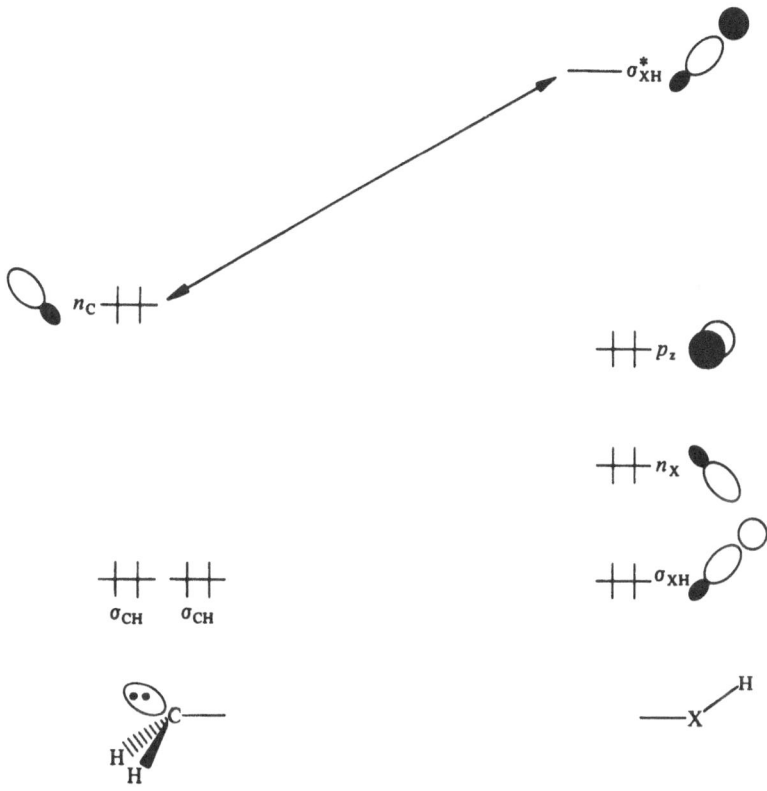

Fig. 50. Interaction diagram of the hybrid MO's in $^-CH_2XH$. The dominant stabilizing interaction involves the carbon lone pair AO and the XH sigma antibonding MO

Next, we consider in greater detail why carbanions adjacent to – SR groups are more stable than those adjacent to –OR groups. The reasons for this can be illustrated by focusing on a fixed conformation of the model $^-CH_2XH$ carbanion, *e. g.* the Y conformation. The following factors are involved:

a) The σ^*_{SH} MO is lower in energy than the σ^*_{OH} MO (see Table 5), *i. e.* –SH is a better electron acceptor than –OH.

b) The overlap integral $S_{n_C\sigma^*_{SH}}$ is larger than $S_{n_C\sigma^*_{OH}}$. This can be understood in a qualitative way by considering the schematic representation of an approximate form of these integrals:

Part IV. Conjugative Interactions

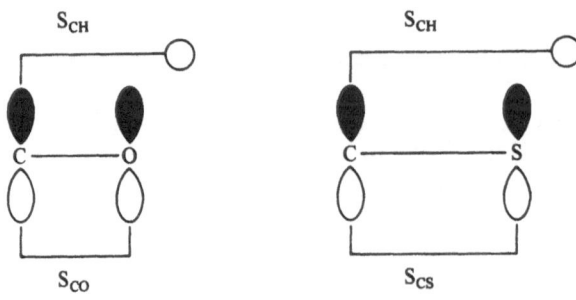

Calculations show that $S_{CO} \approx S_{CS}$. On the other hand, the S_{CH} overlap integral is smaller in the case of sulfur than in the case of oxygen, because the C–O bond is shorter than the C–S bond. This effect can be restated in the following way: the greater n_C–σ^*_{XH} overlap integral in the case of sulfur results from a *smaller* anti-bonding contribution by the group attached to the sulfur. We conclude that the two electron stabilization energy is greater for the case of X=S than X=O.

The results of ab initio calculations are compatible with the conclusion that –SH stabilizes an adjacent carbanion more than –OH[363]. Furthermore, the above discussion represents a viable alternative to the d-orbital model for explaining the enhanced kinetic and equilibrium acidities of molecules containing sulfur groups.

As we have seen, the C–C bond appears to be a better intrinsic acceptor than the C–H bond. A number of experimental observations seem to be consistent with these ideas. For example, the proton in F exchanges 5 times faster than that in G[364, 365].

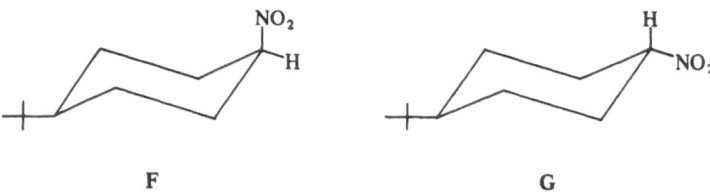

F   G

From the projection drawings below, it is obvious that the lone pair is *anti* to a C–C bond in F and a C–H bond in G. Table 50 lists the relative rates for deuterium exchange in some substituted cyclohexanes. In all cases, the *equatorial* proton exchanges at a much faster rate.

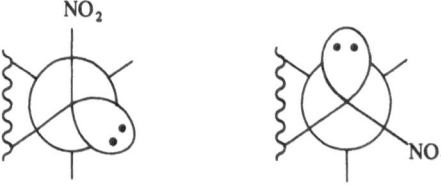

188

Table 50. Relative rates of deuterium exchange[365]

Molecule	Axial	Equatorial
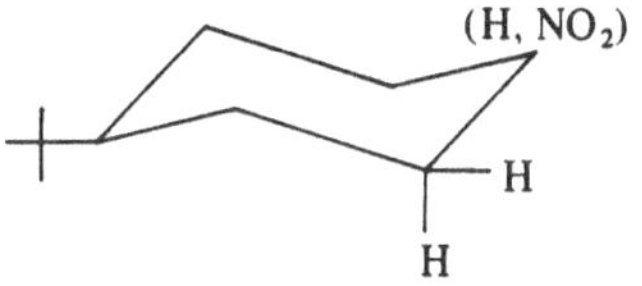 (H, NO$_2$) ... H ... H	2×10^{-1}	1
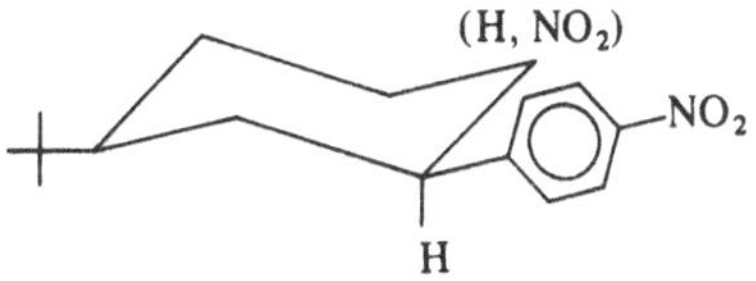 (H, NO$_2$) ... NO$_2$ ... H	4.9×10^{-3}	1
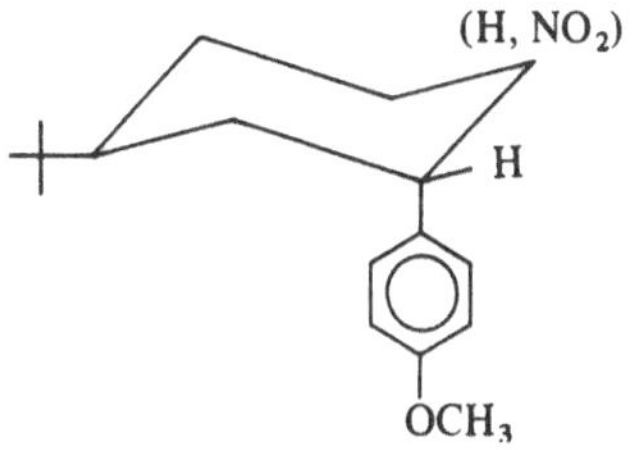 (H, NO$_2$) ... H ... OCH$_3$	4.6×10^{-3}	1

11. Structural Effects of $\sigma-\sigma$ Interactions

In the following sections we shall examine the structural effects of $\sigma-\sigma^*$ interactions on torsional isomerism of the following systems:

A) CH_3-CH_3
B) CH_3-CH_2X molecules
C) XCH_2-CH_2X molecules
D) $XYCH-CH_2X$, $XYCH-CHXY$, and $XYZC-CXYZ$ molecules
E) Rotational Barrier in CH_3-CO-X molecules
F) 1,2-Disubstituted Ring Systems

A. CH_3-CH_3

The prediction of the preferred conformation of substituted ethanes can simply be made by following the recipe:

a) Identify the dominant $\sigma-\sigma^*$ interaction.

b) Select the conformation which maximizes the dominant interaction, *i.e.* the conformation which places the intrinsic donor and acceptor bonds *anti* to each other.

189

The parent system to be considered is ethane itself. Here, the dominant inter-actions, or, better, the only type of interactions are $\sigma_{CH}-\sigma^*_{CH}$. The conformation which places vicinal C—H bonds *anti* with respect to each other, *i.e.* the staggered conformation, is the one which is predicted to be preferred. The same conclusions are reached by considering overlap repulsion and steric effects.

B. CH_3CH_2X Molecules

The simplest type of substituted ethane is the CH_3CH_2X system. When X is more electronegative than H, the dominant interaction is $\sigma_{CH}-\sigma^*_{CX}$ which dictates a staggered conformation. However, the same conformation is also predicted on the basis of overlap repulsion and steric effects. Indeed, irrespective of the nature of X all such molecules exist in staggered form[366].

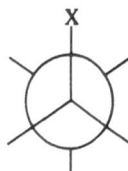

However, an interesting prediction can be made as follows: increasing the electroneg-ativity of X stabilizes the staggered form (minimum) to an increasing extent. Accord-ingly, the rotational barrier should progressively increase in the order $F > OH > NH_2 > CH_3$, assuming that the substituents affect the minimum more than the maximum. If the converse is true, the reverse order is predicted on the basis of conventional steric effects.

The experimental results collected in Table 51 strongly suggest that $\sigma-\sigma^*$ stabi-lization of the staggered (minimum) form of CH_3CH_2X is operative, *i.e.* a purely steric interpretation does not make sense. For example, CH_3CH_2F has roughly the

Table 51. Rotational barriers in CH_3CH_2X molecules

X	Barrier (kcal/mol)[366]
H	2.875
F	3.330
Cl	3.680
Br	3.680
I	3.220
OH	0.770
SH	3.310
SCH$_3$	3.820
CH$_3$	3.400

same barrier as $CH_3CH_2CH_3$ although CH_3 is much bulkier than F. Even more important, CH_3CH_2OH has a barrier which is more than four times less than that of CH_3CH_2F although again OH is bulkier than F.

Replacement of a substituent X by its second row counterpart should lead to better $\sigma-\sigma^*$ interactions. Once more, a substituent effect exerted principally on the minimum will give rise to the orders $Cl > F$, $SR > OR$, $PR_2 > NR_2$, $SiR_3 > CR_3$. Here, unfortunately, the same orders are expected if the principal effect of the substituent is exerted on the maximum. The experimental results collected in Table 51 clearly show that the predicted unequivocal trend is observed.

C. XCH_2CH_2X Molecules

1,2-disubstituted ethanes are extremely interesting molecules since many of these molecules exist in a *gauche* conformation[316)].

The model system FCH_2-CH_2F will illustrate our approach. Here the dominant interaction is $\sigma_{CH}-\sigma^*_{CF}$. The geometry which maximizes this interaction is the *gauche* geometry. Note that the *anti* geometry maximizes the much more inferior $\sigma_{CF}-\sigma^*_{CF}$ and $\sigma_{CH}-\sigma^*_{CH}$ interactions.

$\sigma_{CH}-\sigma^*_{CH}$
$\sigma_{CF}-\sigma^*_{CF}$

$\sigma_{CH}-\sigma^*_{CF}$
$\sigma_{CH}-\sigma^*_{CH}$

We shall now consider what happens if the two fluorines are replaced by OR. In this case, the dominant interactions will be $\sigma_{CH}-\sigma^*_{CO}$ and the tendency for the *gauche* structure will be diminished relative to the case of CH_2F-CH_2F since this interaction is weaker than the $\sigma_{CH}-\sigma^*_{CF}$ interaction. The trend will continue along the series F, OR, NR_2 and CR_3. At this point, it should be mentioned that Pople *et al.* have considered a hyperconjugative mechanism in CH_2F-CH_2F[68)].

Now, consider what happens if the two fluorines are replaced by two chlorines. In this case, the dominant interactions will be $\sigma_{CCl}-\sigma^*_{CCl}$ because the C–Cl bond is both a better donor and a better acceptor than the C–H bond. Hence, it is predicted that 1,2-dichloroethane will have a preferred *trans* conformation which combines in an *anti* orientation the best donor and best acceptor bonds.

The same conclusions are reached for CH_2Br-CH_2Br and CH_2I-CH_2I. Steric effects also, undoubtedly, contribute to the *anti* preference.

Experimentally, these molecules are found to exist in the *anti* conformation[367] and not the *gauche* conformation as does FCH_2CH_2F[368].

D. XYCH–CH₂X, XYCH–CHXY and XYZC–CXYZ Molecules

We now turn our attention to the more complicated system HCF_2-CH_2F. The possible staggered conformations are shown below and the various $\sigma-\sigma^*$ interactions are enumerated.

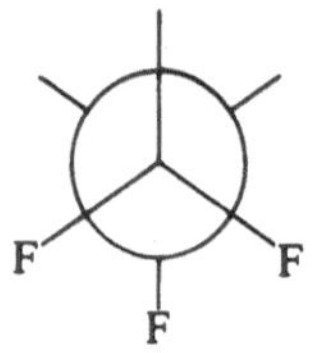

$$\sigma_{CH}-\sigma^*_{CF}$$
A

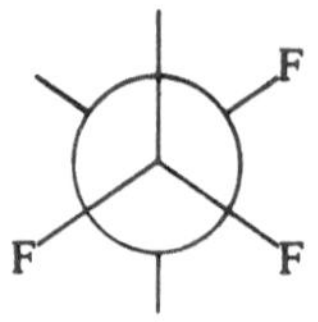

$$\sigma_{CF}-\sigma^*_{CF}, \sigma_{CH}-\sigma^*_{CH}, \sigma_{CH}-\sigma^*_{CF}$$
B

Since the dominant interaction is $\sigma_{CH}-\sigma^*_{CF}$, the predicted preferred conformation is A.

The tetrasubstituted model system HCF_2-CHF_2 constitutes another interesting case. By following familiar reasoning, we predict conformation C to be the most stable conformation. Experimentally, the reverse is found to be true[369]. This can

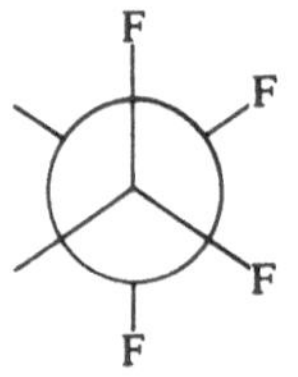

$$\sigma_{CH}-\sigma^*_{CF}$$
$$\sigma_{CF}-\sigma^*_{CF}$$
C

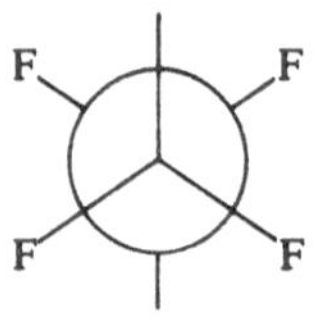

$$\sigma_{CF}-\sigma^*_{CF}$$
$$\sigma_{CH}-\sigma^*_{CH}$$
D

be ascribed to more unfavorable dipolar repulsive interactions in C, which cannot be overcompensated by the $\sigma_{CH}-\sigma^*_{CF}$ conjugative interactions. However, when dipolar interactions are deemphasized and the conjugative interactions are accentuated, a switch in conformational preference is expected. In this sense, an interesting case is CHClBrCHIBr for which two stable conformations have been identified E and F[370]. Furthermore, it was shown that F is more stable than E by 303 cal/mol. This conformational preference can be ascribed to the fact that F places the best intrinsic donor and the best intrinsic acceptor, *i.e.* the C–I and C–Br bonds, in an *anti* relationship.

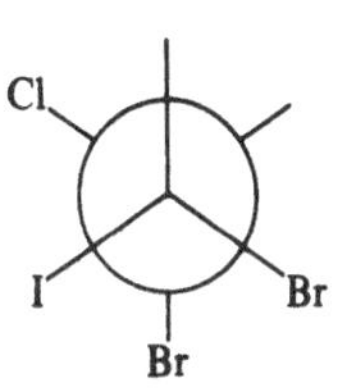

E 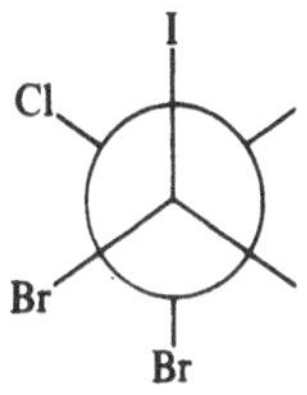   F

We next consider a more complicated problem, that of rotational isomerism in CFClBr–CFClBr. The various conformations which this molecule can exist in are the following:

meso

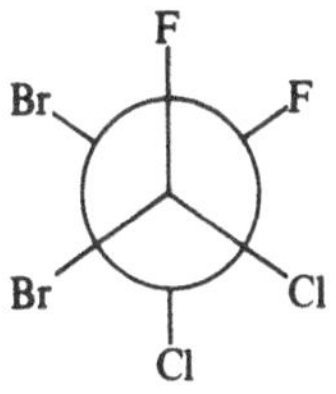

G 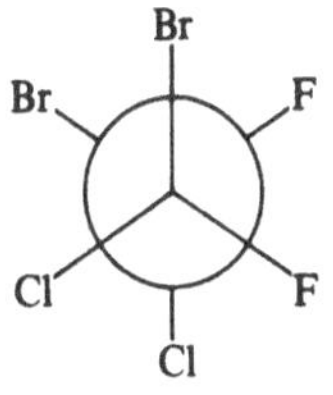   H 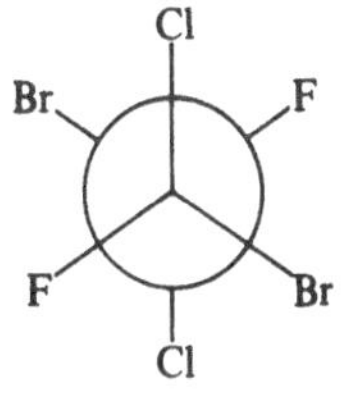   I

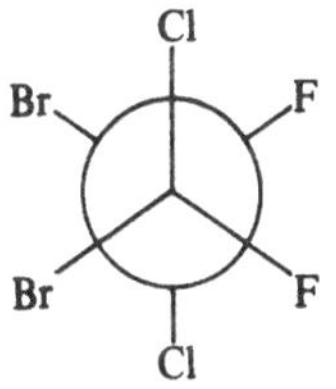

J 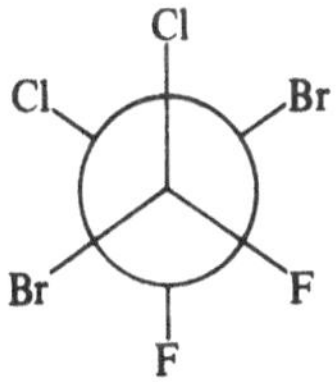   K 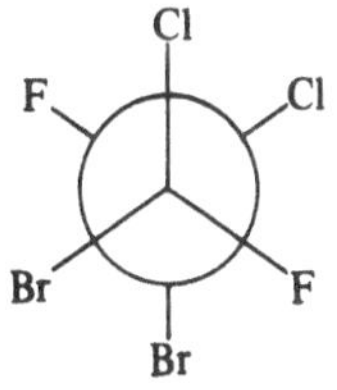   L

racemic

By noting that the best conformation will be the one which allows the interaction of the best donor bond (C–Br) with the best acceptor bond (C–Br), the second best donor bond (C–Cl) with the second best acceptor bond (C–Cl), and the third best donor bond (C–F) with the third best acceptor bond (C–F), we predict the best conformation to be **I** for the *meso* form. The best conformation for the *racemic* form will be **K**, since it is the only one which allows for the interaction of the best donor and best acceptor bonds. We further predict that **L** will be more stable than **J**. Experimentally, **I** and **L** are found to be the most stable conformations of $CClFBrCClFBr$[371].

E. Rotational Barrier in CH_3COX Molecules

We note here that sigma conjugative interactions are also operative in $CH_3–CO–X$ systems and they stabilize the *cis* conformation relative to the *trans* conformation:

193

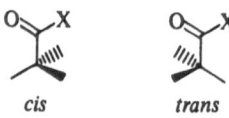

cis   trans

This occurs because the donor bond, C—H, and the best acceptor bond C—X, are in an *anti* geometrical relationship in the *cis* conformation.

As X becomes increasingly electronegative along a row of the Periodic Table, the conformational minimum (*cis*) will be increasingly stabilized relative to the maximum (*trans*) by $\sigma_{CH}-\sigma^*_{CX}$ conjugative interactions. Consequently, the methyl rotational barrier should increase as X becomes more electronegative.

Partial experimental support for this idea is found in the methyl rotational barrier of acetyl fluoride and acetic acid. Specifically, the methyl rotational barrier increases by 561 cal/mol when —OH is replaced by F[131), 132), 137)].

F. Disubstituted Ring Systems

In Part II we discussed conformational isomerism of disubstituted ring systems from the standpoint of nonbonded interactions. Here, we shall examine the importance of $\sigma-\sigma^*$ conjugation in these molecules.

Consider the simplest disubstituted cyclic system, difluorocyclopropane. The two geometrical isomers are shown below and the $\sigma-\sigma^*$ interactions are enumerated, neglecting those involving the central methylene group since they will be identical for both isomers.

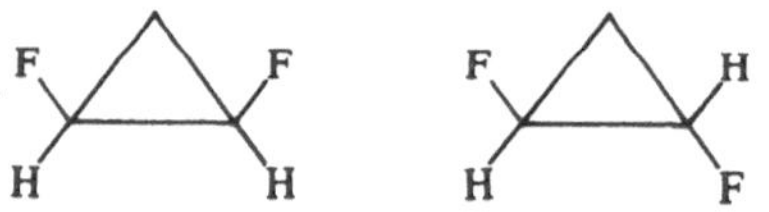

$\sigma_{CF}-\sigma^*_{CF}$ (*syn*) $\sigma_{CH}-\sigma^*_{CF}$ (*syn*)

$\sigma_{CH}-\sigma^*_{CH}$ (*syn*) $\sigma_{CH}-\sigma^*_{CH}$ (*gauche*)

$\sigma_{CH}-\sigma^*_{CF}$ (*gauche*) $\sigma_{CF}-\sigma^*_{CF}$ (*gauche*)

Since the dominant interaction is $\sigma_{CH}-\sigma^*_{CF}$ and following familiar reasoning, it is predicted that the order of stability of the 1,2-disubstituted molecule will be 1,2-*trans* > 1,2-*cis* and that the fluorine atoms will become increasingly negative in the same order.

Experimentally, *trans* 1,2-difluorocyclopropane is more stable than the *cis* isomer[372]. Fluorine and hydrogen charges for *cis* and *trans* 1,2-difluorocyclopropane as calculated by the CNDO/2 method are shown below:

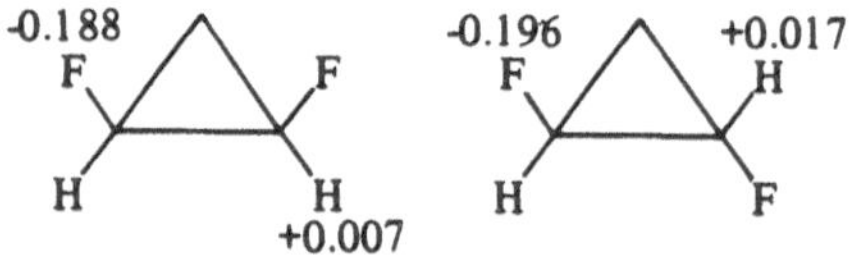

-0.188
F   F
H   H
+0.007

-0.196   +0.017
F   H
H   F

The above results are consistent with either a sigma conjugative effect or electrostatic effect.

Now, consider the effect of replacing the fluorine substituents in the disubstituted cyclopropane discussed above with an atom further down the column in the Periodic Table, e.g. Br. The two geometrical isomers are shown below and the appropriate σ–σ^* interactions are listed.

σ_{CBr}–σ^*_{CBr} (syn)
σ_{CH}–σ^*_{CH} (syn)
σ_{CH}–σ^*_{Br} (gauche)

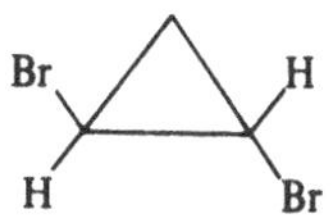

σ_{CH}–σ^*_{CBr} (syn)
σ_{CH}–σ^*_{CH} (gauche)
σ_{CBr}–σ^*_{CBr} (gauche)

Here, the dominant interaction is σ_{CBr}–σ^*_{CBr} and, consequently, the cis-1,2-isomer is predicted to be the most stable isomer barring severe steric effects.

Experimental work in this area is notably lacking. Hence, a systematic study of the relative stabilities of dihalocyclopropanes would be most welcome.

We now focus our attention on conformational isomerism in 1,2-disubstituted cyclohexanes. Our model system will be 1,2-difluorocyclohexane and we first examine the relative stabilities of the two trans 1,2-difluorocyclohexanes, i.e. axial-axial (aa) and equatorial-equatorial (ee). These two molecules are shown below along with a listing of the dominant stabilizing σ–σ^* interactions.

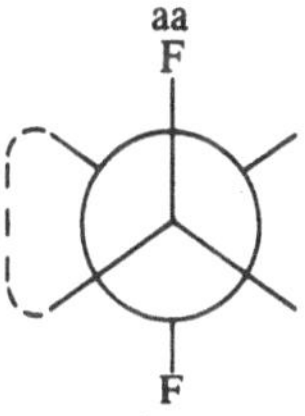

σ_{CC}–σ^*_{CH} (anti)

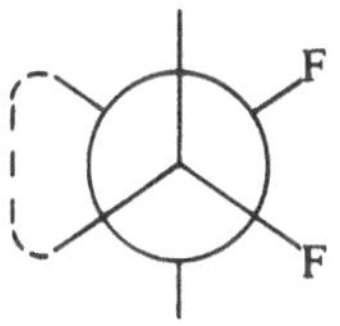

σ_{CC}–σ^*_{CF} (anti)

Clearly, the ee conformation is favored since a C–F bond is a better acceptor than a C–H bond.

Now, we are prepared to compare the relative stabilities of cis and trans-1,2-difluorocyclohexane. The cis conformation, axial-equatorial (ae), is shown below along with the dominant σ–σ^* interactions.

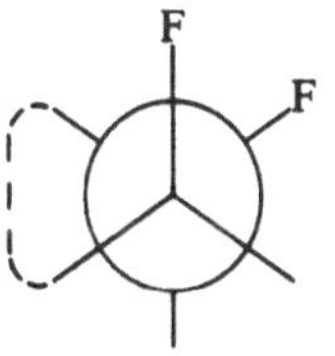

σ_{CC}–σ^*_{CF} (anti)
σ_{CH}–σ^*_{CF} (anti)

We conclude, that the ee conformation is predicted to be more stable than the *ae* conformation since the former conformation enjoys better sigma conjugative interactions. Unfortunately, experimental data about the relative stability of cis and trans difluorocyclohexane is lacking.

We now focus on the effect of replacing fluorine by another halogen of a higher period, *e.g.* Br. The trans and cis conformations of 1,2-dibromo-cyclohexane are shown below along with the dominant orbital interactions:

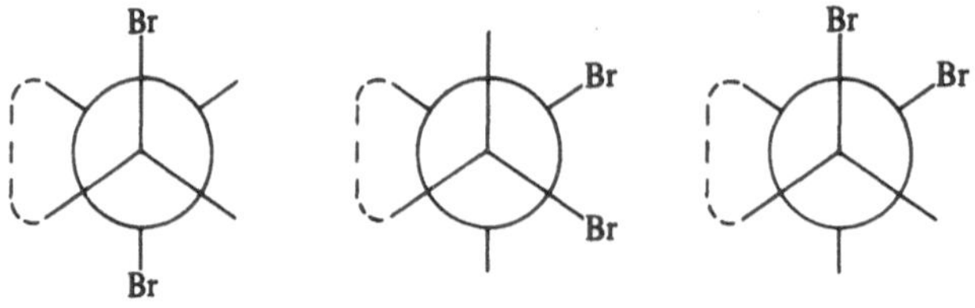

aa
$\sigma_{CBr}-\sigma^*_{CBr}$ *(anti)*

ee
$\sigma_{CC}-\sigma^*_{CBr}$ *(anti)*

ae
$\sigma_{CC}-\sigma^*_{CBr}$ *(anti)*
$\sigma_{CH}-\sigma^*_{CBr}$ *(anti)*

By following familiar arguments, we predict that the order of stability for the 1,2-dibromo-cyclohexanes will be *aa > ee > ae*. This should become increasingly accentuated in the order I > Br > Cl > F.

Typical experimental data are shown in Table 52. As can be seen, the diaxial conformation is preferred for the cases in which adjacent substituent bonds are hermaphroditic, *i.e.* simultaneously very good donor and very good acceptor bonds.

Another interesting example is the problem of conformational isomerism in dihalopyranoses. Our model compound is the 2,3-dichloro-derivative. We first consider the *trans* conformation, *i.e.* the *aa* and *ee* conformations. The two *trans* conformations are shown below and the dominant sigma conjugative interactions are listed.

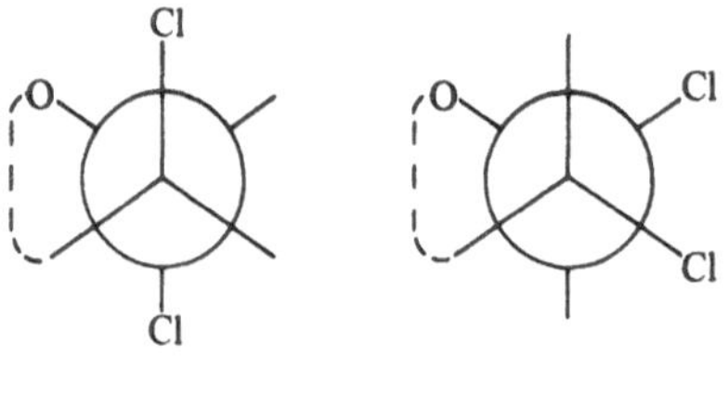

aa
$n_O-\sigma^*_{CCl}$
$\sigma_{CCl}-\sigma^*_{CCl}$ *(anti)*

ee
$\sigma_{CC}-\sigma^*_{CCl}$ *(anti)*

It is clear that a combination of σ–σ* and n–σ*, *i.e.* the anomeric effect[342], interactions favor the *aa* conformation.

The two *cis* conformations are shown below along with a listing of the dominant sigma orbital interactions:

196

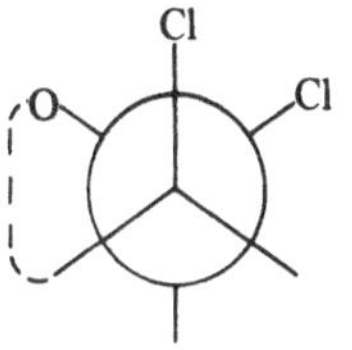

ae

$\sigma_{CC}-\sigma^*_{CCl}$

$\sigma_{CH}-\sigma^*_{CCl}$

$n_O-\sigma^*_{CCl}$

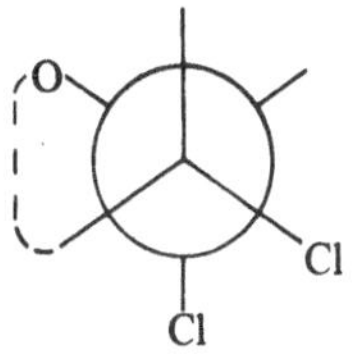

ae'

$\sigma_{CH}-\sigma^*_{CCl}$

Table 52. Relative energies of the conformations of 1,2-disubstituted cyclohexanes, $C_6H_{10}XY$

X	Y	$E_{aa}-E_{ee}$ (kcal/mol)[a]	Ref.
CH$_3$	CH$_3$	3.6	373)
Cl	Cl	.214	374)
Br	Br	−.305	374)
Cl	I	−.242	375)

[a] E_{aa} = Energy of the *axial-axial* conformation.
E_{ee} = Energy of the *equatorial-equatorial* conformation.

Arguing as before, we conclude that the ae conformation will be the most stable *cis* conformation. A comparison of the sigma conjugative interactions which obtain in the most stable *trans* conformation, the *aa* conformation, and the most stable *cis* conformation, the *ae* conformation, leads to the conclusion that the *trans* diaxial isomer is the most stable conformation of this molecule.

Experimentally, the 2,3-dichloro analogue of pyranose exists preferentially in the diaxial form[376]. On the other hand, in the 3,4-dichloro compound, the chlorines, prefer the diequatorial position[377].

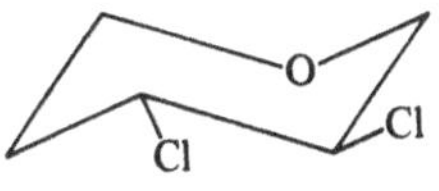

Part V. Bond Ionicity Effects

12. Theory

As we have discussed before, molecules can be dissected into fragments. Unless two fragments are identical, we can define one as the donor fragment D and the other as the acceptor fragment A[j). The electronic states of the composite system can be described in terms of linear combinations of wavefunctions appropriate to the fragments D and A. In order to maximize the simplicity of the theory and yet retain the important features of our analysis, we may choose a minimal basis set of zero order configuration wavefunctions. The zero order configuration wavefunctions forming our basis set are:

1. The no bond configuration DA.
2. The locally excited configurations D*A and DA*.
3. The charge transfer configuration D^+A^- and D^-A^+.

For brevity, we shall use the term configuration as an alternative term for zero order configuration wavefunction. By employing familiar quantum mechanical principles[378], we can write each configuration of the basis set as a linear combination of Slatez determinants. The energies associated with the five configurations can be written as follows:

$$E(DA) \quad = 0 \tag{29}$$

$$E(D^+A^-) \cong I_D - A_A - C \tag{30}$$

$$E(D^-A^+) \cong I_A - A_D - C' \tag{31}$$

$$E(D^*A) \cong G \tag{32}$$

$$E(DA^*) \cong G' \tag{33}$$

In the above equations, I_D is the lowest ionization potential of the donor, I_A that of the acceptor, A_D is the electron affinity of the donor, A_A that of the acceptor, C and C' are the electrostatic attractive interactions of the excess electron of one fragment with the electron hole of the other fragment and G and G' are local fragment excitation energies.

We note at this point that, when the two fragments become identical, the charge transfer configurations and the locally excited configurations will have to be replaced by charge-resonance and exciton-resonance configurations, respectively.

Inspection of Eqs. (29–33) reveals that the quantities I_D, I_A, A_D, A_A, G, and G' can be extracted from experimental data and only C and C' need to be

---

j) When the two fragments or molecules are identical appropriate symmetry adapted wavefunctions should be constructed.

calculated explicitly. Thus, we are in a position to rank the various configurations according to their relative energies as determined by reference to experimental data. These configurations represent the electronic states of the molecule in the hypothetical case of zero configuration interaction. Obviously, the real electronic states result from the interaction of the various configurations and the difference in energy between a configuration and the corresponding electronic state can be thought of as the stabilization or destabilization energy due to configuration interaction.

We can now consider explicitly how configurations interact to produce electronic states. Our first task is to define the Hamiltonian operator. In order to simplify our analysis, we adopt a Hamiltonian which consists of only one electron terms and we set out to develop electronic states which arise from one electron configuration mixing.

$$\hat{H} = \hat{H}_D + \hat{H}_A + \hat{H}' \tag{34}$$

In Eq. (34), $\hat{H}_D$ operates only on fragment D, $\hat{H}_A$ operates only on fragment A and $\hat{H}'$ operates along one or more forming bonds. Next, we assume that the MO's of D and A are orthogonal with respect to $\hat{H}_D$ and $\hat{H}_A$, correspondingly, and that overlap can be neglected. A quantitative determination of the electronic states of the molecule can then be easily accomplished. Specifically, we can construct the energy matrix, where the diagonal elements are the energies of the zero order configurations and the off diagonal elements are the interaction terms between the zero order configurations. The important thing is that many of these matrix elements can be empirically evaluated by reference to experimental data. We can solve the corresponding secular determinant and get the eigenvalues and eigenvectors of the electronic state wavefunctions. The availability of high speed computers and the existence of ionization potential and spectroscopic data which can serve as empirical input make such configuration interaction calculations possible for systems of chemical interest.

While the theoretician might be interested in the subtle features of a problem which can only be revealed by explicit computation, the organic chemist is more interested in a general qualitative theory which can be directly applicable to problems of interest without the need of computer assistance. Hence, we should reduce the quantitative scheme to a qualitative scheme which can still retain the important features of the theory. The various steps which one has to follow in attempting a qualitative configuration interaction analysis are given below:

Step 1. Rank the basis set configurations according to their relative energy. This can be implemented by means of Eqs. (29) to (33).

Step 2. Construct the interaction matrix for the basis set configurations.

Step 3. Generate the electronic states by noting that the energy change of any two nondegenerate basis set configurations arising from their interaction will be inversely proportional to the energy difference between the two basis set configurations and directly proportional to the square of their interaction matrix element. The analogy between MO splitting, which is familiar to the organic chemist, and configuration splitting as a result of interaction is shown in Fig. 51.

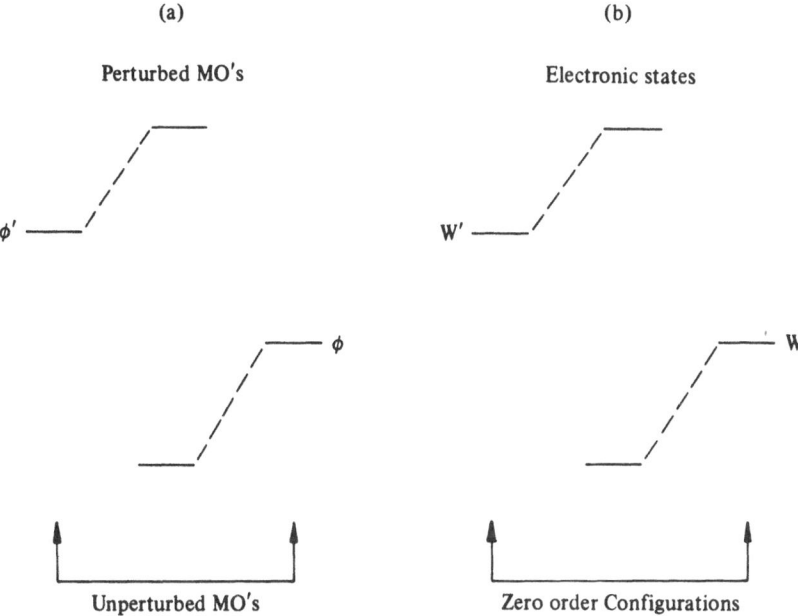

(a)

Perturbed MO's

ϕ'

ϕ

Unperturbed MO's

(b)

Electronic states

W'

W

Zero order Configurations

Fig. 51. (a) MO interaction diagram involving the ϕ and ϕ' MO's. (b) Configuration interaction diagram depicting the interaction of the W and W' configurations

We shall now illustrate how this approach, designated the Linear Combination of Fragment Configurations (LCFC) approach, can be applied to diverse chemical problems.

12.1. The LCFC Approach to Geminal Interactions

The LCFC approach provides the basis for the development of an overview of the shapes of AX_2 and AX_3 molecules. In order to maximize simplicity, we have developed a model which restricts the number of basis set configurations in such a manner that only the key electronic changes accompanying the bending of a linear triatomic molecule or the pyramidalization of a flat tetraatomic molecule are considered. The various features of our model are the following:

a) The electronic configuration of the central atom in AX_2 or AX_3 is taken to be the one which approximates the electronic configuration of A in linear AX_2 and planar AX_3. The same procedure is followed for the ligand X.

b) The basis set configurations are selected in such a manner that they represent the interaction "turned on" by bending or pyramidalization.

c) An equation of the bending or pyramidalization tendency, like Eq. (24) is written where i and j represent now basis set configurations.

We first consider the case of OX_2 as an example. The configuration of the central atom is shown below. Note that two electrons are placed in a $2p_y$ orbital to reproduce the $2p_y$ lone pair of linear OX_2 and one electron is placed in each of the $2p_x$ and 2 s AO's which constitute the two valence AO's of O utilized in bonding in the linear geometry. The energy of the doubly occupied $2p_y$ is approximated by the computed pi ionization potential of linear OX_2 appropriately calibrated with respect to experimental data.

++ $2 p_y$
+ $2 p_x$
+ $2 s$

The configuration of the two ligands is shown below for X=H and X=F.

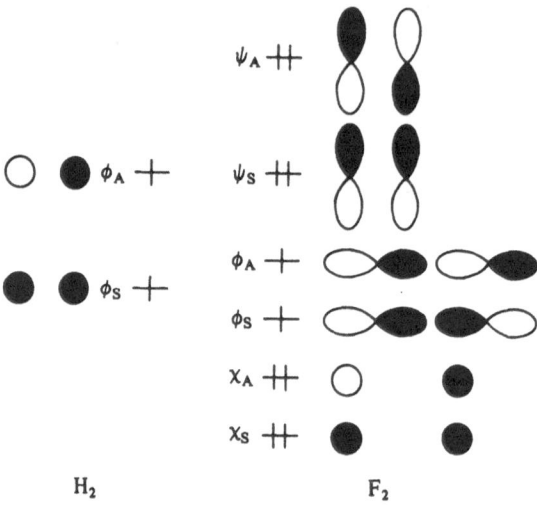

$$H_2 \qquad\qquad\qquad\qquad F_2$$

Note that single occupancy is assigned to AO's which are principally utilized in forming the sigma bonds in linear OX_2.

Let us now identify the MO interaction which is "turned on" upon bending in the case of H_2O. By reference to Fig. 40, we can isolate this interaction as being the one between the $2 p_y$ AO and symmetric MO, ϕ_s, spanning the two hydrogens. The configurations which upon mixing can reproduce this interaction are the ones shown below along with their respective energies.

$2p_y$	++			
$2p_x$	+	+ ϕ_A	+	+
$2s$	+	+ ϕ_S	+	++
			+	
	(O)	(H_2)	(O^+)	(H_2^-)
	E=0		$E = I(2p_y)^O - A(\phi_S)^H + C$	

Proceeding in a similar fashion, we can identify the stabilizing MO interactions which are "turned on" upon bending in F_2O. By reference to Fig. 52 we can isolate four such interactions:

a) The $O2p_y - \phi_S$ interaction
b) The $O2p_y - X_S$ interaction
c) The $O2p_x - \psi_A$ interaction
d) The $O2s - \psi_S$ interaction

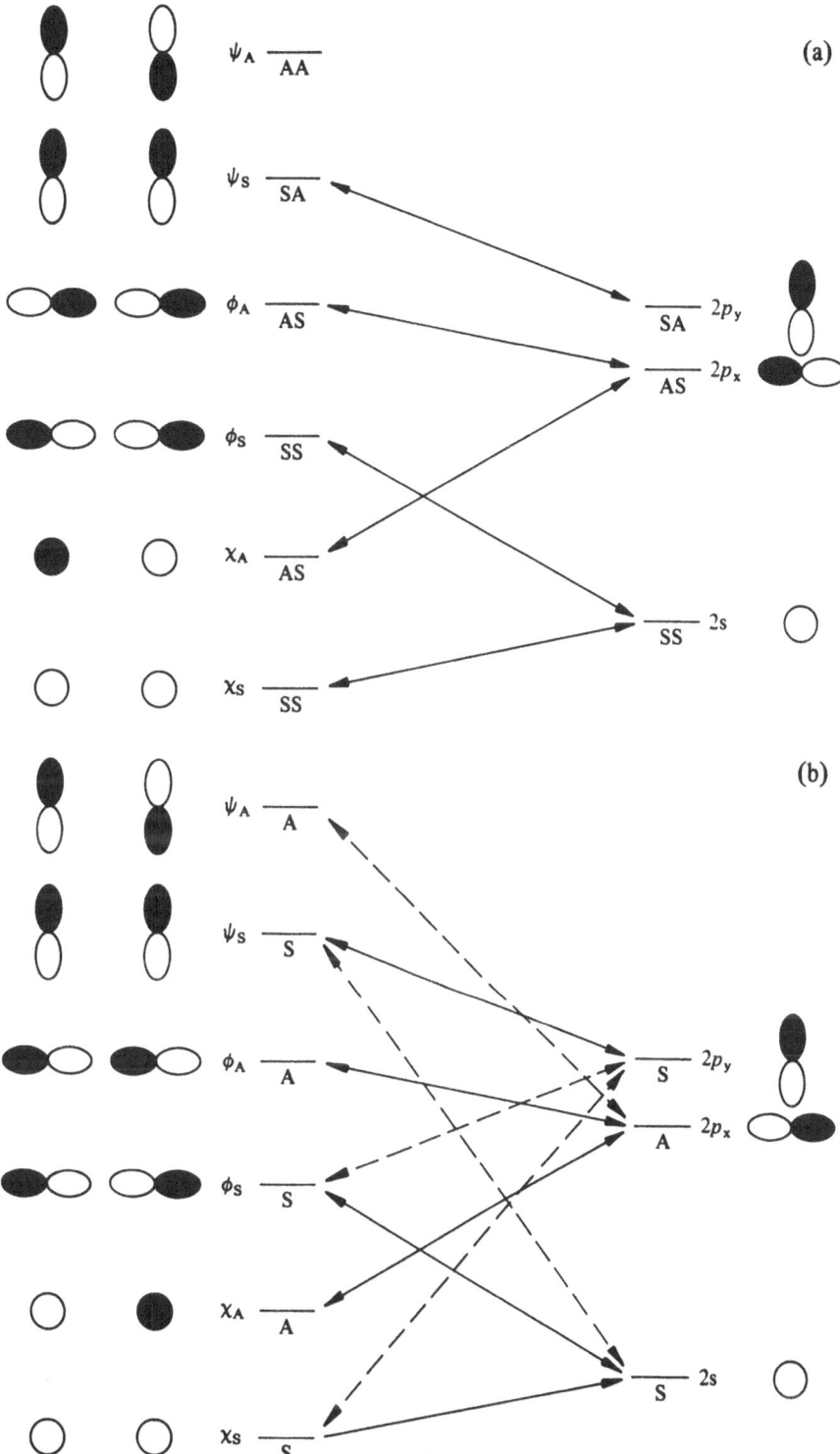

Fig. 52. Interaction diagrams for (a) linear and (b) bent F_2O. The interactions "turned on" upon bending are shown by the dashed lines

Part V. Bond Ionicity Effects

The configuration which upon mixing can reproduce these interactions are given in Scheme 18.

Scheme 18.

a) $02p_y - \phi_s$ interaction.

$2p_y$	$\#$			$+$		
$2p_x$	$+$	$+ \phi_A$			$+$	
		$+ \phi_S$		$+$	$\#$	
$2s$	$+$			$+$		
	(O)	(F_2)		(O^+)	(F_2^-)	
	$E=0$			$E = I(2p_y)^O - A(\phi_S)^F + C$		

b) $02p_y - \chi_s$ interaction is not reproduced. One would have to construct high energy configurations of F_2 in order to achieve that.

c) The $02p_x - \psi_A$ interaction.

$2p_y$	$\#$			$\#$		
$2p_x$	$+$	$\# \psi_A$		$\#$		
		$\# \psi_S$			$+$	
					$\#$	
$2s$	$+$			$+$		
	(O)	(F_2)		(O^-)	(F_2^+)	
	$E=0$			$E = I(\psi_A)^F - A(2p_x)^O + C'$		

d) The $02s - \psi_S$ interaction.

$2p_y$	$\#$			$\#$		
$2p_x$	$+$	$\# \psi_A$		$+$	$\#$	
		$\# \psi_S$			$+$	
$2s$	$+$			$\#$		
	(O)	(F_2)		$(O^-)'$	$(F_2^+)'$	
	$E=0$			$E = I(\psi_S)^F - A(2s)^O + C''$		

The reader will now realize the following trends:

a) The interaction $(O)(F_2) - (O^+)(F_2^-)$ is the one which corresponds to the geminal interaction present also in H_2O.

b) The interaction $(O)(F_2) - (O^-)(F_2^+)$ corresponds to a back-bonding interaction and is identical to the type of interaction discussed in Section 5.1. An additional back-bonding interaction is that between $(O)(F_2)$ and $(O^-)'(F_2^+)'$.

Some comments are now appropriate. Firstly, the ionization potential of a doubly occupied AO is set equal to the ionization potential of the lone pair of an appropriate model system. Secondly, the ionization potential of a singly occupied AO is set equal to the valence state ionization potential[379] of the atom in the appropriate electronic configuration. Thirdly, due to the small splitting of the symmetric and antisymmetric MO's spanning the ligands, e.g. ϕ_S and ϕ_A, the corresponding electron affinities or ionization potentials are set equal to the appropriate valence state electron affinities or ionization potentials. In the cases at hand, the following data have to be used.

$$I(2p_y)^O \simeq I(H_2O, \text{linear}) \qquad \simeq \qquad 12.67 \text{ eV}$$
$$I(\psi_A)^F \simeq I(\psi_S)^F \simeq I(2p)^F \qquad \simeq \qquad 20.88 \text{ eV}$$

$$A\,(\phi_s)^H \simeq 0.747 \text{ eV}$$
$$A\,(2p_x)^O \simeq 2.70 \text{ eV}$$
$$A\,(2s)^O \simeq 35.30 \text{ eV}$$
$$A\,(\phi_s)^F \simeq 3.65 \text{ eV}$$

The diagram shown in Fig. 53 depicts how geminal interactions and back-bonding effects come into play as a result of bending in OF_2 as well as how geminal interactions affect the bending of OH_2. This diagram clearly illustrates that geminal interactions will become increasingly important as the energy gap between $(O)(X_2)$ and $(O^+)(X_2^-)$, $\delta\,E^G$, decreases and the corresponding interaction matrix element, H'_{ij},

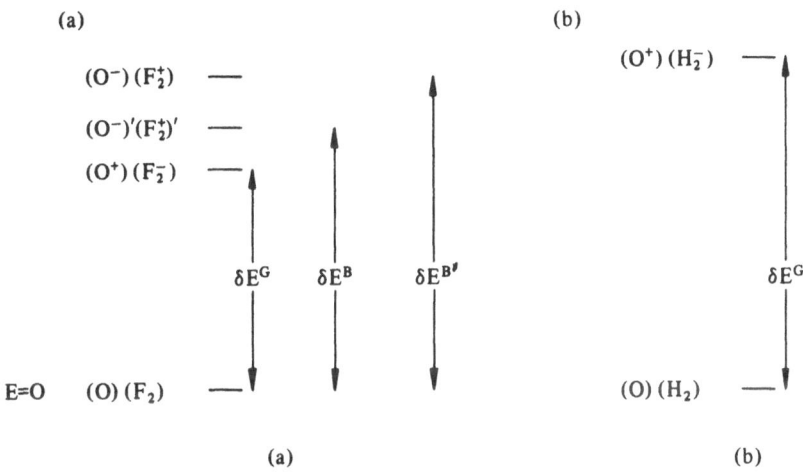

Fig. 53. LCFC interaction diagram for: (a) F_2O and (b) H_2O. $\delta\,E^G$ is the energy change due to the geminal interactions and $\delta\,E^B$ is the energy change due to back bonding of the lone pairs of F to the oxygen

increases in absolute magnitude. It also indicates that back-bonding will become increasingly important as the energy gap between $(O)(X_2)$ and $(O^-)(X_2^+)$, $\delta\,E^B$, decreases and the corresponding interaction matrix element increases.

a) $\overline{N}H_2$, OH_2, $\overset{+}{F}H_2$ series. Here, $\delta\,E^G$ decreases in the direction $\overset{+}{F} \rightarrow \overline{N}$, while the interaction matrix elements do not vary. Hence, we predict angle shrinkage to increase in the direction $\overset{+}{F} \rightarrow \overline{N}$. The same trends are predicted if H is replaced by any group X. Similar results are obtained for the $\overset{-}{P}H_2$, SH_2, $\overset{+}{C}lH_2$ series.

b) H_2O, H_2S series. Here, $\delta\,E^G$ is smaller for H_2S while H'_{ij} varies in an opposite direction. The variation in $\delta\,E^G$ dominates and a smaller angle is predicted for H_2S. The same comparison can be made between $\overline{N}H_2$ and $\overline{P}H_2$ as well as between $\overset{+}{F}H_2$ and $\overset{+}{C}lH_2$. The same trends are predicted if H is replaced by any group X.

c) OF_2, OCl_2, OBr_2, OI_2 series. Here, $\delta\,E^G$ and H'_{ij} both vary in a manner which favors increased angle shrinkage in the direction $OI_2 \rightarrow OF_2$. Similar results are obtained for the SF_2, SCl_2, SBr_2, SI_2 series.

d) Angle shrinkage is expected to be more pronounced in the comparison of H_2O vs. F_2O than in the comparison of H_2S vs F_2S simply because the geminal interactions vary to similar extents while δE^B and H_{ij} favor a much greater back-bonding in OF_2 than in SF_2.

At this point, the reader may reasonably object: is it really appropriate to consider only the orbital interactions which are "turned on" upon bending? This step can be justified on the basis of the observation that as bending occurs, the increase of the overlap integrals appropriate to the interactions which are "turned on" is much more significant than the decrease of the overlap integrals appropriate to the interactions already present in the linear form within a broad range of angles. A typical plot is shown in Fig. 54 for the case of F_2O bending.

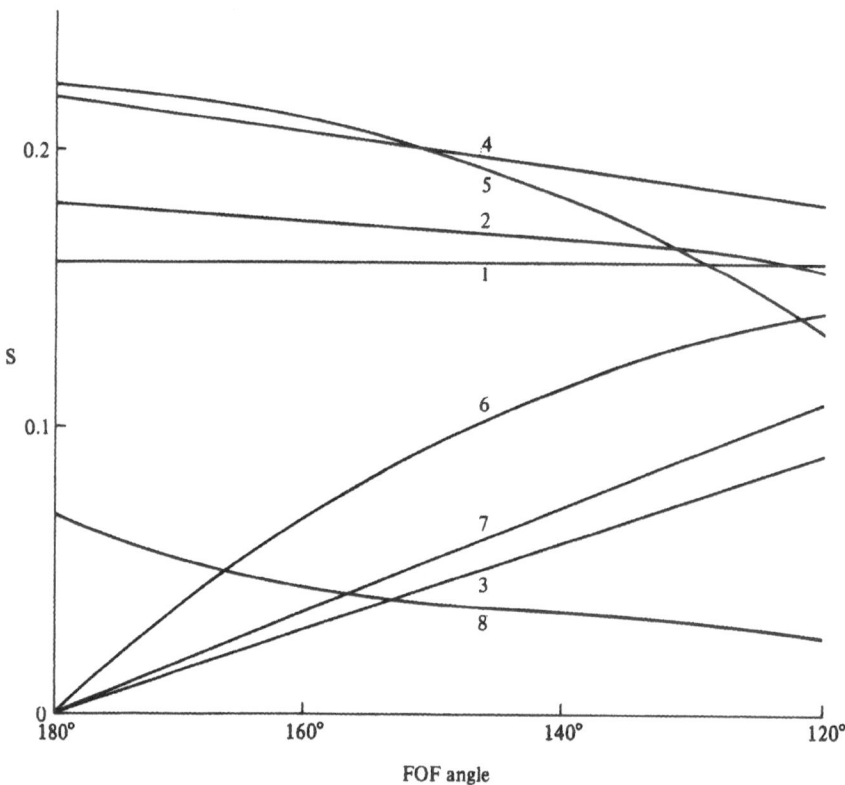

$1 = Os-Fs$	$4 = Op_x-Fs$	$7 = Op_y-Fs$
$2 = Os-Fp_x$	$5 = Op_x-Fp_x$	$8 = Op_y-Fp_y$
$3 = Os-Fp_y$	$6 = Op_x-Fp_y$ and	
	Op_y-Fp_x	

Fig. 54. Plot of AO overlap integrals in F_2O vs. the FOF angle. AO overlap integrals 3, 6 and 7 are involved in the orbital interactions "turned on" by bending

12.2. Structural Isomerism

While great attention has been paid to conformational[380] and, more recently, geometric isomerism, structural isomerism has strangely remained out of the focus of interest. Here, we shall show that structural isomerism may be the type of isomerism which is easiest to understand and will formulate simple predictive rules.

Consider the model systems 1,1- and 1,2-difluoroethane shown below.

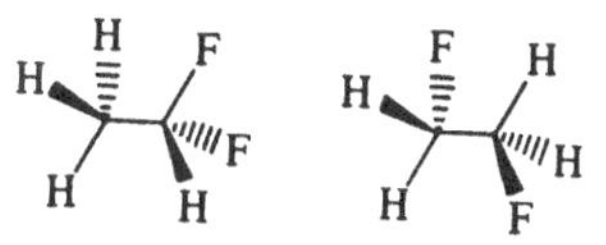

The basis set configurations are shown in Scheme 19 (F_2HC-CH_3) and Scheme 20 (FH_2C-CH_2F). Only the two charge transfer configurations of lowest energy which can mix with the no bond configuration have been included since we are interested in the ground electronic states of the two isomers. The energies of the various basis set configurations have been calculated empirically on the basis of well known equations and the interaction matrix elements have been evaluated with respect to an effective one electron Hamiltonian.

Scheme 19.

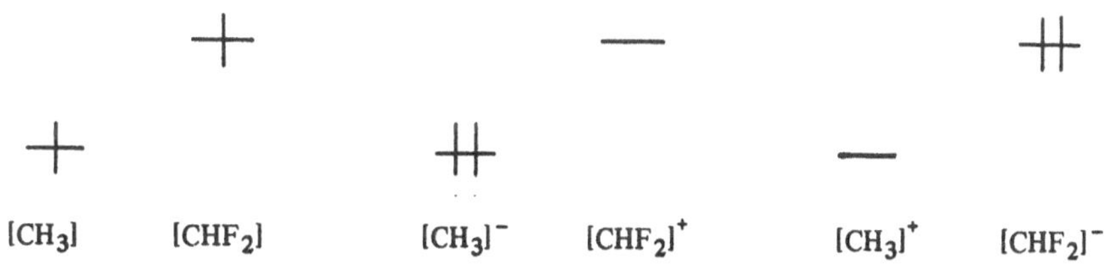

Scheme 20.

The interaction diagram of Fig. 55 shows that Ω_1 and Ω_2 are sandwiched between $[CH_3]^-[CHF_2]^+$ and $[CH_3]^+[CHF_2]^-$. The energetic depression of $[CH_3]^-[CHF_2]^+$ relative to Ω_1 coupled with greater MO overlap in the case of the 1,1-isomer ensure that the interaction of the no bond and charge transfer configurations is greater in CH_3–CHF_2 than in CH_2F–CH_2F giving rise to greater stabilization of the former[k]. Accordingly, we formulate the following rules:

---

[k] In some cases, the relative energy orderings of the A^+B^- and A^-B^+ configurations of the 1,1 isomer depends on the type of computation. However, the A^+A^- configuration of the 1,2 isomer lies always between the A^+B^- and A^-B^+ configurations of the 1,1 isomer.

207

a) A 1,1-disubstituted molecule will always be more stable than its 1,2-isomer, if the two substituents are identical. Furthermore, the difference in their stability will increase as the donor-acceptor properties of the two fragments in the 1,1-isomer are enhanced relative to those in the 1,2-isomer.

b) A 1,1-disubstituted molecule may be more or less stable than its 1,2-isomer, if the two substituents are different. The 1,1-isomer will be more stable if it corresponds to a better donor-acceptor combination and vice versa.

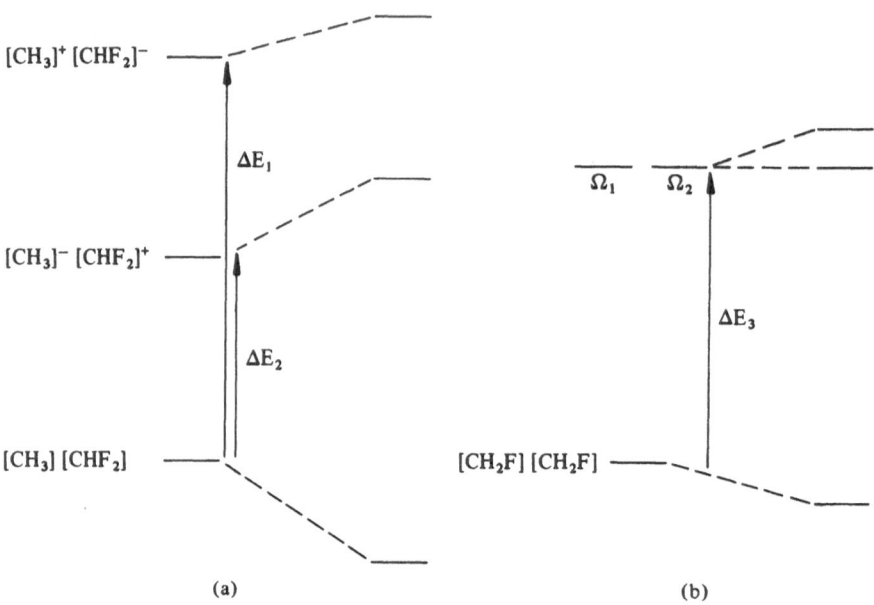

Fig. 55. Configuration interaction diagram for (a) 1,1 difluoroethane and (b) 1,2 difluoroethane. Ω_1 and Ω_2 are charge resonance configurations, *i.e.*

$$\Omega_1 = \frac{1}{\sqrt{2}} \left([CH_2F]^- [CH_2F]^+ + [CH_2F]^+ [CH_2F]^- \right)$$

$$\Omega_2 = \frac{1}{\sqrt{2}} \left([CH_2F]^- [CH_2F]^+ - [CH_2F]^+ [CH_2F]^- \right)$$

A pictorial representation of the interaction matrix elements, *i.e.* a pictorial representation of the key orbital interactions responsible for the greater stability of the 1,1-isomer, is given below.

The greater interaction of the no-bond configuration with the charge transfer configurations in F_2CH-CH_3 than in FCH_2-CH_2F dictates not only a greater stability of the 1,1-isomer but also a shorter $C-C$ bond in F_2CH-CH_3 than in FCH_2-CH_2F. We are then led to the formulation of the following rule: the bond distance between *the atoms which bear the two identical substituents in the 1,2-isomer* will be shorter in the case of the 1,1-isomer unless special effects mitigate this trend.

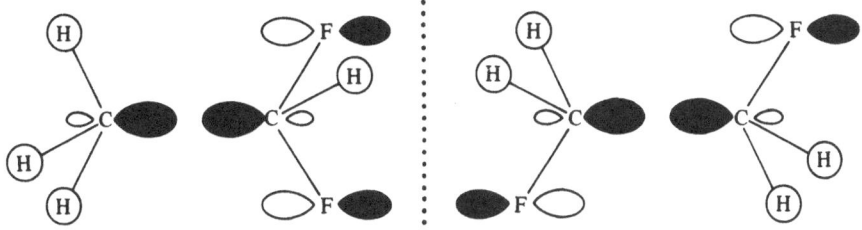

The same type of arguments can be extended to disubstituted olefins. Consider 1,1-difluoroethylene *vs.* 1,2-difluoroethylene. The appropriate basis set configurations are shown in Scheme 21 and 22 for the 1,1-isomer and the 1,2-isomer, respec-

Scheme 21.

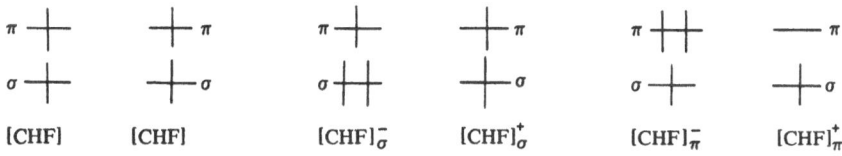

Scheme 22.

tively. Again, only charge transfer configurations which can mix with the no-bond configuration have been considered. The interaction diagram of Fig. 56 shows that the 1,1-isomer will be stabilized relative to the 1,2-isomer for the same reason as in difluoroethane.

An inspection of the interaction matrix elements responsible for the stabilization of the ground electronic state of $F_2C=CH_2$ provides a partial understanding of the angle problem in 1,1-disubstituted olefins. Specifically, the following interaction matrix elements increase as the FCF angle decreases and the HCH angle remains constant because the corresponding MO overlap integrals increase by virtue of relieving long range antibonding interactions. Once more, we have identified a nonbonded interaction effect favoring small angles in $X_2C=CH_2$ molecules[1].

---

[1] Undoubtedly, pi conjugative effects contribute to this trend. See Ref.[1].

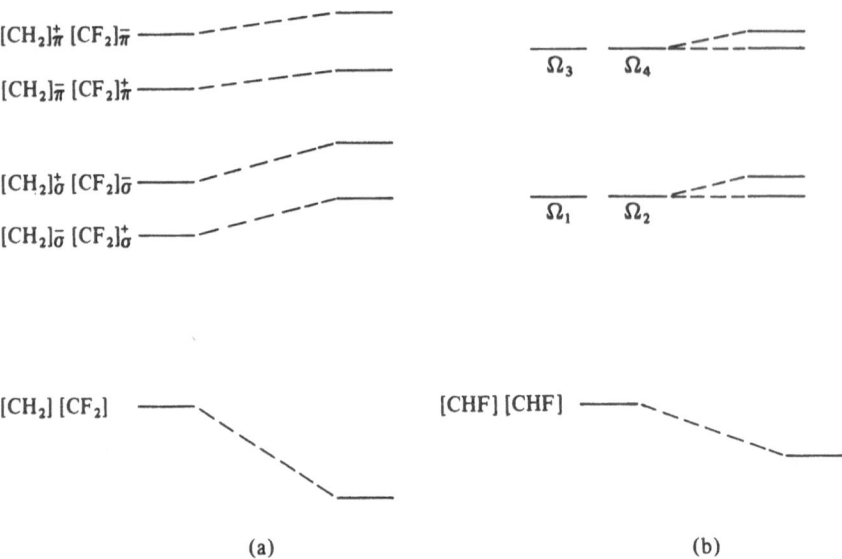

$[CH_2]_\pi^+ [CF_2]_\pi^-$

$[CH_2]_\pi^- [CF_2]_\pi^+$

$\Omega_3 \quad \Omega_4$

$[CH_2]_\sigma^+ [CF_2]_\sigma^-$

$[CH_2]_\sigma^- [CF_2]_\sigma^+$

$\Omega_1 \quad \Omega_2$

$[CH_2] [CF_2]$

$[CHF] [CHF]$

(a)

(b)

Fig. 56. Configuration interaction diagram for (a) 1,1-difluoroethylene and (b) 1,2-difluoro-ethylene. Ω_1 and Ω_2 are sigma charge resonance configurations. Similarly, Ω_3 and Ω_4 are pi charge resonance configurations

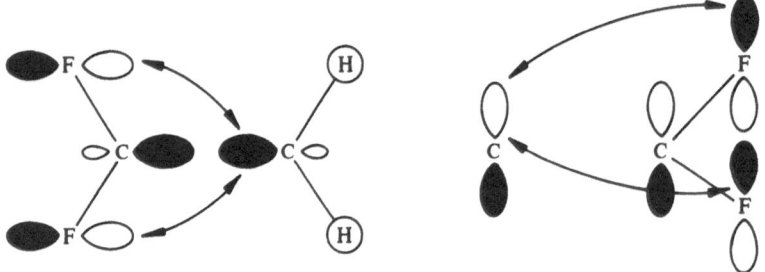

Obviously, as the energy separation of the no bond and charge transfer configurations decreases, an increase in the corresponding matrix element via the angle shrinking mechanism would have an increasingly pronounced effect on the stabilization of the ground state. Thus, as the energy of the lowest charge transfer configuration, *i.e.* $[X_2C]^+[CH_2]^-$ or $[X_2C]^-[CH_2]^+$, of a $X_2C=CH_2$ molecule decreases, the XCX angle will tend to shrink, unless mitigated by a steric effect. Unfortunately, this rule cannot be rigorously tested because precise ionization potentials and electron affinities of the fragments are not known and cannot be estimated in a quantitatively reliable sense. Furthermore, test calculations have led us to suspect that in most such cases electronic effects will run counter to "steric" effects.

The theoretical model which we have proposed is amenable to testing. Thus, we pursued both *ab initio* and CNDO/2 calculations. In the *ab initio* studies, we calculated the three isomers of C_4H_8 at the most stable conformations of the methyl groups employing the Gaussian 70[26] series of computer programs. Extensive geometry optimization was carried out. Since the geometrical parameters for the *cis* and *trans* isomers have been reported before[6], we provide only the data for isobutene[381]

(Table 53). The relative energies of the three structures are given below. Clearly, the predicted greater stability of $X_2C=CH_2$ over $CXH=CHX$ when X is methyl is reproduced well by these computations. Preliminary results[382] indicate that this is also true for the isomers of difluoroethylene and the isomers of dihydroxyethylene.

Table 53. Computed energies and geometric parameters of isobutene

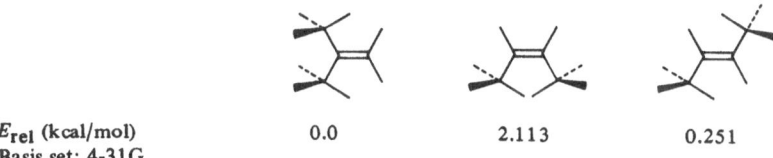

$r(C_1-C_2)$ Å	1.32	1.32	1.32
$r(C_1-C_3)$ Å	1.53	1.53	1.53
$r(C_3-H)$ Å	1.09	1.09	1.09
$r(C_2-H)$ Å	1.07	1.07	1.07
$< C_3C_1C_4$ (deg.)	115.7	116.6	118.4
$< HC_2H$ (deg.)	120	120	120
$< HC_3H$ (deg.)	109.5	109.5	109.5
E_N[a]	119.50258	119.49059	119.47696
E_{el}[a]	−275.38822	−275.37289	−275.35502
E_T[a]	−155.88564	−155.88230	−155.87806

[a] Single 4-31G calculation at the geometries listed above.

E_{rel} (kcal/mol)	0.0	2.113	0.251

Basis set: 4-31G

What kind of evidence is available to support the various rules we have proposed? The relative stability of 1,1- and 1,2-homodisubstituted molecules, where the two substituents are identical, can be assessed in the following manner:

a) By comparing experimentally determined heats of formation. Pertinent data are shown in Table 54.

b) by comparing heats of formation estimated on the basis of group of additivity relationships. Pertinent data are shown in Table 55.

c) Spectroscopic determination of molecular geometries. Pertinent data are shown in Table 56.

An examination of the various tables leads to the following generalizations:

a) In cases where experimental heats of formation are available and *irrespective of the electronegativity of the substituent*, the 1,1-isomer is more stable (Table 54).

211

Part V. Bond Ionicity Effects

Table 54. Experimental heats of formation of isomeric compounds

Compound		ΔH_f° (kcal/mol)	Ref.
MeHC=CHMe	cis	− 1.67	383)
	trans	− 2.67	383)
Me$_2$C=CH$_2$		− 4.04	383)
EtHC=CHMe	cis	− 6.71	383)
	trans	− 7.59	383)
EtMeC=CH$_2$		− 8.68	383)
EtHC=CHEt			
	cis	− 11.38	383)
	trans	− 13.01	383)
Et$_2$C=CH$_2$		− 13.38	383)
CH$_3$CH$_2$CH$_2$CH$_3$		− 30.15	383)
(CH$_3$)$_2$CHCH$_3$		− 32.15	383)
		− 33.05	383)
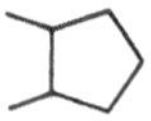	cis	− 30.96	383)
	trans	− 32.67	383)
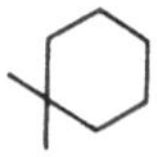		− 43.26	383)
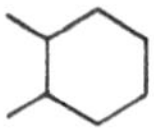	cis	− 41.15	383)
	trans	− 43.02	383)
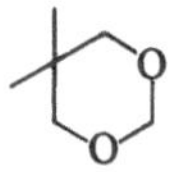		−100.6	383)
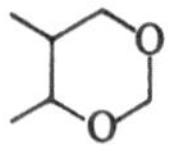		− 98.1	383)
MeHN-NHMe		+ 21.6	383)
Me$_2$N-NH$_2$		+ 20.4	383)
CH$_3$CH(OH)$_2$		−103.1	383)
HOCH$_2$CH$_2$OH		− 92.4	383)
CF$_3$CH$_3$		−178.0	384)
CH$_2$FCF$_2$H		−167	384)
CHCl$_2$CH$_3$		− 30.6	384)
CH$_2$ClCH$_2$Cl		− 29.7	384)
CH$_2$Cl-CHClCH$_3$		− 39.7	384)
CH$_3$CCl$_2$CH$_3$		− 41.4	384)

212

Table 55. Estimated heats of formation for several compounds[383]

		ΔH_f^o (kcal/mol)
FHC=CHF	cis	− 75.2
	trans	− 75.2
$F_2C=CH_2$		− 71.24
ClHC=CHCl	cis	+ 4.2
	trans	+ 4.2
$Cl_2C=CH_2$		+ 4.46
MeHC=CHMe	cis	− 1.98
	trans	− 2.98
$Me_2C=CH_2$		− 3.56
CHF_2CH_3		−119.38
CH_2FCH_2F		−103.6
$CHCl_2CH_3$		− 28.98
CH_2ClCH_2Cl		− 31.2
CF_3CH_3		−168.48
CHF_2CH_2F		−161.1
CCl_3CH_3		− 30.87
$CHCl_2CH_2Cl$		− 34.50

Table 56. The carbon-carbon double bond length in several olefins

Olefin	$r(C=C)$[a]	Ref.
CH_2CH_2	1.337	[385]
cis-CHFCHF	1.324	[386]
CF_2CH_2	1.315	[386]
cis-CHClCHCl	1.354	[387]
trans-CHClCHCl	1.343	[388]
CCl_2CH_2	1.324	[389]
trans-$CH_3CHCHCH_3$	1.347	[390]
$(CH_3)_2CCH_2$	1.331	[391]

[a] In Ångstrom.

b) Heats of formation estimated on the basis of the additivity relationship (Table 55) indicate that, *inter alia*, 1,2-difluoroethylene is more stable than 1,1-difluoroethylene by 4.96 kcal/mol. We suggest that the cases where the additivity relationship indicates that the 1,2-isomer has greater stability than the 1,1-isomer constitute situations where the principle of additivity of thermodynamic quantities breaks down.

c) The variation in the C–C or C=C bond length as a function of the type of substitution is such that it is always shorter in the 1,1-isomer than in the 1,2-isomer *irrespective of the electronegativity of the substituent.*

It is clear that when the two substituents are identical, one does not need any kind of calculation in order to predict relative stabilities. A need for calculation arises only when the two substituents are nonidentical. Even in those cases, qualitative trends have emerged[392]. Exceptions will be found when through space steric repulsion of the substituents in the 1,1-isomer plays a dominant role. This may be the case for pi acceptor substituents such as CN, COOR, C_6H_5, etc.

The approach described above seems to lead to reliable predictions regarding structural isomerism. The principal reasons are the following:

a) The fundamental equation of perturbation theory[16] which describes the energy change of i due to its interaction with j involves an interaction matrix element H_{ij} and an energy gap $\epsilon_i-\epsilon_j$. In comparisons of 1,1- vs 1,2-isomers the interaction matrix element remains relatively constant and the energy gap determines relative stability. This situation is not always met in discussions of reactivity problems where the H_{ij} term can be found to vary faster than the $\epsilon_i-\epsilon_j$ terms in certain comparisons.

b) A factor which undoubtedly contributes to a greater stability of the 1,1 structural isomers is intrafragment n–σ* conjugative interactions of the type shown below[314].

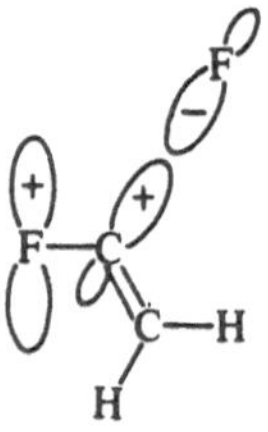

These interactions are expected to be important when the substituents are F, Cl, OR, SR, etc. On the other hand, this type of mechanism will not significantly contribute towards a greater stability of the 1,1-isomer when the substituents are alkyl groups.

Before departing the topic of structural isomerism, we wish to point out that the key ideas presented here fit within a broad scheme of chemical reasoning. Thus, the combination of two different types of fragments, namely, two A and two B fragments to yield A_2 plus B_2 is less stable than the combination of the same fragments to yield 2 AB, the latter being superior because it involves a donor-acceptor pairing. The above statement is supported by ample experimental evidence. Representative examples are shown below[383]:

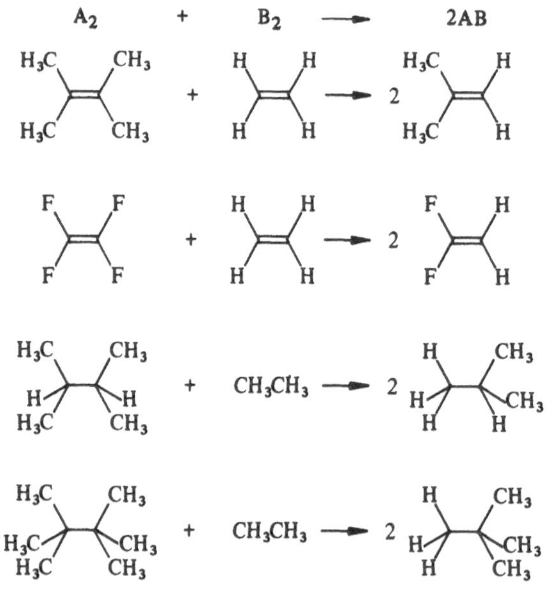

A_2	+	B_2	⟶	2AB	$2\,\Delta H_f(AB)-\Delta H_f(A_2+B_2)$ (kcal/mol)
					−16.16
					−19.10
					− 1.57
					− 5.11

Similarly, the combination of three different types of fragments, namely, A, B and two C fragments to yield AB plus C_2 may be more or less stable than the combination of the same fragments to yield AC plus BC depending upon the donor-acceptor interrelationships. In all cases, however, the success of the approach is guaranteed only if matrix elements remain relatively constant.

12.3. Bond Strengths

We shall now examine the effect of bond ionicity upon bond strength. For example, consider the bond strength of H–X as the electronegativity of X increases while the interatomic distance, r_{HX}, as well as the spatial overlap between the AO's of H and X remain constant. The LCFC interaction diagram of Fig. 57 indicates that increased mixing of the HX and H^+X^- configurations should lead to stronger HX bonding.

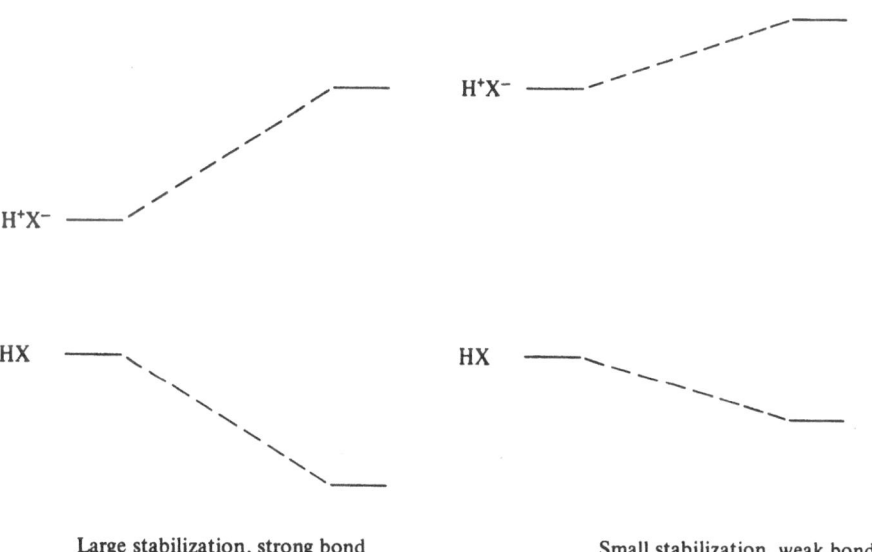

Large stabilization, strong bond   Small stabilization, weak bond

Fig. 57. LCFC interaction diagram for (a) highly electronegative X atom and (b) moderately electronegative X atom

The degree of mixing of the no bond and lowest charge transfer configuration is expected to increase as the ionization potential of one singly occupied AO decreases and the electron affinity of the second singly occupied AO entering bonding increases while the AO coefficients remain unity. However, in situations where an increase of the polarity of the system, *i.e.* a decrease of the quantity $I_D - A_A$, is counteracted by a drastic shrinkage of the AO coefficients of the uniting centers,

Part V. Bond Ionicity Effects

the reduction of the absolute magnitude of the interaction matrix element may dominate the diminution of the energy gap separating the interacting DA and D^+A^- configurations. The smaller C–H bond dissociation energy of $CH_2=CH–CH_2–H$ as compared with that of $CH_3–CH_2–CH_2–H$ is an example of matrix element control effect.

(a) Group MO's

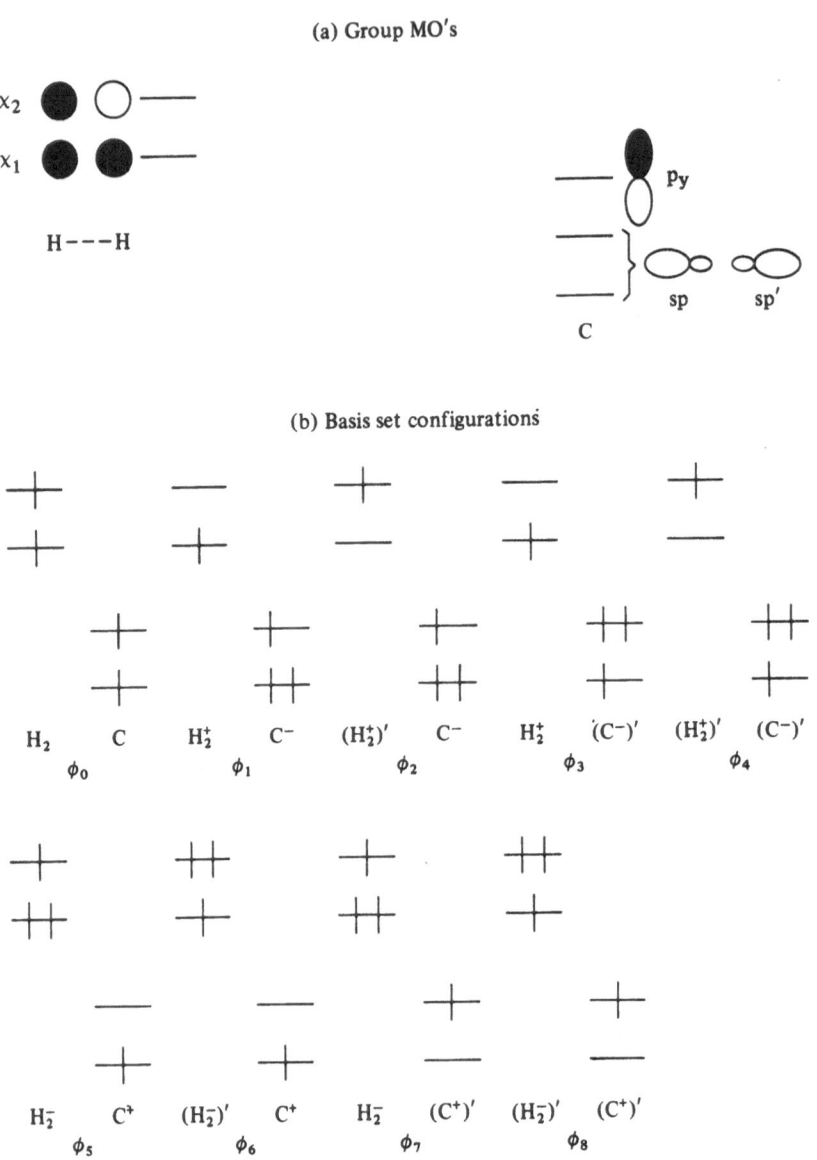

(b) Basis set configurations

Fig. 58. (a) Hydrogen group MO's and carbon sigma AO's for H_2C. (b) The important basis set configurations for linear H_2C

216

A well recognized effect, often referred to as the fluorine effect, is related to bond shortening upon progressive substitution of one center by electronegative groups such as fluorine[316]. The LCFC approach leads to a satisfactory understanding of this trend.

Consider the simplest system where successive substitution of hydrogens by fluorine may obtain, i.e. H_2A where $A = C$. The sigma group MO's of the ligand and the sigma AO's of the central atom are shown in Fig. 58 along with the most important configurations for a fixed geometry, e.g. a linear geometry. The latter geometry is chosen in order to simplify the problem. Thus, only the two sp AO's of the central carbon atom need be considered since the p_y AO cannot mix with either ϕ_1 or ϕ_2. This restricts the number of configurations which are necessary for an appropriate description of the bonding.

The interaction of the various configurations leads to sigma bond formation and this can be pictorialized by reference to the one electron interaction matrix elements responsible for mixing of the various configurations. Typical examples are shown below.

H   C   H   H   C   H

$<H_2C|\hat{H}'|H_2^-C^+>$ $<H_2C|\hat{H}'|(H_2^-)'C^+>$

Clearly, successive replacement of a hydrogen by a more electronegative atom will shrink the energy gap between the no bond and charge transfer configuration of the type $C^+H_2^-$ leading to stronger bonding of the central atom and each of the ligands (see Fig. 59). The effect will further be accentuated by substitution of the second hydrogen by a more electronegative atom. Accordingly, the LCFC approach predicts that successive replacement of hydrogens by more electronegative atoms should lead to increasing shortening of all sigma bonds emanating from the central atom, other things being equal.

The effect of successive fluorination of hydrocarbons on bond lengths has been extensively discussed in a recent review article which summarizes the most important experimental trends[316]. These trends are nicely accounted for by the LCFC approach.

12.4. Valence Isomerization

The effects of substituents on valence tautomerization can be partially understood on the basis of the LCFC approach. Consider for example, the two valence isomerizations shown below:

$$\text{(cyclobutene, positions 1,4,2,3)} \longrightarrow \text{(butadiene, positions 1,2,3,4)} \qquad \Delta H = -8 \text{ kcal/mol}[394]$$

$$\text{(F, F}_2\text{, F, F}_2\text{ cyclobutene)} \longrightarrow \text{(F}_6\text{ butadiene)} \qquad \Delta H = 11.7 \text{ kcal/mol}[395]$$

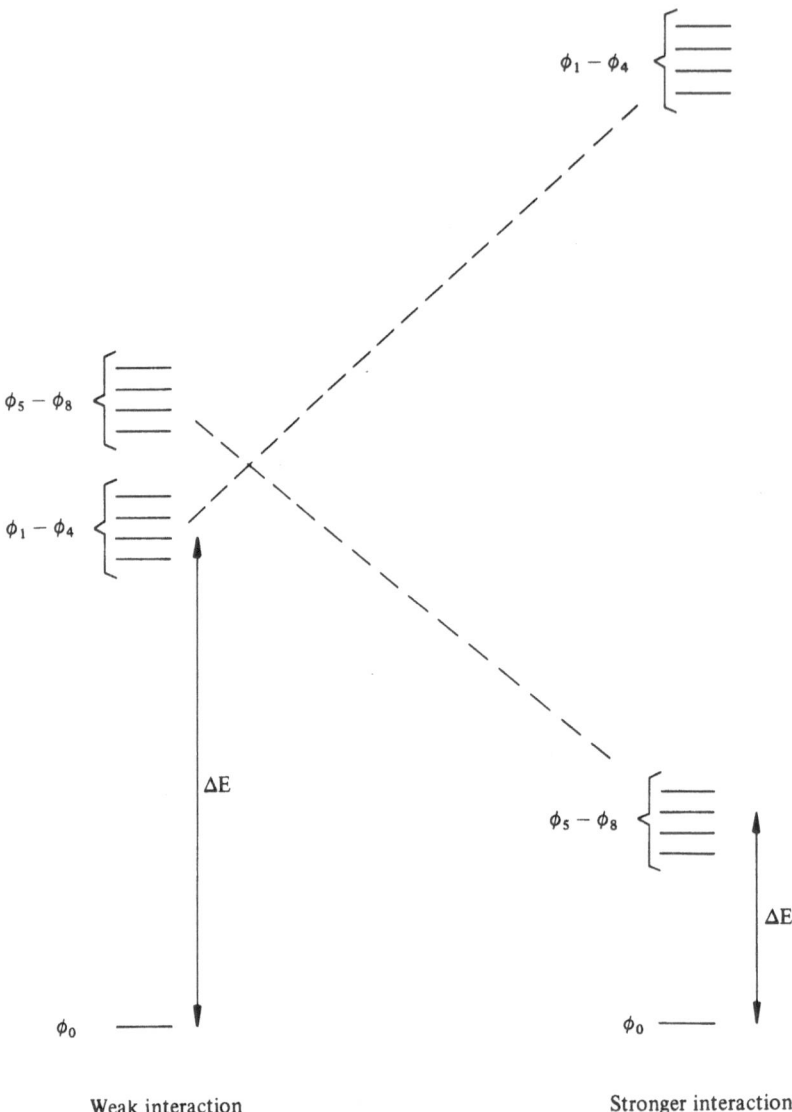

Fig. 59. Dependence of the energy gap between no bond and charge transfer configurations as a function of ligand electronegativity. Increasing ligand electronegativity results into stronger bonding between the central atom and each of the ligands

One could say that the major change in this transformation amounts to a change in the hybridization of C_3 and C_4 from sp^3 to sp^2 and an accompanying change in the bond C_3-X (or C_4-X) bond strengths. The LCFC diagrams shown in Fig. 60 clearly demonstrate that the rate of change in C–X bond strengths will be greater for the fluoro derivative. Other effects may well contribute towards the preference for attachment to saturated rather than unsaturated carbon exhibited by fluorine.

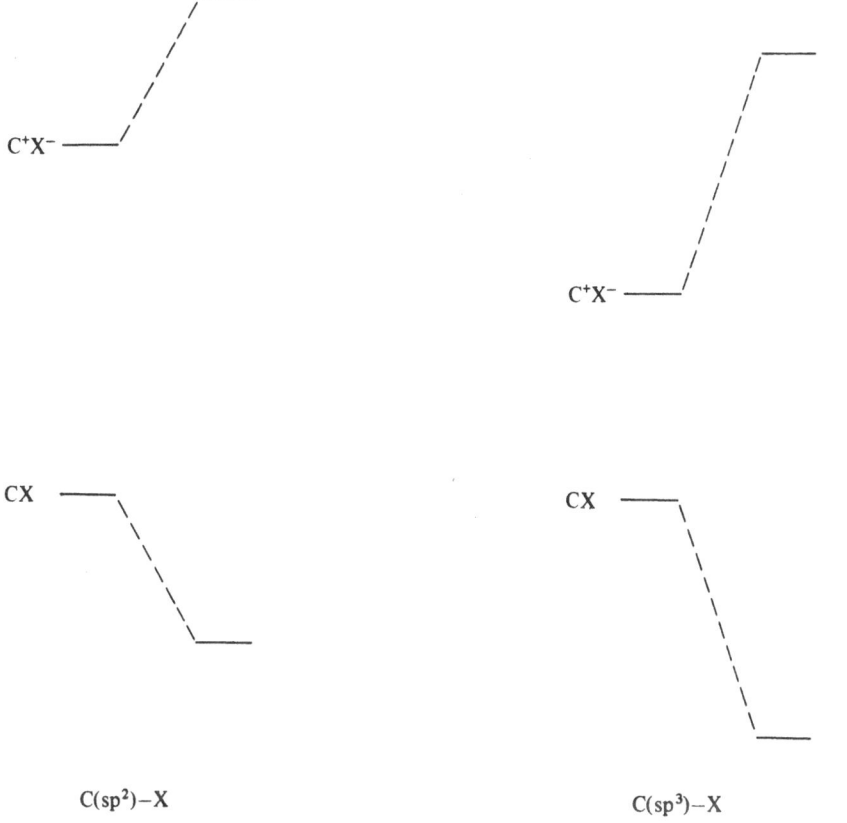

C^+X^- ————

C^+X^- ————

CX ————

CX ————

$C(sp^2)-X$ $C(sp^3)-X$

Fig. 60. The differential *stabilization* of C−X bonds as a function of hybridization

In short, all the problems discussed in Part V involve a variation of bond strengths accompanying a structural change. The LCFC approach is well suited for monitoring this effect.

13. Other Approaches

General theories of molecular structure have been advocated by different groups in the past but their applicability has been restricted to the shape, *i.e.* bond lengths and bond angles, of "small" molecules. Two types of approaches deserve special mention:

a) The "hybridization" theories based upon valence bond type arguments[396].

b) The Valence Shell Electron Pair Repulsion (VSEPR) model based upon an interpretation of the Pauli Exclusion Principle[397].

Firstly, we shall point out that the aforementioned theories have been extremely useful in organizing our thinking regardless of "how theoretically valid" they actually

219

were. Secondly, we shall argue that these models have now served their purpose and should be replaced by the more general OEMO model advocated in this work. This latter model is applicable to important problems, such as conformational, geometrical, and structural isomerism, whereas the "hybridization" and VSEPR models are not. Furthermore, we have been able to explain trends in molecular shape previously unintelligible on the basis of the "hybridization" or VSEPR models. These trends are, in summary, the following:

a) The fact that perfluorination shrinks the angles of NH_3 but not of PH_3, etc.

b) The fact that replacement of Y=CH_2 by Y=O shrinks the XCX angle in CX_2=Y.

c) The fact that PH_3 has smaller angles than NH_3, etc.

We shall next discuss other approaches to structural problems which have been aired in the literature and which have led to conclusions which are incompatible with ours. In one case, the conclusions are accidentally compatible.

Disheartening as it is to criticize the efforts of other research groups, it is necessary to pinpoint such approaches which, to the best of our knowledge, are deficient, in order to serve the interests of the experimentalist who is not a theoretical expert. Indeed, very often the reader of scientific journals is exposed to conflicting theoretical viewpoints. As a result, a confusion arises as to what is right and what is wrong. In view of these considerations, we have endeavored to provide a detailed critique of other approaches and demonstrate in some detail their shortcomings.

Wolfe and collaborators have stated the following three rules[332, 398–400]:

1. Electron pair-electron pair, electron pair-polar bond, or polar bond-polar bond interactions cause a significant increase in rotation-inversion barriers of atoms bearing these substituents.

2. When electron pairs or polar bonds are placed or generated on adjacent pyramidal atoms, *syn* or *anti* periplanar orientations are disfavored energetically with respect to that structure which contains the maximum number of *gauche* interactions.

3. The relative importance of the "*gauche* effects" associated with polar bonds and lone electron pairs is polar bond-polar bond > polar bond-lone pair > lone pair-lone pair.

The first rule apparently pertains to inversion problems. In **B**, as we already have

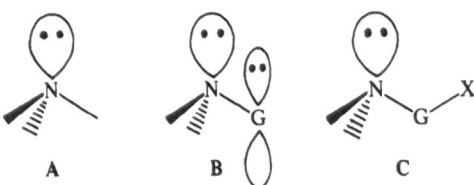

A   B   C

discussed, electron pair-electron pair interaction will increase the inversion barrier relative to that in **A**. However, in **C**, where the G—X bond is assumed to be polar, the inversion barrier may be reduced (G—X= good acceptor bond) or increased (G—X= poor acceptor bond) relative to **A**. Clearly, rule 1 will lead to erroneous predictions when the polar G—X bond becomes a good acceptor.

The second rule is directly relevant to conformational isomerism. In Part IV of this work we have provided the theoretical justification of this rule and also identified the situation where this rule may break down. A typical exception is shown below and involves a molecule where two vicinal polar bonds constitute the best donor-acceptor fragment combination and, thus, lead to a conformational preference placing them in an *anti*-relationship.

The third rule leads to the inference that decreasing the polarity of the C−X and/or O−Y bond in $CXH_2 - OY$ will reduce the magnitude of the anomeric effect[333]. Following the reasoning exemplified in Part II, we can produce many cases which contradict this inference. Thus, the anomeric effect gets stronger along the series $X = I > Br > Cl > F$ where C−X bond polarity varies in the opposite direction.

Up until this point, the contribution of the aforementioned workers could be characterized as the formulation of purely empirical rules based upon consideration of some, but not all, experimental evidence. The rules of Wolfe and collaborators were devoid of any theoretical basis and, thus, could never do anything more than act as basis for extrapolation from known experimental data. Thus, the variation of the anomeric effect as a function of halogen substituent alluded to before never attracted the attention of these authors simply because their rules were founded on an intuitive concept of bond polarity which deceptively seemed to be consistent with the great majority of the experimental facts encompassed by the three rules as stated.

Wolfe and collaborators attempted to understand the origin of the "*gauche* effect" by using an intuitive interpretation of good quality *ab initio* calculations. The result of these attempts has been the generation of a model which is confusing.

For instance, Wolfe and collaborators concluded that the origin of a barrier can be understood principally in terms of the interactions of the bonded electron pairs. The authors further stated: "The nonbonded electron pairs, until now considered to be directed ligands with stereochemical character of their own, must be exhibiting nondirectional behavior, *i.e.*, they create a quasispherical potential field in which the true ligands (or bonding pairs) move. Such an averaged interaction between nonbonded electron pairs would not significantly influence the shape of a resulting potential barrier"[332]. This interpretation of the "*gauche* effect" is theoretically unclear. Part of the problem lies in the confusion associated with the Canonical MO (CMO) and localized MO (LMO) representations. For example, from inspection of the various CMO's of H_2O one can immediately recognize two types of lone pairs, an sp^2 and a p lone pair, something confirmed by photoelectron spectroscopic results[35]. This point is discussed by Salem and co-workers in a recent article[401], while the difference between CMO's and LMO's is discussed in many texts, *e.g.* Ref.[402]. When CMO's are properly analyzed, one arrives at the conventional picture of directed hybrid AO's, *e.g.* see the excellent text by Salem and Jørgensen[20]. By transforming CMO's into LMO's, one arrives at a different, yet equivalent picture of molecular structure, which, however, suffers from a distinct disadvantage, *i.e.* LMO's are not connected to experiment by a Koopmans' type theorem[28] as are the CMO's. Many times an organic chemist will ask: how is it possible for methane to have four equivalent C–H bonds and, yet, two different ionization potentials[35] implying two different sets of MO's? This question indicates a failure to realize that the four equivalent bonds of methane refer to a localized description, where the LMO's are not connected to Koopmans' theorem, while the alternative delocalized description utilizes CMO's, *i.e.* three degenerate CMO's and one lower energy CMO, which are connected to Koopmans' theorem.

Finally, it should be mentioned that Wolfe and his collaborators, and, more recently, Liberles and collaborators[403], emphasized the fact that conformational preferences arise as a balance of attractive and repulsive components of the total energy of the system. It should be pointed out that such discussions, based on Allen's component analysis[404], do not have any predictive value since they require the results of the full quantum mechanical calculation of the system under investigation.

We now proceed to discuss in detail two other approaches which are directly relevant to the ideas presented in this book. Thus, in a recent paper, Kollman[405] has questioned the role of "nonbonded attraction" in determining the stereochemical properties of difluoroethylenes and related molecules. He has suggested instead that the "attraction" is due to changes in the nature of the C–X bonding orbital as X becomes more electronegative.

The criticism focuses on the following points:

a) In our original approach[1], we treated the fluoroethylenes and other halo-ethylene type molecules by neglecting the s lone pairs and focusing attention only on the interactions of the p_x and p_z lone pairs (see Scheme 23). This simplification was based on the fact that s lone pairs lie so low in energy that they cannot interact efficiently through bond. In other words, their contribution to nonbonded attraction

Scheme 23. Axis convention for $X_2C=CH_2$ and X_2Y

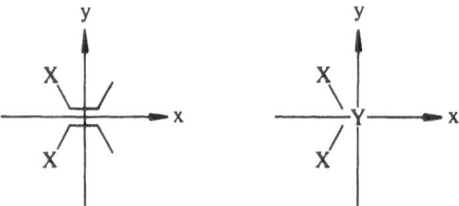

is small or negligible within the framework of the model we employed. Of course, this is borne out by *ab initio* calculations which show that the F2s–F2s overlap population in 1,1 difluoroethylene remains virtually constant when the FCF angle varies from 108° to 120°.

b) In our original approach, we approximated the in-plane lone pair of a halogen as a p_x lone pair. Kollman pointed out that the in-plane lone pairs have contributions from all the in-plane basis AO's. However, the approximation used in our paper is known to have great practical utility. Thus, in photoelectron spectroscopy a lone pair ionization potential often refers to the energy required to remove an electron from an orbital which is not a "pure" lone pair orbital. Nonetheless, the concept of the "impure" lone pair is a useful one. It should be said that the results of our analysis would remain unaffected if instead of a p_x lone pair we assumed a hybrid lone pair.

In the computational section, Kollman reports the results of *ab initio* calculations on model "pseudoethylenes", $H_1'H_2'C_1'=C_2'H_3H_4$, where H_1', H_2', C_1' and C_2' separately, or in combinations, bear modified nuclear charges in order to simulate the polarity of C–F bonds due to the difference in the electronegativities of the atoms. He finds that as the simulated electronegativity of H_1' or H_1' and H_2' increases, the corresponding HCH angle tends to shrink. Since the key calculations are those simulating 1,1-difluoroethylene, where the additional lone pair interactions can obtain, we shall focus, as the author did, on them.

On the basis of the results mentioned above, Kollman draws the conclusion that "the results presented. . .show clearly that the attractive interaction in fluoroethylenes is an electronegativity effect and not due to "lone pair" interactions since there are no lone pairs in our model system". The limitations of this argument can best be realized by reference to the work of Mislow *et al*[52] who used the electronegativity simulation approach for studying the effect of substituents on the inversion barrier of ammonia. These workers found that pseudoammonia, NH_2H', displays an increased barrier relative to ammonia. Thus, the higher barrier of NH_2F relative to NH_3 can be *partially* ascribed to an "electronegativity effect". This "electronegativity effect" is the resultant of geminal interactions. But Mislow *et al.* were careful to consider the additional influence of lone pair-lone pair interaction and, in fact, argued that this latter interaction is more important than the "electronegativity effect". In short, contrary to his claim, Kollman's *ab initio* results do not rule out that non-bonded interactions between lone pairs or molecular fragments can be the dominant effect determining the XCX angle in $X_2C=CH_2$. As we have already discussed, the small angles in such molecules result from nonbonded *and* geminal interactions. The importance of the nonbonded attraction component can be realized easily by con-trasting systems where pi nonbonded attraction, due to pi stabilizing as well as de-stabilizing orbital interactions, plus geminal interaction obtain to systems where no pi nonbonded attraction due to pi stabilizing orbital interactions is possible. Thus, for example, the oxygen series shown in Scheme 24 suggest that only fluorine groups

Scheme 24. Impact of nonbonded attraction due to pi stabilizing interactions on molecular shape[a]

F F	Cl Cl	H$_3$C CH$_3$	H H
109.3 ± 0.4°	114.5 ± 1.0°	115.8 ± 0.6°	116.2 ± 0.8°

F O F	Cl O Cl	H$_3$C O CH$_3$	H O H
103.3°	111°	111°	104.9°

[a] Pi stabilization on angle shrinkage is possibie only in the ethylene series.

are capable of shrinking the HOH angle. By contrast, in the ethylene series, the XCX angle is smaller than the HCH angle of ethylene irrespective of the electronegativity of the group X. A further interesting comparison is that between the series H_2O, HOF, F_2O and the series $CH_2=CH_2$, $CHF=CH_2$, $CF_2=CH_2$. Here, two fluorines cause a greater angle shrinkage than one fluorine in the ethylene systems where the

additional influence of pi nonbonded attraction due to pi stabilizing orbital interactions can be felt while the situation is strikingly different in the oxygen systems. Next, Kollman assumes that the small FCF angle in $F_2C=CH_2$ can be understood in terms of the F_2C fragment, i. e. this implies that transferring a fragment X_2C from vacuum to within a molecule should have a small or no effect on the X_2C angle. This argument is sound if one assumes that other interactions which arise within the molecule are unimportant relative to any effects already present in the CX_2 fragment. Of course, this is not correct since we have already seen that the angle of a fragment XCX depends on the heteroatom which is attached on the carbon atom (see Table 32).

Kollman chose to attempt to understand the angle shrinkage of 1,1 difluoroethylene by comparing this molecule to ethylene itself. He then noted that, in $F_2C=CH_2$ there is a greater withdrawal of charge from the p_y than from the p_x or s carbon AO's relative to ethylene. The same trend was found for the corresponding pseudoethylene. The author proposed and intuitive interpreation: "The carbon s and p_x orbitals contribute relatively more to the C−F bonding than the p_y and the fluorines move in toward the x axis to take advantage of the location of higher electron density". However, we have argued that C−F bonding involves more p_y than p_x carbon AO character and calculations show exactly that. Pertinent computational data is collected in Table 57 and the overlap populations make it unambiguously clear that C−F bonding has a dominant contribution by the carbon $2p_y$ AO. Furthermore, the noted trend that there is a greater withdrawal of charge from the p_y than the p_x or s carbon AO's in $F_2C=CH_2$ than in $CH_2=CH_2$ is not general. Thus, for example, in the triatomic fragments H_2O or F_2O and H_2C or F_2C, where the angle is kept constant, there is greater withdrawal of charge from p_x rather than p_y (see Table 58). Here, we deal with the absurd situation of a mechanism for angle shrinkage based upon fragment consideration supported by calculation of the total molecule and denied by calculation of the appropriate fragment. The key factor influencing these charge redistribution effects is the back donation from the occupied n_A lone pair MO to the vacant π_g and π_u MO's of the ethylenic unit or the p_y AO of the central atom. This effect is greater for the oxygen system and tends to counteract the normal charge depletion expected simply on the basis of geminal interactions (see Section 5.0).

Table 57. Sigma overlap populations in AX_2 and $CH_2=AX_2$ systems

Molecule	Calculation type	A	X	Geometry employed[a)	$A_{2s}-X$	$A_{2p_x}-X$	$A_{2p_y}-X$
H_2O	EH	O	H	Experimental	.1244	.0557	.1630
F_2O	EH	O	F	Experimental OF bond length FÔF angle is as in H_2O	.0459	.0744	.1221
H_2C	EH	C	H	Bond length and angle are taken from experimental C_2H_4 geometry	.1318	.0648	.1314
F_2C	EH	C	F	Experimental C–F bond length, FĈF angle same as HĈH angle in CH_2	.0163	.0529	.0711
$H_2C=CH_2$	EH	C	H	Experimental	.1475	.0703	.2110
$F_2C=CH_2$	EH	C	F	C_2H_4 geometry; experimental C–F bond length	.0535	.0712	.1450
$H_2C=CH_2$	Ab initio– STO–3G basis set	C	H	STO–3G optimized	.1313	.0820	.1813
$F_2C=CH_2$	Ab initio STO–3G basis set	C	F	STO–3G optimized	.0530	.0557	.1041

a) The x axis bisects the XAX, XCX and HCH angles and the molecules lie on the xy plane.

Part V. Bond Ionicity Effects

Table 58. AO occupation numbers in diverse AX_2 and $CH_2{=}CX_2$ systems

Molecule	Calculation type	Geometry[a]	2s	$2p_x$	$2p_y$
H_2O	EH	Experimental	1.482	1.204	1.779
F_2O	EH	Experimental OF bond length; FOF angle same as in H_2O	1.702	.806	1.460
Difference ($F_2O{-}H_2O$)		--	.220	−.398	−.319
H_2C	EH	Bond lengths and angles taken from experimental C_2H_4 geometry.	1.040	1.680	.369
F_2C	EH	Experimental C−F bond length; FCF angle same as HCH angle.	1.363	1.188	.199
Difference ($F_2C{-}H_2C$)			.323	−.492	−.170
$H_2C{=}CH_2$	EH	Experimental	1.080	.971	1.045
$F_2C{=}CH_2$	EH	C_2H_4 geometry; experimental C−F bond length.	1.099	.859	.623
Difference ($C_2H_2F_2{-}C_2H_2$)			.019	−.112	−.422
$H_2C{=}CH$	Ab initio	STO−3G optimized	.756	.590	.655
$F_2C{=}CH_2$	Ab initio	STO−3G optimized	.891	.604	.487
Difference ($C_2H_2F_2{-}C_2H_2$)			.135	.014	−.168

[a] The x axis bisects the XAX, XCX and HCH angles and the molecules lie in the xy plane.

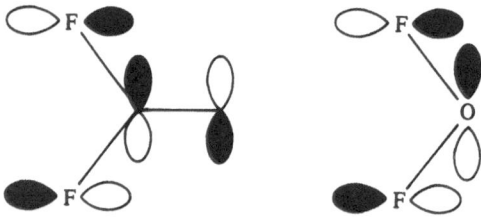

We now come to other aspects of Kollman's work. Specifically, in our original work, we commented upon the fact that greater nonbonded attraction between lone pairs is obtained in U or Y molecules. The relative stability of isomeric U and Y systems, obviously, depends on much more than lone pair interactions. Accordingly, our original work did not deal with this topic. Thus, the statement of Kollman,

"MO symmetry arguments ... lead to incorrect conclusions on the relative energies of 'Y' and 'U' systems" is misrepresentation of our work.

As we have discussed, the LCFC approach predicts that all 1,1-disubstituted molecules, where the two substituents are identical, are more stable than 1,2-isomers, regardless of the electronegativity of the substituent. On the other hand, Kollman proposed that the greater stability of 1,1-difluoroethylene relative to the 1,2-*cis*-isomer is a "charge" effect. An electrostatic depiction was provided in order to exemplify the argument, *i. e.* Kollman implied that better charge dispersal obtains in **A** than in **B**. This "charge" model would not work for substituents which are not highly electronegative because in such an event, "charge" effects would be small. Furthermore, one might suspect that even in the case of electronegative substituents, such as F, the model is actually wrong, *i. e.* 1,1-difluoroethylene is more destabilized relative to *cis*-1,2-difluoroethylene because of the proximity of the negatively charged fluorines.

To test our suspicion, we used a point charge model and *ab initio* calculated atomic charges in order to evaluate the electrostatic energy of 1,2-*cis* and 1,1-difluoroethylene.

	A	**B**
Electrostatic Energy (kcal/mol) (STO–3G)	13.61	−2.92

In a recent article Bingham questioned the existence of long range attractive interactions and cited the Kollman work and certain experimental cases as evidence against our proposal[406]. The experimental data which Bingham cites are consistent with our ideas. Thus, the geometries of the isomeric 1,2-difluoroethylenes and the conformational preference exhibited by diaminomaleonitrile have already been discussed in previous sections. The structural features of these molecules are understandable in terms of our concepts. The preferred conformation of the pentadienyl anion and that of *cis*-hexacyanobutadiene anion are probably dictated by conventional steric and electrostatic effects. On the other hand, acetylacetonate adopts a conformation where nonbonded attraction between the methyl groups can occur.

We believe that the criticism of Bingham is without a real basis. Accidentally, the final conclusion, *i. e.* overlap repulsions favoring "crowded" structures, is, in some cases, correct.

We shall now enumerate the various theoretical aspects of Bingham's work which we hold to be unsatisfactory.

a) Bingham states that pi electrons must delocalize preferentially through trans conformations. This fact is expressed by the comment that "electrons hate to go around corners" (e. e. electrons delocalize preferentially in straight lines)[407] and may be interpreted quantum mechanically as a "kinetic energy effect". Normally, the term delocalization is connected to a comparison, i. e. we say that the pi system of *trans* butadiene is more delocalized than the pi system of *cis* butadiene meaning that the former pi system is more stabilized than the latter, the converse being true for the corresponding dianions. In this sense, the preference for delocalization depends on the number of π electrons. Furthermore while initially Bingham tells us that delocalization is a kinetic energy effect, he subsequently uses an interaction diagram in order to ascertain the extent of delocalization which he associates with a net sum of stabilizing and destabilizing orbital interactions. These, of course, are evaluated with respect to an effective one electron Hamiltonian which contains both kinetic energy and potential energy terms!

b) Bingham considers the interaction of two orbitals and reiterates the well known result of perturbation theory that a two electron interaction is stabilizing and a four electron interaction is destabilizing. In the former case only a bonding MO (BMO) is occupied while in the other case a bonding MO and an antibonding MO (ABMO) are occupied, the ABMO being more antibonding than the BMO is bonding. The author proposed that "When only bonding (or nonbonding) molecular orbitals are occupied, electron delocalization will be energetically favorable. When antibonding molecular orbitals are also occupied, such delocalization becomes energetically unfavorable." In this proposal the terms bonding, nonbonding, and antibonding are not properly clarified. Specifically, a given MO can be classified as bonding, nonbonding, or antibonding only by reference to appropriate fragment orbitals which combine to give rise to the MO in question. Depending on whether the reference fragments are atoms or molecular subunits one may arrive at different conclusions.

For example, by combining the previous statements, Bingham formulates the following rule. "Electron delocalization will, therefore, stabilize *trans* conformations relative to the corresponding *cis* structures when only bonding or nonbonding molecular orbitals are occupied. *Cis* conformations will be stabilized (less destabilized) relative to *trans* when antibonding molecular orbitals are also filled." Let us pursue the implications for a simple case, 1,3-butadiene, choosing as reference fragments molecular subunits. This molecule has one BMO and one ABMO by reference to the noninteracting ethylenic fragments. However, Bingham chose to regard this molecule as having two BMO's using therefore the atoms as the reference fragments, but no statement is made to justify this choice. Rather, the author proceeds to predict that *trans*-1,3-butadiene will be more stable than *cis* 1,3-butadiene! Further predictions are made on a similar basis.

Other fallacies can be found. Thus, ionization potentials are cited as evidence of the extent of delocalization while it can be easily shown that there is no such connection.[m]

---

[m] For example see: Epiotis, N. D., Cherry, W.: J. Amer. Chem. Soc. *98*, 4365 (1976).

In other papers[408], the discussion of conjugative destabilization has been based on an incorrect assumption, *i. e.* that the destabilizing interaction of filled orbitals increases as their energy separation decreases, while, as we have seen in Eq. (5′), four electron overlap repulsions depend not on the energy separation of the two interacting MO's but rather on the sum of them. Thus, situation **A** is less destabilizing than situation **B** (constant S_{ij}).

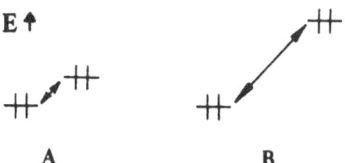

The mishandling of basic theoretical principles is a very general and, perhaps, unavoidable trend. Specifically, the following approach seems to be taken by various authors:

a) A survey of certain experimental facts is made.

b) A theoretical parameter which correlates these facts is suggested. Since one can correlate a restricted number of facts by virtually anything imaginable, these approaches are purely correlative and constitute restatement of facts in an alternative language. For example, a well read chemist knows that increased electronegativity of X along a row of the Periodic Table makes AX_2 progressively more bent. Rediscovering an already known fact by calculation is nothing significant. Here, a valuable contribution would be to develop a theory which would lead to the anticipation of reversals, if that is possible, as we have done in this work.

In concluding this section, we wish to emphasize that all general theories are, by necessity, not perfect. Thus, our own approaches to problems of molecular structure have been refined with time and will certainly be improved further.

14. Conclusion

In this work, we have identified the following three types of interactions:

a) Pi or sigma nonbonded interactions.

b) Pi or sigma conjugative interactions.

c) Geminal interactions.

d) Bond ionicity effects.

These constitute the alphabet for the construction of rationalizations and formulation of predictions regarding structural variations of organic molecules.

Pi and sigma nonbonded interactions as well as geminal interactions have been formulated by us. Pi and sigma conjugative ("hyperconjugative") interactions have been recognized for a long time but a systematic study has never appeared before. Part of this work constitutes such a study. Finally, bond ionicity effects have also been recognized for a long time. However, their applicability to problems of structural isomerism has never been realized before.

We have seen that in many cases unambiguous predictions can be made regarding the preferred geometry of a molecule. In other cases, conflicting effects demand some quantitative assessement of the dominant effect, *i. e.* an unambiguous qualitative prediction cannot be made. Finally, we have seen that whenever unambiguous electronic predictions fail, "steric effects" seem to be the culprit.

While the predictive aspect of one electron MO theory is impressive, we believe that the greatest contribution this approach makes is that it allows for the deliberate, rational design of molecular systems which will exhibit the geometrical preference desired by the designer. In other words, once we have isolated the key electronic factors present in a molecular system, we can manipulate them in order to achieve a certain goal. We hold this manipulation aspect to be the greatest virtue of the theoretical approach we have described. We believe that theory becomes a formidable weapon when it can suggest possibilities unfathomed by ordinary intuition or extrapolation from the available experimental data.

While the manipulations involved in the practice of one electron MO theory are simple, it is clear that, unless someone is well familiar with the intricacies involved, mistakes can easily be made. We hope that in Part II we have provided sufficient warning of the pitfalls which await the careless and/or inexperienced worker who tries to apply MO methodology to a chemical problem. Needless to say, the proliferation of canned computer programs capable of performing quantum mechanical calculations of varied degrees of sophistication, makes forays into the theoretical arena irresistible to nonexperts. Whether this will turn out to be a panacea or a source of confusion for the experimentalists remains to be seen.

At this point, a few words about chemical "effects" are in order. Various trends regarding molecular structure have been noted in the literature. Some of them are listed below:

a) The "fluorine" effect controlling the bond lengths in $CF_n H_{4-n}$.
b) The "*cis*" effect in geometric isomerism.
c) The "*gauche*" effect in conformational analysis.
d) The "rabbit ear" effect in conformational analysis.
e) The "anomeric" effect in conformational analysis.

In this work, we have seen that all these effects are understandable in terms of MO theory. They arise as a result of dominance of one or more MO interactions of the type discussed in this work.

The methods delineated in this work can be easily extended to any problem of molecular structure. Thus, for example, it can be easily shown that our ideas can be successfully applied to various types of organic chemical problems which have not yet been considered systematically using OEMO theory. Some of the problems currently under investigation in our laboratories are the following:

a) The conformation and stability of radicals.
b) The participation, or absence thereof, of d orbitals in bonding.
c) Intermolecular interactions involving macromolecules and "small" organic molecules.
d) The effect of substituents upon the absorption of radiation by organic molecules.
e) Hydrogen bonding and the stability of charge transfer complexes.

Finally, we consider certain basic problems that remain to be tackled. These can be formulated as follows:

a) What is the operational significance of correlation energy? For example, consider *cis*- and *trans*-1,2-difluoroethylene. The greater stability of the *cis* isomer is definitely due to a great extent to the factors discussed in this work. Another contributor may be correlation energy favoring the more crowded, *i.e. cis,* form. Accordingly, calculation of the Hartree-Fock limit of both isomers can tell us to what extent the geometrical preference of 1,2-difluoroethylene is due to correlation energy and to what extent is due to the factors described here.

b) Why do the bonding predictions of the OEMO method match so well the results of explicit ab initio calculations involving geometry optimization? In our work, we have demonstrated that the trends in overlap populations accompanying a geometrical change can be predicted very accurately by the OEMO method and that *ab initio* calculations bear out these predictions to an astonishing degree. We are worried by this success of our approach! Could it be that orbital symmetry expresses itself not only in monoelectronic matrix elements but also in bielectronic ones and both effects go in the same direction? We are investigating this problem, which, by the way, is not as simple as it may superficially appear. We note that a commendable effort in this direction has been made by Van Catledge[410].

The problems of *why* certain theoretical models seem to work, henceforth designated the justification problem, is a constant preoccupation of theoretical chemists. The papers written on justification are numerous. We note here the most recent attempt by Buenker and Peyerimhoff to summarize arguments and data relevant to the justification of the old Mulliken-Walsh model of molecular structure which proved particularly successful in the interpretation of the dependence of bond angle and bond length distortions on the number of electrons (*i.e.* MO occupancy) in "small" molecules[265].

c) Can we develop rules for evaluating the dominant effect when more than one important effects conflict in dictating a geometrical preference? Progress in this direction will be very welcome.

In addition to these questions, there are a number of additional specific problems which need further probing. Thus, the nature of lone pair sigma nonbonded interactions in diverse systems and its impact on molecular shape awaits a more detailed analysis using high quality *ab initio* wavefunctions. The problem of cooperative effects is another central problem which needs systematic scrutiny. Thus, the rotational barrier in $CH_3CX=Y$ depends on the nature of X and Y in a complex fashion. Specifically, the $CH_3\cdots X$ pi nonbonded interaction is a function of the nature of X and Y but also of the distance between CH_3 and X which is modulated by the sigma hyperconjugative interaction between Y and the CH_3-C-X fragment.

In the present treatise of structural chemistry, we have described a complex of ideas which were developed in the period of the last five years. However, in some cases, reasoning similar to ours has been employed by other workers in the course of their independent studies of various types of structural problems. We feel that the following works deserve special mention, although we hasten to add that the list is by no means exhaustive:

a) The investigation of the general topic of through space and through bond coupling by Hoffmann and his collaborators[60-64] provided the stimulation for our work on nonbonded attraction. The same concept was developed independently by Hoffmann's school[412].

b) The theoretical analysis of the effect of hyperconjugative interactions on conformational preferences. Here, the list of the main contributors is long but the names of Romers and Altona[335-342], Pople[295, 297, 314, 413], Hoffmann[344, 413], Salem[344, 401, 413], Schleyer[295, 297, 413] and Hehre[314, 344, 413] deserve special mention.

c) The independent investigation of the effect of electronegativity on the shape and inversion barriers of molecules by C. C. Levin[264] and B. M. Gimarc[414].

d) The interpretation of a variety of conformational effects by focusing on repulsive nonbonded interactions offered by Lowe[66], Salem and Hehre[98].

e) The work of Pople and Hehre on the conformational preferences of diverse systems such as ROR'[149], C_4[214] and diene molecules[199].

f) The recognition of the stereoelectronic control of conformational preference and chemical reactivity[415-420].

Note Added in Proof:

The following is a list of references to previous relevant works which came to our attention after submission of the manuscript as well as publications which appeared after the completion of our treatise: (a) *Nonbonded Interactions:* Fraser, R. R., Grindley, T. B., Passannanti, S.: Can. J. Chem. *53*, 2473 (1975); Spencer, T. A., Leong, C. W.: Tetrahedron Lett. *1975*, 3889; Fraser, R. R., Dhawan, K. L.: Chem. Commun. *1976*, 674; Skaarup, S., Skancke, P. N., Boggs, J. E.: J. Amer. Chem. Soc. *98*, 6106 (1976); Schander, J., Russell, B. R.: J. Amer. Chem. Soc. *98*, 6900 (1976); Hehre, W. J., Pople, J. A., Devaquet, A.: J. Amer. Chem. Soc. *98*, 664 (1976); Durig, J. R., Griffin, M. G.: J. Mol. Spect. *64*, 252 (1977); Newton, M. D., Jeffrey, G. A.: J. Amer. Chem. Soc. *99*, 2413 (1977); Vishveshwara, S., Pople, J. A.: J. Amer. Chem. Soc. *99*, 2422 (1977); Binkley, J. S., Pople, J. A.: Chem. Phys. Lett. *45*, 197 (1977); Dondoni, A., Gilli, G., Sacerdoti, M.: J. Chem. Soc. Perk. *II*, 1036 (1976). (b) *Geminal Interactions:* Cherry, W., Epiotis, N., Borden, W. T.: Accounts Chem. Res. *10*, 167 (1977); Gordon, M. S., Fischer, H.: J. Amer. Chem. Soc. *91*, 2471 (1968). (c) *Sigma Conjugation:* Hamlow, H. P., Okada, S., Nakagawa, N.: Tetrahedron Lett. *1964*, 2553; Krueger, P. J., Jan, J., Wieser, H.: J. Molec. Struct. *5*, 375 (1970); Hanstein, W., Traylor, T.G.: Tetrahedron Lett. *1967*, 4451; Hanstein, W., Berwin, H. J., Traylor, T. G.: J. Amer. Chem. Soc. *92*, 7476 (1970); Traylor, T. G., Hanstein, W., Berwin, H. J., Clinton, N. A., Brown, R. S.: J. Amer. Chem. Soc. *93*, 5715 (1971); Pitt, C. G.: J. Organometal. Chem. *61*, 49 (1973); Baddeley, G.: Tetrahedron Lett. *1973*, 1645; Klein, J.: Tetrahedron *30*, 3349 (1974); Bingham, R. C.: J. Amer. Chem. Soc. *97*, 6743 (1975); Abe, A.: J. Amer. Chem. Soc. *98*, 6477 (1976); Lehn, J.-M., Wipff, G.: J. Amer. Chem. Soc. *98*, 7498 (1976). (d) *Bond Ionicity:* Pickard, J. M., Rodgers, A. S.: J. Amer. Chem. Soc. *98*, 6115 (1976); Wu, E., Rodgers, A. S.: J. Amer. Chem. Soc. *98*, 6112 (1976).

References

[1] Epiotis, N. D.: J. Amer. Chem. Soc. *95*, 3087 (1973)

[2] Epiotis, N. D., Cherry, W.: Chem. Commun. 278 (1973)

[3] Epiotis, N. D., Bjorkquist, D., Bjorkquist, L., Sarkanen, S.: J. Amer. Chem. Soc. *95*, 7558 (1973)

[4] Epiotis, N. D., Sarkanen, S., Bjorkquist, D., Bjorkquist, L., Yates, R.: J. Amer. Chem. Soc. *96*, 4075 (1974)

[5] Epiotis, N. D., Yates, R. L., Bernardi, F.: J. Amer. Chem. Soc. *97*, 5961 (1975)

[6] Epiotis, N. D., Yates, R. L.: J. Amer. Chem. Soc. *98*, 461 (1976)

[7] Yates, R. L., Epiotis, N. D., Bernardi, F.: J. Amer Chem. Soc. *97*, 6615 (1975)

[8] Epiotis, N. D., Yates, R. L., Bernardi, F., Schlegel, H. B.: J. Amer. Chem. Soc. *98*, 2385 (1976)

[9] Cherry, W. R., Epiotis, N. D.: J. Amer. Chem. Soc. *98*, 1135 (1976)

[10] Epiotis, N. D.: J. Amer. Chem. Soc. *95*, 1191 (1973)

[11] Epiotis, N. D.: J. Amer. Chem. Soc. *95*, 1200 (1973)

[12] Epiotis, N. D.: J. Amer. Chem. Soc. *95*, 1206 (1973)

[13] Epiotis, N. D.: J. Amer. Chem. Soc. *95*, 1214 (1973)

[14] Epiotis, N. D.: Angew. Chem. Int. Ed. *13*, 751 (1974)

[15] Epiotis, N. D., Bernardi, F., Yates, R. L., Cherry, W., Larson, J. R., Shaik, S.: to be published

[16] Dewar, M. J. S.: The molecular orbital theory of organic chemistry. New York: McGraw Hill 1969

[17] Heilbronner, E., Bock, H.: Das HMO-Modell und seine Anwendung. Weinheim/Bergstr.: Verlag Chemie 1968

[18] Salem, L.: J. Amer. Chem. Soc. *90*, 543 (1968)

[19] Imamura, A.: Mol. Phys. *15*, 225 (1968); Fukui, K., Fujimoto, H.: Bull. Chem. Soc. Jap. *41*, 1989 (1968)

[20] Jorgensen, W. L., Salem, L.: The organic chemist's book of orbitals. New York: Academic Press 1973

[21] Baird, N. C., West, R. M.: J. Am. Chem. Soc. *93*, 4427 (1971)

[22] Müller, K.: Helv. Chim. Acta *53*, 1112 (1970)

[23] Longuet-Higgins, H. C., de. V. Roberts, M.: Proc. Roy. Soc. *A 224*, 336 (1954); *A 230*, 110 (1955)

[24] Pople, J. A., Beveridge, D. L.: Approximate molecular orbital theory. New York: McGraw Hill 1970

[25] Murrell, J. N.: The theory of the electronic spectra of organic molecules. New York: John Wiley 1963.

[26] Hehre, W. J., Stewart, R. F., Pople, J. A.: J. Chem. Phys. *51*, 2657 (1969); Hehre, W. J., Lathan, W. A., Ditchfield, R., Newton, M. D., Pople, J. A.: Quantum chemistry program exchange. Bloomington Ind.: Indiana University; Ditchfield, R., Hehre, W. J., Pople, J. A.: J. Chem. Phys. *54*, 724 (1971)

[27] Mulliken, R. S.: Acc. Chem. Res. *9*, 7 (1976)

[28] Koopmans, T.: Physica *1*, 104 (1934)

[29] Streitwieser, Jr., A.: Molecular orbital theory for organic chemists. New York: John Wiley 1961

[30a] Turner, D. W., Baker, C., Baker, A. D., Brundle, C. R.: Molecular photoelectron spectroscopy. New York: John Wiley 1970

[b] Guimon, C., Gombeau, D., Pfister-Guillouzo, G., Asbrink, L., Sandström, J.: J. Electron. Spectr. *4*, 49 (1974)

[31] Ledwith, A., Woods, H. J.: J. Chem. Soc. (B) *1970*, 310

[32] Watanabe, K., Motte, J.: J. Quant. Spect. Radiat. Transf. *2*, 369 (1962)

[33] Estimated from the relation I + A = const.; const. = 7.7. eV. See Ref.[25]

[34] Houk, K. N., Munchausen, L. L.: J. Amer. Chem. Soc. *98*, 937 (1976)

[35] Baker, A. D., Betteridge, D.: Photoelectron spectroscopy. New York: Pergamon Press 1972

36) Pritchard, H. O.: Chem. Revs. *52*, 529 (1953)

37) Eland, J. H. D.: Photoelectron spectroscopy. London: Butterworths 1974

38) Page, F. M.: Advan. Chem. Ser. *36*, 68 (1962)

39) Fukui, K., Morokuma, K., Kato, H., Yonezawa, T.: Bull. Chem. Soc. Japan *36*, 217 (1963)

40) March, J.: Advanced organic chemistry. New York: McGraw Hill 1968

41) Fendler, E. J., Fendler, J. H.: Adv. Phys. Org. Chem. *8*, 271 (1970)

42) Lund, H.: Acta. Chem. Scand. *14*, 1927 (1960)

43 a) Potts, A. W., Lempka, H. J., Streets, D. G., Price, W. C.: Phylos. Trans. R. Soc. London *A 268*, 59 (1970)

b) Katsumata, s., Iwai, T., Kimura, K.: Bull. Chem. Soc. Japan *46*, 3391 (1973)

c) Cradock, S., Whiteford, R. A.: J. Chem. Soc. Faraday II *68*, 281 (1972)

d) Elbel, S.: Ph. D. Dissertation, University of Frankfurt, 1974.

e) Potts, A. W., Price, W. C.: Proc. R. Soc. London *A 326*, 165 (1972)

f) Dewar, M. J. S., Worley, S. D.: J. Chem. Phys. *50*, 654 (1969)

44) Potts, A. W., Price, W. C.: Proc. Roy. Soc. *326A*, 181 (1972)

45) Neurt, H., Clasen, H.: Z. Naturforsch. a, *7*, 410 (1952)

46) Hoffmann, R.: J. Chem. Phys. *39*, 1397 (1963)

47) Hoffmann, R., Lipscomb, W. N.: J. Chem. Phys. *36*, 2179, 3489 (1962)

48) Hoffmann, R., Lipscomb, W. N.: J. Chem. Phys. *37*, 2872 (1962)

49) Bingham, R. C., Dewar, M. J. S., Lo, D. H.: J. Amer. Chem. Soc. *97*, 1285 (1975)

50) Baird, N. C., Dewar, M. J. S.: J. Chem. Phys. *50*, 1262 (1969)

51) Dewar, M. J. S., Haselbach, E.: J. Amer. Chem. Soc. *92*, 590 (1970)

52) Rauk, A., Allen, L. C., Mislow, K.: Angew. Chem. Int. Edit. Engl. *9*, 400 (1970)

53) Hückel, E.: Z. Physik *70*, 204 (1931)

54) Hückel, W.: Theoretical principles of organic chemistry, 7th edit. Houston: Elsevier Pub. Co. 1955

55) Bergman, E. D., Pullman, B., (ed.): Aromaticity, pseudoaromaticity, and antiaromaticity. Jerusalem: The Israel Academy of Sciences and Humanities 1971

56) Lewis, D.: Facts and theories of aromaticity. London: MacMillan Press 1975

57) Garratt, P. J.: Aromaticity. New York: McGraw Hill 1971

58) Clar, E.: The aromatic sextet. New York: Wiley 1972

59) Zimmerman, H. E.: Accounts Chem. Res. *4*, 272 (1971)

60) Hoffmann, R.: Accounts Chem. Res. *4*, 1 (1971)

61) Hoffmann, R., Heilbronner, E., Gleiter, R.: J. Amer. Chem. Soc. *92*, 706 (1970)

62) Hoffmann, R., Imamura, A., Zeiss, G. D.: J. Amer. Chem. Soc. *89*, 5215 (1967)

63) Hoffmann, R., Imamura, A., Hehre, W. J.: J. Amer. Chem. Soc. *90*, 1499 (1968)

64) Hoffmann, R., Swaminathan, S., Odell, B. G., Gleiter, R.: J. Amer. Chem. Soc. *92*, 7091 (1970)

65) Pilar, F. L.: Elementary quantum chemistry. New York: McGraw-Hill 1968

66) Lowe, J. P.: J. Amer. Chem. Soc. *96*, 3759 (1974); Lowe, J. P.: J. Amer Chem. Soc. *92*, 3799 (1970)

67) Fitzgerald, W. E., Jahz, G. J.: J. Mol. Spect. *1*, 49 (1957)

68) Radom, L., Lathan, W. A., Hehre, W. J., Pople, J. A.: J. Amer. Chem. Soc. *95*, 693 (1973)

69) Hirota, E.: J. Chem. Phys. *37*, 283 (1962)

70) Morino, Y., Kuchitsu, K.: J. Chem. Phys. *28*, 175 (1958)

71) Sarachman, T. N.: j. Chem. Phys. *39*, 469 (1963)

72) Komaki, C., Ichishuma, I., Kuratani, K., Miyazawa, T., Shimanouchi, T., Mizushuma, S.: Bull. Chem. Soc. Japan *28*, 330 (1955)

73) Armstrong, S.: Appl. Spect. *23*, 575 (1969)

74) Abdurahmanov, A. A., Rahimova, R. A., Imanov, L. M.: Phys. Lett. *A 32*, 123 (1970)

75) Krueger, P. J., Mettee, H. D.: Can. J. Chem. *42*, 326 (1964)

76) Hagen, K., Hedberg, K.: J. Amer. Chem. Soc. *95*, 8263 (1973)

77) Iagarashi, M., Yamada, M.: Bull. Chem. Soc. Japan *29*, 871 (1956)

78) Almenningen, A., Bastiansen, O., Fernholt, L., Hedberg, K.: Acta. Chem. Scand. *25*, 1946 (1971)

79) Smith, D. W., Hedberg, K.: J. Chem. Phys. *25*, 1282 (1956)
80) Fateley, W. G., Bent, H. A., Crawford, Jr., B.: J. Chem. Phys. *31*, 204 (1959)
81) Groth, P.: Nature *198*, 1081 (1963)
82) Cartwright, B. S., Robertson, J. H.: Chem. Commun. *1966*, 82
83) Howell, J. M., Van Wazer, J. R.: J. Amer. Chem. Soc. *96*, 7902 (1974)
84) Nimon, L. A., Seshadri, K. S., Taylor, R. C., White, D.: J. Chem. Phys. *53*, 2416 (1970)
85) Gayles, J. N., Self, J.: J. Chem. Phys. *40*, 3530 (1964)
86) Finch, A., Hyams, I., Steele, D.: Spectrochim. Acta *21*, 1423 (1965)
87) Armstrong, G. T., Marantz, S.: J. Chem. Phys. *38*, 169 (1963)
88) Binenboym, J., Burcat, A., Lifshitz, A., Shamir, J.: J. Amer. Chem. Soc. *88*, 5039 (1966)
89) Eysel, H. H.: J. Mol. Struct. *5*, 275 (1970)
90) Private communication given in Ref.[89)]
91) Hirota, E.: J. Chem. Phys. *45*, 1984 (1966)
92) Fateley, W. G., Miller, F. A.: Spectrochim. Acta. *19*, 611 (1963)
93) Möller, K. D., DeMeo, A. R., Smith, D. R., London, L. H.: J. Chem. Phys. *47*, 2609 (1967)
94) Kilb, R. W., Lin, C. C., Wilson, E. B.: J. Chem. Phys. *26*, 1695 (1957); Verdier, P. H., Wilson, E. B.: J. Chem. Phys. *29*, 340 (1958)
95) Fateley, W. G., Miller, F. A.: Spectrochim. Acta *17*, 857 (1961)
96) Kilb, R. W., Lin, C. C.: Bull. Am. Phys. Soc. *9*, 198 (1956)
97) See Ref.[66)]
98) Hehre, W. J., Salem, L.:Chem. Commun. *1973*, 754
99) Fieser, L. F., Fieser, M.: Reagents for organic synthesis. New York: John Wiley 1967
100) Whipple, E. B., Stewart, W. E., Reddy, G. S., Goldstein, J. H.: J. Chem. Phys. *34*, 2136 (1961)
101) Wood, R. E., Stevenson, D. P.: J. Amer. Chem. Soc. *63*, 1650 (1941)
102) Gardner, D. V., McGreer, D. E.: Can. J. Chem. *48*, 2104 (1970)
103) Noyes, R. M., Dickinson, R. G.: J. Amer. Chem. Soc. *65*, 1427 (1943)
104) Furuyama, S., Golden, D. M., Benson, S. W.: J. Phys. Chem. *72*, 3204 (1968)
105) Olson, A. R., Maroney, W.: J. Amer. Chem. Soc. *56*, 1320 (1934)
106) Noyes, R. M., Dickinson, R. G., Schomaker, V.: J. Amer. Chem. Soc. *67*, 1319 (1945)
107) Miller, S. I., Weber, A., Cleveland, F. F.: J. Chem. Phys. *23*, 44 (1955)
108) Craig, N. C., Lo, Y. S., Piper, L. G., Wheeler, J. C.: J. Phys. Chem. *74*, 1712 (1970)
109) Waldron, J. T., Snyder, W. H.: J. Amer. Chem. Soc. *95*, 5491 (1973)
110) Okuyama, T., Fueno, T., Furukawa, J.: Tet. *25*, 5409 (1969
111) Crump, J. W.: J. Org. Chem. *28*, 953 (1963)
112) Viehe, H. G.: Angew. Chem. *75*, 793 (1963)
113) Sarachman, T. N.: J. Chem. Phys. *49*, 3146 (1968)
114) Kilpatrick, J. E., Pitzer, K. S.: J. Res. Natl. Bur. Stand. *37*, 163 (1946)
115) Harwell, K. E., Hatch, L. F.: J. Amer. Chem. Soc. *77*, 1682 (1955)
116) Salomaa, P., Nissi, P.: Acta. Chem. Scand. *21*, 1386 (1967)
117) Rhoads, S. J., Chattopadhyay, J. K., Waali, E. E.: J. Org. Chem. *35*, 3352 (1970)
118) Price, C. C., Snyder, W. H.: J. Amer. Chem. Soc. *83*, 1773 (1961)
119) Beaudet, R. A., Wilson, E. B.: J. Chem. Phys. *37*, 1133 (1962)
120) Siegel, S.: J. Chem. Phys. *27*, 989 (1957)
121) Beaudet, R. A.: J. Chem. Phys. *40*, 2705 (1964)
122) Beaudet, R. A.: J. Chem. Phys. *37*, 2398 (1962)
123) Beaudet, R. A.: J. Chem. Phys. *38*, 2548 (1963)
124) Laurie, V. W.: J. Chem. Phys. *32*, 1588 (1960)
125) Scarzafava, E., Allen, L. C.: J. Amer. Chem. Soc. *93*, 311 (1971)
126) English, A. D., Palke, W. E.: J. Amer. Chem. Soc. *95*, 8536 (1973)
127) Simmons, H. E., Blomstrom, D. C., Vest, R. D.: J. Amer. Chem. Soc. *84*, 4756 (1962)
128) Orgel, L., Lohrmann, R.: Accts. Chem. Res. 7, 368 (1974)
129) Bredereck, H., Schmötzer, G., Oehler, E.: Ann. Chem. *600*, 81 (1956)
130) Penfold, B. R., Lipscomb, W. N.: Acta Cryst. *14*, 589 (1961)
131) Pierce, L., Krisher, L. C.: J. Chem. Phys. *31*, 875 (1959)

References

132) Pierce, L.: Bull. Amer. Phys. Soc. *1*, 198 (1956)
133) Sinnott, K. M.: J. Chem. Phys. *34*, 851 (1961)
134) Sinnott, K. M.: Bull. Amer. Phys. Soc. *1*, 198 (1956)
135) Krisher, L. C.: J. Chem. Phys. *33*, 1237 (1960)
136) Moloney, M. J., Krisher, L. C.: J. Chem. Phys. *45*, 3277 (1966)
137) Tabor, W. J.: J. Chem. Phys. *27*, 974 (1957)
138) Dellepiane, G., Overand, J.: Spectrochim. Acta. *22*, 593 (1966)
139) Buric, Z., Krueger, P. J.: Spectrochim. Acta. *30 A*, 2069 (1974)
140) Krisher, L. C., Wilson, E. B.: J. Chem. Phys. *31*, 882 (1959)
141) Krisher, L. C.: J. Chem. Phys. *33*, 304 (1960)
142) Steifvater, O. L., Sheridan, J.: Proc. Chem. Soc. *1963*, 368
143) Nakagawa, I., Ichishima, I., Kuratani, K., Miyazawa, T., Shimanouchi, T., Mizushima, S. I.: J. Chem. Phys. *20*, 1720 (1952)
144) Karabatsos, G. J., Fenoglio, D. J., Lande, S. S.: J. Amer. Chem. Soc. *91*, 3572 (1969); Karabatsos, G. J., Fenoglio, D. J.: Top. Stereochem. *5*, 167 (1970)
145) Crowder, G. A., Northam, F.: J. Mol. Spect. *26*, 98 (1968)
146) Wieser, H., Laidlaw, W. G., Krueger, P. J., Fuhrer, H.: Spectrochim. Acta. *24A*, 1055 (1968)
147) Hayashi, M., Kuwada, K.: J. Mol. Struct. *28*, 147 (1975)
148) Steinmetz, W. E.: J. Amer. Chem. Soc. *95*, 2777 (1973)
149) Cremer, D., Binkley, J. S., Pople, J. A., Hehre, W. J.: J. Amer. Chem. Soc. *96*, 6900 (1974)
150) Laurie, V. W., Wollrab, J.: Bull. Amer. Phys. Soc. 327 (1963)
151) Itoh, T.: J. Phys. Soc. Japan *11*, 264 (1956)
152) Kasai, P. H., Myers, R. J.: J. Chem. Phys. *30*, 1096 (1959)
153) Ivash, E. V., Dennison, D. M.: J. Chem. Phys. *21*, 1804 (1953)
154) Hirota, E., Matsumura, C., Morino, Y.: Bull. Chem. Soc. Japan *40*, 1124 (1967)
155) Weiss, S., Leroi, G. E.: J. Chem. Phys. *48*, 962 (1968)
156) Laurie, V. W.: J. Chem. Phys. *34*, 1516 (1961)
157) Hirota, E.: J. Chem. Phys. *45*, 1984 (1966)
158) Nelson, R., Pierce, L.: J. Mol. Struct. *18*, 344 (1965)
159) Herschbach, D. R.: J. Chem. Phys. *31*, 91 (1959)
160) Jean, Y., Salem, L.: Chem. Comm. 1971, 382
161) Owen, N. L., Sheppard, N.: Proc. Chem. Soc. (London) *1963*, 264
162) Owen, N. L., Sheppard, N.: Trans. Faraday Soc. *60*, 634 (1964)
163) Samdel, S., Seip, H. M.: J. Mol. Struct. *28*, 193 (1975)
164) Riveros, J. M., Wilson, E. B.: J. Chem. Phys. *46*, 4605 (1967)
165) Miyazawa, T.: Bull. Chem. Soc. Japan *34*, 691 (1961)
166) Quade, C. R., Lin, C. C.: J. Chem. Phys. *38*, 540 (1963)
167) Curl, Jr., R. F.: J. Chem. Phys. *30*, 1529 (1959)
168) Piercy, I. E., Subrahmanyam, S. V.: J. Chem. Phys. *42*, 1475 (1965)
169) Subrahmanyam, S. V., Piercy, J. E.: J. Acoust. Soc. Amer. *37*, 340 (1965)
170) Bailey, J., North, A. M. Trans. Faraday Soc. *64*, 1499 (1968)
171) Slie, W. M., Litovitz, R. A.: J. Chem. Phys. *39*, 1538 (1963)
172) Wyn-Jones, E., Pehtrick, R. A.: Topics Stereochem., *5*, 205 (1970)
173) Suzuki, J., Tsuboi, M., Shimanouchi, T., Mizushima, S.: Spectrochim. Acta *16*, 471 (1960)
174) Gutowsky, H. S., Holm, C. M.: J. Chem. Phys. *25*, 1228 (1956)
175) Mizushima, S. I., Shimanouchi, T., Nagakura, S., Kouratani, K., Tsuboi, T., Babs, H., Fujioka, O.: J. Amer. Chem. Soc. *72*, 3490 (1950)
176) See Ref. [8]
177) Cahill, P., Gold, L. P., Owen, N. L.: J. Chem. Phys. *48*, 1620 (1968)
178) Radom, L., Pople, J. A., Schleyer, P. v. R.: J. Amer. Chem. Soc. *94*, 5935 (1972)
179) Young, W. G., Sharman, S. H., Winstein, S.: J. Amer. Chem. Soc. *82*, 1376 (1960)
180) Bank, S.: J. Amer. Chem. Soc. *87*, 3245 (1965)
181) Bank, S., Schriesheim, A., Rowe, C. A.: J. Amer. Chem. Soc. *87*, 3244 (1965)

182) Schriesheim, A., Rowe, C. A.: Tet. Lett. 405 (1962)
183) Haag, W. O., Pines, H.: J. Amer. Chem. Soc. *82*, 387 (1960)
184) Price, C. C., Snyder, W. H.: Tet. Lett. 69 (1962)
185) Price, C. C.; Snyder, W. H., J. Amer. Chem. Soc. *83*, 1773 (1961)
186) Prosser, T. J.: J. Amer. Chem. Soc. *83*, 1701 (1961)
187) Strand, T. G., Cox, Jr., H. L.: J. Chem. Phys. *44*, 2426 (1966)
188) Strand, T. G.: J. Chem. Phys. *44*, 1611 (1966)
189) Streltsova, I. N., Struchkov, Y. T.: Chem. Abstr. *56*, 8112c (1962)
190) Ballester, M., Olivella, S.: Polychloroaromatic Compounds. Suschitzky, H. ed. New York: Plenum Press 1974, p. 20
191) Hush, N. S., Pople, J. A.: Trans. Faraday Soc. *51*, 600 (1955)
192) Woolfenden, W. R., Grant, D. M.: J. Amer. Chem. Soc. *88*, 1496 (1966)
193) Almenningen, A., Bastianen, O., Tratteberg, M.: Acta. Chem. Scand. *12*, 1221 (1958)
194) Marais, D. J., Sheppard, N., Stoicheff, P.: Tet. *17*, 163 (1962)
195) Radom, L., Pople, J. A.: J. Amer. Chem. Soc. *92*, 4786 (1970)
196) Dumbacher, B.: Theor. Chim. Acta. *23*, 346 (1972)
197) Skancke, P. N., Boggs, J. E.: J. Mol. Struct. *16*, 179 (1973)
198) Pincelli, U., Cadioli, B., Levy, B.: Chem. Phys. Lett. *13*, 249 (1972)
199) Devaquet, A. J. P., Townshend, R. E., Hehre, W. J.: J. Amer. Chem. Soc. *98*, 4068 (1976)
200) Kuchitsu, K., Fukuyama, J., Morino, Y.: J. Mol. Struct. *1*, 463 (1968); Currie, G. N., Ramsay, D. A.: Can. J. Phys. *49*, 317 (1971)
201) Cherniak, E. A., Costain, C. C.: J. Chem. Phys. *45*, 104 (1966); Durig, J. R., Tong, C. C., Li, Y. S.: J. Chem. Phys. *57*, 4425 (1972)
202) Ha, T. K.: J. Mol. Struct. *12*, 171 (1972)
203) Skancke, P. N., Saebo, S.: J. Mol. Struct. *28*, 279 (1975)
204) Dykstra, C. E., Schaefer III, H. F.: J. Amer. Chem. Soc. *97*, 7210 (1975)
205) Szasz, G. J., Sheppard, N.: Trans. Far. Soc. *49*, 358 (1953)
206) Hagen, K., Hedberg, K.: J. Amer. Chem. Soc. *95*, 1003 (1973)
207) Hagen, K., Hedberg, K.: J. Amer. Chem. Soc. *95*, 4796 (1973)
208) Marais, O. J., Sheppard, N., Stoicheff, B. P.: Tet. *17*, 163 (1962)
209) Beaudet, R. A.: J. Chem. Phys. *42*, 3758 (1965)
210) Beaudet, R. A.: J. Amer. Chem. Soc. *87*, 1390 (1965)
211) Bothner-By, A. A., Jung, D.: J. Amer. Chem. Soc. *90*, 2342 (1968)
212) See Ref.[205)
213) Chang, C. H., Andreassen, A. L., Bauer, S. H.: J. Org. Chem. *36*, 920 (1971)
214) Hehre, W. J., Pople, J. A.: J. Amer. Chem. Soc. *97*, 6941 (1975)
215) Watson, H. C.: Prog. in Stereochem. *4*, 299 (1969)
216) Phillips, D. C.: Nat. Acad. Sci. USA *57*, 484 (1967)
217) Blake, C. C. F., Mair, G. A., North, A. C. T., Phillips, D. C., Sharma, V. R.: Proc. Royal Soc. (London) *B167*, 365 (1967)
218) Principles and techniques of protein chemistry. Leach, S. J. (ed.) New York: Academic Press 1969
219) Birktoft, J. J., Matthews, B. M., Blow, D. M.: Biochem. Biophys. Res. Commun. *36*, 131 (1969)
220) Lipscomb, W. N., Hartsuck, J. A., Reeke, G. N, Quiocho, F. A., Bethge, P. H., Ludwig, M. L., Steitz, T. A., Muirhead, H., Coppola, J. C.: Brookhaven Symposia in Biology *21*, 24 (1968)
221) Goodman, M., Verdini, A. S., Choi, N. S., Masuda, Y.: Top. Stereochem. *5*, 69 (1970)
222) McDiarmid, R. S.: Ph. D. Dissertation, Harvard University 1965
223) Fraser, R. D. B., Harrap, B. S., MacRae, T. P., Stewart, F. H. C., Suzuki, E.: Biopolymers *5*, 251 (1967)
224) Bamford, C. H., Brown, L., Cant, E. M., Elliot, A., Hanby, W. E., Malcolm, B. R.: Nature *176*, 396 (1955)
225) Rich, A., Crick, F. H. C.: Nature *176*, 915 (1955)
226) Shipman, L. L., Christoffersen, R. E.: J. Amer. Chem. Soc. *95*, 4733 (1973)
227) Shipman, L. L., Christoffersen, R. E.: J. Amer. Chem. Soc. *95*, 1408 (1973)

228) Allinger, N. L., Freiberg, L. A.: J. Amer. Chem. Soc. *83*, 5028 (1961)

229) Groth, P., Hassel, O.: Proc. Chem. Soc. 1963, 218

230) Frey, H., Martin, H., Hekman, M.: Chem. Commun. 1975, 204

231) Norman, Jr., J. G., Kolari, H. J.: J. Amer. Chem. Soc. *97*, 33 (1975)

232) Norman, Jr., J. G., Kolari, H. J.: Chem. Commun. 1975, 649

233) Jonathan, N., Ross, K., Tomlinson, V.: Int. J. Mass Spect. and Ion Phys. *4*, 51 (1970)

234) Lake, R. F., Thompson, H.: Proc. Roy. Soc. *A 315*, 323 (1970)

235) Momigny, J.: Nature *199*, 1179 (1963)

236) Momigny, J.: Nature *191*, 1089 (1961)

237) Smith, D. B., Gilbert, A., Orger, B., Tyrrel, H.: Chem. Commun. 1974, 334

238) Bralsford, R., Harris, P. V., Price, W. C.,: Proc. Roy. Soc. *A258*, 459 (1960)

239) Worley, S. D.: Chem. Rev. *71*, 295 (1971)

240) Bock, H., Stafast, H.: Chem. Ber. *105*, 1158 (1972)

241) Farragher, A. L., Page, F. M.: Trans. Faraday Soc. *63*, 2369 (1967)

242) Briegleb, G.: Angew. Chem. Internat. Ed. *3*, 617 (1964)

243) Matsen, F. A.: J. Chem. Phys. *24*, 602 (1956)

244) Fry, A. J.: Fort. Chem. Forsch. *34*, 1 (1972)

245) Rosenthal, I., Hayes, J. R., Martin, A. J., Elving, P. J.: J. Amer. Chem. Soc. *80*, 3050 (1958)

246) Elving, P. J., Teitelbaum, C.: J. Amer. Chem. Soc. *71*, 3916 (1949)

247) Pasternak, R.: Helv. Chim. Acta. *31*, 753 (1948)

248) Mairanovskii, S. G., Bergel, L. D.: Russ. J. Phys. Chem. *34*, 112 (1960)

249) Jones, L. C., Taylor, L. W.: Anal. Chem. *27*, 228 (1955)

250) Salahub, D. R.: Theort. Chim. Acta. *22*, 330 (1971)

251) Lacher, J. R., Hummel, L. E., Bohmfalk, E. F., Park, J. D.: J. Amer. Chem. Soc. *72*, 5486 (1950)

252) Colthrup, N. B., Daly, L. H., Wiberly, S. E.: Introduction to infrared and raman spectroscopy. New York: Academic Press 1964

253) Levin, I. W., Pearce, R. A. R.: J. Mol. Spect. *49*, 91 (1974)

254) Levin, I. W. Pearce, R. A. R., Harris, W. C.: J. Chem. Phys. *59*, 3048 (1973)

255) Craig, N. C., Overend, J.: J. Chem. Phys. *51*, 1127 (1969)

256) Wasserman, A.: Diels-alder reactions. New York: Elsevier 1965

257) Onishchenko, A. S.: Diene synthesis. Israel, Jerusalem: S. Monson Binding 1964

258) Sauer, J., Wiest, H., Mielert, A.: Z. Naturforsch. *17B*, 203 (1962)

259) Stork, G., White, W. N.: J. Amer. Chem. Soc. *78*, 4609 (1956)

260) Stork, G., Clarke, F. H.: J. Amer. Chem. Soc. *78*, 4619 (1956)

261) Bordwell, F. G., Pagani, G.: J. Amer. Chem. Soc. *97*, 118 (1975)

262) Bordwell, F. G., Mecca, T. G.: J. Amer. Chem. Soc. *97*, 123, 127 (1975)

263) Bordwell, F. G., Wiley, P. F., Mecca, T. G.: J. Amer. Chem. Soc. *97*, 132 (1975)

264) Levin, C. C.: J. Amer. Chem. Soc. *97*, 5649 (1975)

265) Buenker, R. J., Peyerimhoff, S. D.: Chem. Revs. *74*, 127 (1974)

266) Kim, H., Pearson, E. F., Appelman, E. H. J.: J. Chem. Phys. *56*, 1 (1972)

267) Tables of Interatomic Distances and Configurations in Molecules and Ions. Special Pub. 18, Chem. Soc. (London), Oxford Press 1965

268) Pierce, L., Jackson, R. H., DiCianni, N.: J. Chem. Phys. *35*, 2240 (1961)

269) The values were obtained from Extended Hückel calculations

270) Herzberg, G.: Science *177*, 123 (1972)

271) Mirri, A. M., Scappini, F., Cazzoli, G.: J. Mol. Spect. *38*, 218 (1971)

272) Beagley, B., Clark, A. H., Hewitt, T. G.: J. Chem. Soc. A. 1968, 658

273) Johnson, D. R., Powell, F. X.: Science *164*, 950 (1969)

274) Palmer, K. J.: J. Amer. Chem. Soc. *60*, 2360 (1938)

275) Burrus, C. A., Gordy, W.: Phys. Revs. *92*, 274 (1953)

276) Stenkamp, L. Z., Davidson, E. R.: Theoret. Chim. Acta *30*, 283 (1973)

277) Frost, A. A.: J. Phys. Chem. *72*, 1289 (1968)

278) Almenningen, A., Bastiansen, O.: Acta Chem. Scand. *9*, 815 (1955)

279) Lide, D. R.: J. Chem. Phys. *38*, 456 (1963)

280) Hirota, E., Morino, Y.: J. Mol. Spect. *33*, 460 (1970)

281) Lambert, J.: Topics Stereochem. *6*, 19 (1971)

282) Lambert, J., Oliver, W. L., Packard, B. S.: J. Amer. Chem. Soc. *93*, 933 (1971)

283) Lehn, J.-M., Wagner, J.: Chem. Comm. *1968*, 148

284) Lehn, J. M.: Fort. Chem. Forsch. *15*, 311 (1970)

285) Riddell, F. G., Lehn, J. M., Wagner, J.: Chem. Comm. *1968*, 1403

286) Lehn, J. M., Wagner, J.: Tet. *26*, 4227 (1970)

287) Harrison, J. F.: Accts. Chem. Res. *7*, 378 (1974)

288) Laurie, V. W., Pence, D. T.: J. Chem. Phys. *38*, 2693 (1963)

289) Powell, F. X., Lide, R. D.: J. Chem. Phys. *45*, 1067 (1966)

290) Dill, J. D., v. R. Schleyer, P., Pople, J. A.: J. Amer. Chem. Soc. *97*, 3402 (1975)

291) Allen, H. C., Plyler, E. K.: J. Chem. Phys. *31*, 1062 (1959)

292) Kauzmann, W.: Quantum chemistry. New York: Academic Press 1957

293) Hirota, E.: J. Chem. Phys. *42*, 2071 (1965)

294) Bothner-By, A. A., Günther, H.: Discus. Faraday Soc. *34*, 127 (1962)

295) Radom, L., Pople, J. A., Schleyer, P. v. R.: J. Amer. Chem. Soc. *94*, 5935 (1972)

296) For a discussion of group electronegativities see Wells, P. R.: Prog. Phys. Org. Chem. *6*, 111 (1968)

297) Radom, L., Pople, J. A., Buss, V., Schleyer, P. v. R.: J. Amer. Chem. Soc. *92*, 6380, 6987 (1970)

298) Baker, J. W., Nathan, W. S.: J. Chem. Soc. 1935, 1844

299) Gould, E. S.: Mechanism and structure in organic chemistry. New York: Holt, Rinehart and Winston 1959

300) Schubert, W. M., Gurka, D. F.: J. Amer. Chem. Soc. *91*, 1443 (1969)

301) Schubert, W. M., Sweeney, W. A.: J. Org. Chem. *21*, 119 (1956)

302) Schubert, W. M., Robins, J.: J. Amer. Chem. Soc. *80*, 559 (1958)

303) Schubert, W. M., Murphy, R. B., Robins, J.: Tet. *17*, 199 (1962)

304) Schubert, W. M., Minton, R. G.: J. Amer. Chem. Soc. *82*, 6188 (1960)

305) Hehre, W. J., McIver, R. T., Pople, I. A., Schleyer, P. v. R.: J. Amer. Chem. Soc. *96*, 7162 (1974)

306) Baker, A. D., May, D. P., Turner, D. W.: J. Chem. Soc. B *1968*, 22

307) Jaffe, H. H., Orchin, M.: Theory and applications of ultraviolet spectroscopy. New York: Wiley-Interscience 1962

308) Freeman, F.: Chem. Revs. *75*, 439 (1975)

309) Trombetti, A., John, J. W. C., cited in: J. W. Nibler and Bondybey, V. E.: J. Chem. Phys. *60*, 1307 (1974)

310) Pople, J. A.: Tet. *30*, 1605 (1974)

311) Chang, C. H., Porter, R. F., Bauer, S. H.: J. Amer. Chem. Soc. *92*, 5313 (1970)

312) Almenningen, A., Anfinsen, I. M., Haaland, A.: Acta. Chem. Scand. *24*, 1230 (1970)

313) Veillard, A.: Chem. Phys. Lett. *4*, 51 (1969)

314) Radom, L., Hehre, W. J., Pople, J. A.: J. Amer. Chem. Soc. *94*, 2371 (1972)

315) Pedersen, L.: J. Mol. Struct. *3*, 510 (1969)

316) Yokozeki, A., Bauer, S. H.: Fortschr. Chem. Forsch. *53*, 71 (1975)

317) Hirota, E., Sugisaki, R., Nielsen, C. J., Sorensen, G. O.: J. Mol. Spect. *49*, 251 (1974)

318) Almenningen, A., Bastiansen, O., Motzfeldt, T.: Acta. Chem. Scand. *23*, 2848 (1969)

319) Miller, R. F., Curl, Jr., R. F.: J. Chem. Phys. *34*, 1847 (1961)

320) Kitano, N., Kuchitsu, K.: Bull. Chem. Soc. Japan *46*, 3048 (1973)

321) Derissen, J. L.: J. Mol. Struct. *7*, 67 (1971)

322) Pierce, L., Krisher, L. C.: J. Chem. Phys. *31*, 875 (1959)

323) Tsuchiya, S., Kimura, M.: Bull. Chem. Soc. Japan *45*, 736 (1972)

324) Foote, C. S.: J. Amer. Chem. Soc. *86*, 1853 (1964)

325) Schleyer, P. v. R.: J. Amer. Chem. Soc. *86*, 1854, 1856 (1964)

326) Kuchitsu, K.: J. Chem. Phys. *44*, 906 (1966)

327) Takagi, K., Oka, T.: J. Phys. Soc. Japan *18*, 1174 (1963)

328) Carlos, Jr., J. L., Karl, Jr., R. R., Bauer, S. H.: J. Chem. Soc. Faraday Trans. II *70*, 177 (1974)

References

329) Laurie, V. W., Pence, D. T., Jackson, R. H.: J. Chem. Phys. *37*, 2959 (1962)
330) Livingston, R. L., Rao, C. N. R., Kaplan, L. H., Rocks, L.: J. Amer. Chem. Soc. *80*, 5368 (1958)
331) Robinson, G. W.: J. Chem. Phys. *21*, 1741 (1953)
332) Wolfe, S., Rauk, A., Tel, L. M., Csizmadia, I. G.: J. Chem. Soc. *B*, 136 (1971)
333) Romers, C., Altona, C., Buys, H. R., Havinga, E.: Topics Stereochem. *4*, 39 (1969)
334) Altona, C., Romers, C., Havinga, E.: Tetrahedron Lett. *1959*, 16
335) Altona, C., Knobler, C., Romers, C.: Rec. Trav. Chim. *82*, 1089 (1963)
336) Altona, C., Romers, C.: Rec. Trav. Chim. *82*, 1080 (1963)
337) Altona, C., Knobler, C., Romers, C.: Acta Cryst. *16*, 1217 (1963)
338) Rutten, E. W. M., Nibbering, N., MacGillavry, C. H., Romers, C.: Rec. Trav. Chim. *87*, 888 (1968)
339) De Wolf, N., Romers, C., Altona, C.: Acta Cryst. *22*, 715 (1967)
340) Kalff, H. T., Romers, C.: Rec. Trav. Chim. *85*, 198 (1966)
341) Kalff, H., Romers, C.: Acta Cryst. *18*, 164 (1965)
342) Altona, C., Romers, C.: Acta Cryst. *16*, 1225 (1963)
343) Jeffrey, G. A., Pople, J. A., Radom, L.: Carbohydrate Res. *25*, 117 (1972)
344) David, S., Eisenstein,.O., Hehre, W. J., Salem, L., Hoffmann, R.: J. Amer. Chem. Soc. *95*, 3806 (1973)
345) Jones, R. A. Y., Katrizky, A. R., Richards, A. C., Wyatt, R. J., Bishop, R. J., Sutton, L. E.: J. Chem. Soc. *B*, 127 (1970)
346) Lambert, J. B., Oliver Jr., W. L., Jackson III, G. F.: Tetrahedron Lett. *1969*, 2027
347) Lambert, J. B., Oliver, Jr., W. L.: Tet. *27*, 4245 (1971)
348) Blackburne, I. D., Katritzky, A. R., Takeuchi, T.: Accounts Chem. Res. *8*, 300 (1975)
349) Cook, M. J., Jones, R. A. Y., Katritzky, A. R., Mañas, M. M., Richards, A. C., Sparrow, A. J., Trepanier, D. L.: J. Chem. Soc. Perkin Trans. II, *4*, 325 (1973)
350) Booth, H., Lemieux, R. U.: Can. J. Chem. *49*, 777 (1971)
351) Angiolini, L., Duke, R. P., Jones, R. A. Y., Katritzky, A. R.: J. Chem. Soc. Perkin Trans. *2*, 674 (1972)
352) Strietwieser, Jr., A., Holtz, D.: J. Amer. Chem. Soc. *89*, 692 (1967)
353) Strietwieser, Jr., A., Marchand, A. P., Pudjaatmaka, A. H.: J. Amer. Chem. Soc. *89*, 693 (1967)
354) Hollyhead, W. B., Stephens, R., Tatlow, J. C., Westwood, W. T.: Tet. *25*, 1777 (1969); Campbell, S. F., Stephens, R. S., Tatlow, J. C.: ibid. *21*, 2997 (1965)
355) Heilbronner, E., Hornung, V., Bock, H., Alt, H.: Angew. Chem. Internat. Edit. *8*, 524 (1969)
356) Brown, R. S.: Can. J. Chem. *53*, 2446 (1976)
357) Krueger, P. J., Jan, J.: Can. J. Chem. *48*, 3229 (1970)
358) Krueger, P. J., Jan, J.: Can. J. Chem. *48*, 3236 (1970)
359) Cilento, G.: Chem. Revs. *60*, 146 (1960)
360) Seebach, D.: Synthesis *1*, 17 (1969)
361) Eliel, E. L., Hartman, A. A., Abatjoglou, A. G.: J. Amer. Chem. Soc. *96*, 1807 (1974)
362) Oae, S., Tajaki, W., Ohno, A.: J. Amer. Chem. Soc. *83*, 5036 (1961)
363) Epiotis, N. D., Yates, R. L., Bernardi, F., Wolfe, S.: J. Amer. Chem. Soc., *98*, 5435 (1976)
364) Bordwell, F. G., Vestling, M. M.: J. Amer. Chem. Soc. *89*, 3906 (1967)
365) Bordwell, F. G., Yee, K. C.: J. Amer. Chem. Soc. *92*, 5933 (1970)
366) Lowe, J. P.: Prog. Phys. Org. Chem. *6*, 1 (1968)
367) Tokuhiro, T.: J. Chem. Phys. *41*, 438 (1964); Mizushima, S.: Structure of molecules and internal rotation. New York: Academic Press 1954
368) Klaboe, P., Nielsen, J. R.: J. Chem. Phys. *33*, 1764 (1960)
369) Klaboe, P., Nielsen, J. R.: J. Chem. Phys. *32*, 899 (1960)
370) Thomas, W. A.: NMR spectroscopy as an aide in conformational analysis. In: Annual repors in NMR spectroscopy *3*, 91 (1971)
371) Newmak, R. A., Sederholm, C. H.: J. Chem. Phys. *43*, 602 (1965)
372) Craig, N. C., Piper, L. G., Wheeler, V. L.: J. Phys. Chem. *75*, 1453 (1971)
373) Prosen, E. J., Johnson, W. H., Rossini, F. D.: J. Res. Natl. Bur. Std. *39*, 173 (1947)

374) Reeves, L. W., Stromme, K. D.: Trans. Far. Soc. *57*, 390 (1961)

375) Premuzic, E., Reeves, L. W., Can. J. Chem. *40*, 1870 (1962)

376) Lemieux, R. U., Farser, B.: Can. J. Chem. *43*, 1460 (1965)

377) Planje, M. C.: Ph. D. Thesis, University of Leiden, 1966.

378) McGlynn, S. P., Vanquickenborne, L. G., Kinoshita, M., Carroll, D. G.: Introduction to applied quantum chemistry. New York: Holt, Rinehart and Winston 1972

379) Basch, H., Viste, A., Gray, H. B.: J. Chem. Phys. *44*, 10 (1966)

380) Orville-Thomas, W. J.: Internal rotation in molecules. New York: John Wiley 1974

381) Radom, L., Pople, J. A.: J. Amer. Chem. Soc. *92*, 4786 (1970)

382) Bernardi, F.: unpublished results

383) Benson, S. W., Cruickshank, F. R., Golden, D. M., Haugen, G. R., O'Neal, H. E., Rodgers, A. S., Shaw, R., Walsh, R.: Chem. Revs. *69*, 279 (1969)

384) Cox, J. P., Pilcher, G.: Thermochemistry of organic and organometallic compounds. New York: Academic Press 1970

385) Bartell, L. S., Roth, E. A., Hollowell, C. P., Kuchitsu, K., Young, J. E., Jr.: J. Chem. Phys. *42*, 2683 (1965)

386) Laurie, V. W., Pence, P. T.: J. Chem. Phys. *38*, 2693 (1963)

387) Davis, M. I., Hanson, H. P.: J. Phys. Chem. *69*, 4091 (1965)

388) Table of Interatomic Distance and Configuration in Molecules and Ions. Special Pub. 18, Supplement 1956–1959. Chem. Soc. (London): Oxford Press 1965

389) Livingston, R. L., Rao, C. N. R., Kaplan, L. M., Rocks, L.: J. Amer. Chem. Soc. *80*, 5368 (1958)

390) Almenningen, A., Anfinsen, I. M., Haaland, A.: Acta Chem. Scand. *24*, 43 (1970)

391) Bartell, L. S., Bonham, R. A.: J. Chem. Phys. *32*, 824 (1960)

392) Epiotis, N. D.: in preparation

393) The relevant bond dissociation energies are as follows: H-F: ~135 kcal/mol; H-I: ~71 kcal/mol; $CH_2 = CHCH_2 - H$: ~77 kcal/mol; $CH_3CH_2CH_2 - H$: ~ 71 kcal/mol;

394) Hauser, W. P., Walters, W. P.: J. Phys. Chem. *67*, 1328 (1963)

395) Schlag, E. W., Peatman, W. B.: J. Amer. Chem. Soc. *86*, 1676 (1964)

396) Walsh, A. D.: Discussions Faraday Soc. *2*, 18 (1947); Walsh stated the following rule: if a group X attached to carbon is replaced by a more electronegative group Y, the carbon valency toward Y has more p character than it had toward X. For a review, see: Bent, H. A.: Chem. Revs. *61*, 275 (1961)

397) Gillespie, R. J.: J. Chem. Ed. *51*, 367 (1974); Gillespie, R. J.: Molecular geometry. London: Van Nostrand Reinhold Co. 1972; Bent, H. A.: J. Chem. Ed. *40*, 446 (1963); Linnet, J. W.: The electronic structure of molecules. London: Methuen 1964

398) Wolfe, S., Tel, L. M., Liang, J. H., Csizmadia, I. G.: J. Amer. Chem. Soc. *94*, 1361 (1972)

399) Wolfe, S.: Accounts Chem. Res. *5*, 102 (1972)

400) Wolfe, S., Tel, L. M., Haines, W. J., Robb, M. A., Csizmadia, I. G.: J. Amer. Chem. Soc. *95*, 4863 (1973)

401) Eisenstein, O., Anh, N. T., Jean, Y., Devaquet, A., Cantacuzene, J., Salem, L.: Tetrahedron *30*, 1717 (1974)

402) Levin, I. N.: Quantum chemistry. Boston: Allyn Bacon Inc. 1968

403) Eilers, J. E., Liberles, A.: J. Amer. Chem. Soc. *97*, 4183 (1975)

404) Davidson, R. B., Allen, L. C.: J. Chem. Phys. *54*, 2828 (1971)

405) Kollman, P.: J. Amer. Chem. Soc. *96*, 4362 (1974); See, also: Kollman, P., McKelvey, J., Gund, P.: J. Amer. Chem. Soc. *97*, 1640 (1975)

406) Bingham, R. C.: J. Amer. Chem. Soc. *98*, 535 (1976)

407) Eyring, H., Stewart, G. H., Smith, R. F.: Proc. Nat. Acad. Sci. USA *44*, 259 (1958); Stewart, G. H., Eyring, H.: J. Chem. Ed. *35*, 550 (1958)

408) Bingham, R. C., Dewar, M. J. S.: J. Amer. Chem. Soc. *95*, 7182 (1973); Krusic, P. J., Bingham, R. C.: J. Amer. Chem. Soc. *98*, 230 (1976)

409) Clementi, E., Raimondi, D. L.: J. Chem. Phys. *38*, 2686 (1963)

410) Van Catledge, F. A.: J. Amer. Chem. Soc. *96*, 5693 (1974)

411) Howell, J. M.: J. Amer. Chem. Soc. *98*, 886 (1976)

References

412) Hoffmann, R., Olofson, R. A.: J. Amer. Chem. Soc. *88*, 943 (1966) ; Hoffmann, R.,
Levin, C. C., Moss, R. A.: J. Amer. Chem. Soc. *95*, 629 (1973)

413) Hoffmann, R., Radom, L., Pople, J. A., v. R. Schleyer, P., Hehre, W. J., Salem, L.:
J. Amer. Chem. Soc. *94*, 6221 (1972)

414) Gimarc, B. M.: Accounts Chem. Res. *7*, 384 (1974)

415) Deslongchamps, P.: Tetrahedron. *31*, 2463 (1975)

416) Eliel, E. L., Nader, F. W.: J. Amer. Chem. Soc. *92*, 584 (1970)

417) Lehn, J. M., Wipff, G., Bürgi, H. B.: Helv. Chim. Acta. *57*, 493 (1974)

418) Bürgi, H. B., Lehn, J. M., Wipff, G.: J. Amer. Chem. Soc. *96*, 1956 (1974)

419) Lehn, J. M., Wipff, G.: J. Amer. Chem. Soc. *96*, 4048 (1974)

420) Bürgi, H. B., Dunitz, J. D., Lehn, J. M., Wipff, G.: Tetrahedron *30*, 1563 (1974)

Author Index Volumes 26 - 70

The volume numbers are printed in italics

Albini, A., and Kisch, H.: Complexation and Activation of Diazenes and Diazo Compounds by Transition Metals. *65*, 105–145 (1976).

Altona, C., and Faber, D. H.: Empirical Force Field Calculations. A Tool in Structural Organic Chemistry. *45*, 1–38 (1974).

Anderson, J. E.: Chair-Chair Interconversion of Six-Membered Rings. *45*, 139–167 (1974).

Anet, F. A. L.: Dynamics of Eight-Membered Rings in Cyclooctane Class. *45*, 169–220 (1974).

Ariëns, E. J., and Simonis, A.-M.: Design of Bioactive Compounds. *52*, 1–61 (1974).

Aurich, H. G., and Weiss, W.: Formation and Reactions of Aminyloxides. *59*, 65–111 (1975).

Bardos, T. J.: Antimetabolites: Molecular Design and Mode of Action. *52*, 63–98 (1974).

Barnes, D. S., see Pettit, L. D.: *28*, 85–139 (1972).

Bauer, S. H., and Yokozeki, A.: The Geometric and Dynamic Structures of Fluorocarbons and Related Compounds. *53*, 71–119 (1974).

Baumgärtner, F., and Wiles, D. R.: Radiochemical Transformations and Rearrangements in Organometallic Compounds. *32*, 63–108 (1972).

Bernardi, F., see Epiotis, N. D.: *70*, 1–242 (1977).

Bernauer, K.: Diastereoisomerism and Diastereoselectivity in Metal Complexes. *65*, 1–35 (1976).

Boettcher, R. J., see Mislow, K.: *47*, 1–22 (1974).

Brandmüller, J., and Schrötter, H. W.: Laser Raman Spectroscopy of the Solid State. *36*, 85–127 (1973).

Bremser, W.: X-Ray Photoelectron Spectroscopy. *36*, 1–37 (1973).

Breuer, H.-D., see Winnewisser, G.: *44*, 1–81 (1974).

Brewster, J. H.: On the Helicity of Variously Twisted Chains of Atoms. *47*, 29–71 (1974).

Brocas, J.: Some Formal Properties of the Kinetics of Pentacoordinate Stereoisomerizations. *32*, 43–61 (1972).

Brunner, H.: Stereochemistry of the Reactions of Optically Active Organometallic Transition Metal Compounds. *56*, 67–90 (1975).

Buchs, A., see Delfino, A. B.: *39*, 109–137 (1973).

Bürger, H., and Eujen, R.: Low-Valent Silicon. *50*, 1–41 (1974).

Burgermeister, W., and Winkler-Oswatitsch, R.: Complexformation of Monovalent Cations with Biofunctional Ligands. *69*, 91–196 (1977).

Butler, R. S., and deMaine, A. D.: CRAMS – An Automatic Chemical Reaction Analysis and Modeling System. *58*, 39–72 (1975).

Caesar, F.: Computer-Gas Chromatography. *39*, 139–167 (1973).

Čársky, P., and Zahradník, R.: MO Approach to Electronic Spectra of Radicals. *43*, 1–55 (1973).

Chandra, P.: Molecular Approaches for Designing Antiviral and Antitumor Compounds. *52*, 99–139 (1974).

Chapuisat, X., and Jean, Y.: Theoretical Chemical Dynamics: A Tool in Organic Chemistry. *68*, 1–57 (1976).

Cherry, W.R., see Epiotis, N.D.: *70*, 1–242 (1977).

Christian, G. D.: Atomic Absorption Spectroscopy for the Determination of Elements in Medical Biological Samples. *26*, 77–112 (1972).

Clark, G. C., see Wasserman, H. H.: *47*, 73–156 (1974).

Clerc, T., and Erni, F.: Identification of Organic Compounds by Computer-Aided Interpretation of Spectra. *39*, 91–107 (1973).

Clever, H.: Der Analysenautomat DSA-560. *29*, 29–43 (1972).

Connors, T. A.: Alkylating Agents. *52*, 141–171 (1974).

Craig, D. P., and Mellor, D. P.: Discriminating Interactions Between Chiral Molecules. *63*, 1–48 (1976).

Cram, D. J., and Cram, J. M.: Stereochemical Reaction Cycles. *31*, 1–43 (1972).

Gresp, T. M., see Sargent, M. V.: *57*, 111–143 (1975).

Dauben, W. G., Lodder, G., and Ipaktschi, J:: Photochemistry of β, γ-Unsatuared Ketones. *54*, 73–114 (1974).

DeClercq, E.: Synthetic Interferon Inducers. *52*, 173–198 (1974).

Degens, E. T.: Molecular Mechanisms on Carbonate, Phosphate, and Silica Deposition in the Living Cell. *64*, 1–112 (1976).

Delfino, A. B., and Buchs, A.: Mass Spectra and Computers. *39*, 109–137 (1973).

DeMaine, A. D., see Butler, R. S.: *58*, 39–72 (1975).

DePuy, C. H.: Stereochemistry and Reactivity in Cyclopropane Ring-Cleavage by Electrophiles. *40*, 73–101 (1973).

Devaquet, A.: Quantum-Mechanical Calculations of the Potential Energy Surface of Triplet States. *54*, 1–71 (1974).

Dimroth, K.: Delocalized Phosphorus-Carbon Double Bonds. Phosphamethincyanines, λ^3-Phosphorins and λ^5-Phosphorins. *38*, 1–150 (1973).

Döpp, D.: Reactions of Aromatic Nitro Compounds *via* Excited Triplet States. *55*, 49–85 (1975).

Dougherty, R. C.: The Relationship Between Mass Spectrometric, Thermolytic and Photolytic Reactivity. *45*, 93–138 (1974).

Dryhurst, G.: Electrochemical Oxidation of Biologically-Important Purines at the Pyrolytic Graphite Electrode. Relationship to the Biological Oxidation of Purines. *34*, 47–85 (1972).

Dürr, H.: Reactivity of Cycloalkene-carbenes. *40*, 103–142 (1973).

Dürr, H.: Triplet-Intermediates from Diazo-Compounds (Carbenes). *55*, 87–135 (1975).

Dürr, H., and Kober, H.: Triplet States from Azides. *66*, 89–114 (1976).

Dürr, H., and Ruge, B.: Triplet States from Azo Compounds. *66*, 53–87 (1976).

Dugundji, J., and Ugi, I.: An Algebraic Model of Constitutional Chemistry as a Basis for Chemical Computer Programs. *39*, 19–64 (1973).

Eglinton, G., Maxwell, J. R., and Pillinger, C. T.: Carbon Chemistry of the Apollo Lunar Samples. *44*, 83–113 (1974).

Eicher, T., and Weber, J. L.: Structure and Reactivity of Cyclopropenones and Triafulvenes. *57*, 1–109 (1975).

Epiotis, N. D., Cherry, W. R., Shaik, S., Yates, R. L., and Bernardi, F.: Structural Theory of Organic Chemistry. *70*, 1–242 (1977).

Erni, F., see Clerc, T.: *39*, 139–167 (1973).

Eujen, R., see Bürger, H.: *50*, 1–41 (1974).

Faber, D. H., see Altona, C.: *45*, 1–38 (1974).

Fietzek, P. P., and Kühn, K.: Automation of the Sequence Analysis by Edman Degradation of Proteins and Peptides. *29*, 1–28 (1972).

Finocchiaro, P., see Mislow, K.: *47*, 1–22 (1974).

Fischer, G.: Spectroscopic Implications of Line Broadening in Large Molecules. *66*, 115–147 (1976).

Fluck, E.: The Chemistry of Phosphine. *35*, 1–64 (1973).

Flygare, W. H., see Sutter, D. H.: *63*, 89–196 (1976).

Fowler, F. W., see Gelernter, H.: *41*, 113–150 (1973).

Freed, K. F.: The Theory of Raditionless Processes in Polyatomic Molecules. *31*, 105–139 (1972).

Fritz, G.: Organometallic Synthesis of Carbosilanes. *50*, 43–127 (1974).

Fry, A. J.: Stereochemistry of Electrochemical Reductions. *34*, 1–46 (1972).

Ganter, C.: Dihetero-tricycloadecanes. *67*, 15–106 (1976).

Gasteiger, J., Gillespie, P., Marquarding, D., and Ugi, I.: From van't Hoff to Unified Perspectives in Molecular Structure and Computer-Oriented Representation. *48*, 1–37 (1974).

Geick, R.: IR Fourier Transform Spectroscopy. *58*, 73–186 (1975).

Geist, W., and Ripota, P.: Computer-Assisted Instruction in Chemistry. *39*, 169–195 (1973).

Gelernter, H., Sridharan, N. S., Hart, A. J., Yen, S. C., Fowler, F. W., and Shue, H.-J.: The Discovery of Organic Synthetic Routes by Computer. *41*, 113–150 (1973).

Gerischer, H., and Willig, F.: Reaction of Excited Dye Molecules at Electrodes. *61*, 31–84 (1976).

Gillespie, P., see Gasteiger, J.: *48*, 1–37 (1974).

Gleiter, R., and Gygax, R.: No-Bond-Resonance Compounds, Structure, Bonding and Properties. *63*, 49–88 (1976).

Guibé, L.: Nitrogen Quadrupole Resonance Spectroscopy. *30*, 77–102 (1972).

Gundermann, K.-D.: Recent Advances in Research on the Chemiluminescence of Organic Compounds. *46*, 61–139 (1974).

Gust, D., see Mislow, K.: *47*, 1–22 (1974).

Gutman, I., and Trinajstić, N.: Graph Theory and Molecular Orbitals. *42*, 49–93 (1973).

Gutmann, V.: Ionic and Redox Equilibria in Donor Solvents. *27*, 59–115 (1972).

Gygax, R., see Gleiter, R.: *63*, 49–88 (1976).

Haaland, A.: Organometallic Compounds Studied by Gas-Phase Electron Diffraction. *53*, 1–23 (1974).

Häfelinger, G.: Theoretical Considerations for Cyclic (pd) π Systems. *28*, 1–39 (1972).

Hariharan, P. C., see Lathan, W. A.: *40*, 1–45 (1973).

Hart, A. J., see Gelernter, H.: *41*, 113–150 (1973).

Hartmann, H., Lebert, K.-H., and Wanczek, K.-P.: Ion Cyclotron Resonance Spectroscopy. *43*, 57–115 (1973).

Hehre, W. J., see Lathan, W. A.: *40*, 1–45 (1973).

Hendrickson, J. B.: A General Protocol for Systematic Synthesis Design. *62*, 49–172 (1976).

Hengge, E.: Properties and Preparations of Si-Si Linkages. *51*, 1–127 (1974).

Henrici-Olivé, G., and Olivé, S.: Olefin Insertion in Transition Metal Catalysis. *67*, 107–127 (1976).

Herndon, W. C.: Substituent Effects in Photochemical Cycloaddition Reactions. *46*, 141–179 (1974).

Höfler, F.: The Chemistry of Silicon-Transition-Metal Compounds. *50*, 129–165 (1974).

Ipaktschi, J., see Dauben, W. G.: *54*, 73–114 (1974).

Jacobs, P., see Stohrer, W.-D.: *46*, 181–236 (1974).

Jahnke, H., Schönborn, M., and Zimmermann, G.: Organic Dyestuffs as Catalysts for Fuel Cells. *61*, 131–181 (1976).

Jakubetz, W., see Schuster, P.: *60*, 1–107 (1975).

Jean, Y., see Chapuisat, X.: *68*, 1–57 (1976).

Jørgensen, C. K.: Continuum Effects Indicated by Hard and Soft Antibases (Lewis Acids) and Bases. *56*, 1–66 (1975).

Julg, A.: On the Description of Molecules Using Point Charges and Electric Moments. *58*, 1–37 (1975).

Kaiser, K. H., see Stohrer, W.-D.: *46*, 181–236 (1974).

Khaikin, L. S., see Vilkow, L.: *53*, 25–70 (1974).

Kisch, H., see Albini, A.: *65*, 105–145 (1976).

Kober, H., see Dürr, H.: *66*, 89–114 (1976).

Kompa, K. L.: Chemical Lasers. *37*, 1–92 (1973).

Kratochvil, B., and Yeager, H. L.: Conductance of Electrolytes in Organic Solvents. *27*, 1–58 (1972).

Krech, H.: Ein Analysenautomat aus Bausteinen, die Braun-Systematic. *29*, 45–54 (1972).

Kühn, K., see Fietzek, P. P.: *29*, 1–28 (1972).

Kustin, K., and McLeod, G.C.: Interactions Between Metal Ions and Living Organisms in Sea Water. *69*, 1–37 (1977).

Kutzelnigg, W.: Electron Correlation and Electron Pair Theories. *40*, 31–73 (1973).

Lathan, W. A., Radom, L., Hariharan, P. C., Hehre, W. J., and Pople, J. A.: Structures and Stabilities of Three-Membered Rings from *ab initio* Molecular Orbital Theory. *40*, 1–45 (1973).

Lebert, K.-H., see Hartmann, H.: *43*, 57–115 (1973).

Lodder, G., see Dauben, W. G.: *54*, 73–114 (1974).

Luck, W. A. P.: Water in Biologic Systems. *64*, 113–179 (1976).

Lucken, E. A. C.: Nuclear Quadrupole Resonance. Theoretical Interpretation. *30*, 155–171 (1972).

Mango, F. D.: The Removal of Orbital Symmetry Restrictions to Organic Reactions. *45*, 39–91 (1974).

Maki, A. H., and Zuclich, J. A.: Protein Triplet States. *54*, 115–163 (1974).

Margrave, J. L., Sharp., K. G., and Wilson, P. W.: The Dihalides of Group IVB Elements. *26*, 1–35 (1972).

Marius, W., see Schuster, P.: *60*, 1–107 (1975).

Marks, W.: Der Technicon Autoanalyzer. *29*, 55–71 (1972).

Marquarding, D., see Gasteiger, J.: *48*, 1–37 (1974).

Maxwell, J. R., see Eglinton, G.: *44*, 83–113 (1974).

McLeod, G.C., see Kustin, K.: *69*, 1–37 (1977).

Mead, C. A.: Permutation Group Symmetry and Chirality in Molecules. *49*, 1–86. (1974).

Meier, H.: Application of the Semiconductor Properties of Dyes Possibilities and Problems. *61*, 85–131 (1976).

Meller, A.: The Chemistry of Iminoboranes. *26*, 37–76 (1972).

Mellor, D. P., see Craig, D. P.: *63*, 1–48 (1976).

Michl, J.: Physical Basis of Qualitative MO Arguments in Organic Photochemistry. *46*, 1–59 (1974).

Minisci, F.: Recent Aspects of Homolytic Aromatic Substitutions. *62*, 1–48 (1976).

Mislow, K., Gust, D., Finocchiaro, P., and Boettcher, R. J.: Stereochemical Correspondence Among Molecular Propellers. *47*, 1–22 (1974).

Nakajima, T.: Quantum Chemistry of Nonbenzenoid Cyclic Conjugated Hydrocarbons. *32*, 1–42 (1972).

Nakajima, T.: Errata. *45*, 221 (1974).

Neumann, P., see Vögtle, F.: *48*, 67–129 (1974).

Oehme, F.: Titrierautomaten zur Betriebskontrolle. *29*, 73–103 (1972).

Olivé, S., see Henrici-Olivé, G.: *67*, 107–127 (1976).

Papoušek, D., and Špirko, V.: A New Theoretical Look at the Inversion Problem in Molecules. *68*, 59–102 (1976).

Pearson, R. G.: Orbital Symmetry Rules for Inorganic Reactions from Perturbation Theory. *41*, 75–112 (1973).

Perrin, D. D.: Inorganic Medicinal Chemistry. *64*, 181–216 (1976).

Pettit, L. D., and Barnes, D. S.: The Stability and Structure of Olefin and Acetylene Complexes of Transition Metals. *28*, 85–139 (1972).

Pignolet, L. H.: Dynamics of Intramolecular Metal-Centered Rearrangement Reactions of Tris-Chelate Complexes. *56*, 91–137 (1975).

Pillinger, C. T., see Eglinton, G.: *44*, 83–113 (1974).

Pople, J. A., see Lathan, W. A.: *40*, 1–45 (1973).

Puchelt, H.: Advances in Inorganic Geochemistry. *44*, 155–176 (1974).

Pullman, A.: Quantum Biochemistry at the All- or Quasi-All-Electrons Level. *31*, 45–103 (1972).

Quinkert, G., see Stohrer, W.-D.: *46*, 181–236 (1974).

Radom, L., see Lathan, W. A.: *40*, 1–45 (1973).

Renger, G.: Inorganic Metabolic Gas Exchange in Biochemistry. *69*, 39–90 (1977).

Rice, S. A.: Conjectures on the Structure of Amorphous Solid and Liquid Water. *60*, 109–200. (1975).

Rieke, R. D.: Use of Activated Metals in Organic and Organometallic Synthesis. *59*, 1–31 (1975).

Ripota, P., see Geist, W.: *39*, 169–195 (1973).

Rüssel, H. and Tölg, G.: Anwendung der Gaschromatographie zur Trennung und Bestimmung anorganischer Stoffe/Gas Chromatography of Inorganic Compounds. *33*, 1–74 (1972).

Ruge, B., see Dürr, H.: *66*, 53–87 (1976).

Sargent, M. V., and Cresp, T. M.: The Higher Annulenones. *57*, 111–143 (1975).

Schäfer, F. P.: Organic Dyes in Laser Technology. *61*, 1–30 (1976).

Schneider, H.: Ion Solvation in Mixed Solvents. *68*, 103–148 (1976).

Schönborn, M., see Jahnke, H.: *61*, 133–181 (1976).

Schrötter, H. W., see Brandmüller, J.: *36*, 85–127 (1973).

Schuster, P., Jakubetz, W., and Marius, W.: Molecular Models for the Solvation of Small Ions and Polar Molecules. *60*, 1–107 (1975).

Schutte, C. J. H.: The Infra-Red Spectra of Crystalline Solids. *36*, 57–84 (1973).

Scrocco, E., and Tomasi, J.: The Electrostatic Molecular Potential as a Tool for the Interpretation of Molecular Properties. *42*, 95–170 (1973).

Shaik, S., see Epiotis, N.D.: *70*, 1–242 (1977).

Sharp, K. G., see Margrave, J. L.: *26*, 1–35 (1972).

Shue, H.-J., see Gelernter, H.: *41*, 113–150 (1973).

Simonetta, M.: Qualitative and Semiquantitative Evaluation of Reaction Paths. *42*, 1–47 (1973).

Simonis, A.-M., see Ariëns, E. J.: *52*, 1–61 (1974).

Smith, S. L.: Solvent Effects and NMR Coupling Constants. *27*, 117–187 (1972).

Špirko, V., see Papoušek, D.: *68, 59*–102 (1976).

Sridharan, N. S., see Gelernter, H.: *41*, 113–150 (1973).

Stohrer, W.-D., Jacobs, P., Kaiser, K. H., Wiech, G., and Quinkert, G.: Das sonderbare Verhalten elektronen-angeregter 4-Ringe-Ketone. – The Peculiar Behavior of Electronically Exited 4-Membered Ring Ketones. *46*, 181–236 (1974).

Stoklosa, H. J., see Wasson, J. R.: *35*, 65–129 (1973).

Suhr, H.: Synthesis of Organic Compounds in Glow and Corona Discharges. *36*, 39–56 (1973).

Sutter, D. H., and Flygare, W. H.: The Molecular Zeeman Effect. *63*, 89–196 (1976).

Thakkar, A. J.: The Coming of the Computer Age to Organic Chemistry. Recent Approaches to Systematic Synthesis Analysis. *39*, 3–18 (1973).

Tölg, G., see Rüssel, H.: *33*, 1–74 (1972).

Tomasi, J., see Scrocco, E.: *42*, 95–170 (1973).

Trinajstić, N., see Gutman, I.: *42*, 49–93 (1973).

Trost, B. M.: Sulfuranes in Organic Reactions and Synthesis. *41*, 1–29 (1973).

Tsuji, J.: Organic Synthesis by Means of Transition Metal Complexes: Some General Patterns. *28*, 41–84 (1972).

Turley, P. C., see Wasserman, H. H.: *47*, 73–156 (1974).

Ugi, I., see Dugundji, J.: *39*, 19–64 (1973).

Ugi, I., see Gasteiger, J.: *48*, 1–37 (1974).

Veal, D. C.: Computer Techniques for Retrieval of Information from the Chemical Literature. *39*, 65–89 (1973).

Vennesland, B.: Stereospecifity in Biology. *48*, 39–65 (1974).

Vepřek, S.: A Theoretical Approach to Heterogeneous Reactions in Non-Isothermal Low Pressure Plasma. *56*, 139–159 (1975).

Vilkov, L., and Khainkin, L. S.: Stereochemistry of Compounds Containing Bonds Between Si, P, S, Cl, and N or O. *53*, 25–70 (1974).

Vögtle, F., and Neumann, P.: [2.2] Paracyclophanes, Structure and Dynamics. *48*, 67–129 (1974).

Vollhardt, P.: Cyclobutadienoids. *59*, 113–135 (1975).

Wänke, H.: Chemistry of the Moon. *44*, 1–81 (1974).

Wagner, P. J.: Chemistry of Excited Triplet Organic Carbonyl Compounds. *66*, 1–52 (1976).

Wanczek, K.-P., see Hartmann, K.: *43*, 57–115 (1973).

Wasserman, H. H., Clark, G. C., and Turley, P. C.: Recent Aspects of Cyclopropanone Chemistry. *47*, 73–156 (1974).

Wasson, J. R., Woltermann, G. M., and Stoklosa, H. J.: Transition Metal Dithio- and Diselenophosphate Complexes. *35*, 65–129 (1973).

Weber, J. L., see Eicher, T.: *57*, 1–109 (1975).

Weiss, A.: Crystal Field Effects in Nuclear Quadrupole Resonance. *30*, 1–76 (1972).

Weiss, W., see Aurich, H. G.: *59*, 65–111 (1975).

Wentrup, C.: Rearrangements and Interconversion of Carbenes and Nitrenes. *62*, 173–251 (1976).

Werner, H.: Ringliganden-Verdrängungsreaktionen von Aromaten-Metall-Komplexen. *28*, 141–181 (1972).

Wiech, G., see Stohrer, W.-D.: *46*, 181–236 (1974).

Wild, U. P.: Characterization of Triplet States by Optical Spectrocsopy. *55*, 1–47 (1975).

Wiles, D. R., see Baumgärtner, F.: *32*, 63–108 (1972).

Willig, F., see Gerischer, H.: *61*, 31–84 (1976).

Wilson, P. W., see Margrave, J. L.: *26*, 1–35 (1972).

Winkler-Oswatitsch, R., see Burgermeister, W.: *69*, 91–196 (1977).

Winnewisser, G., Mezger, P. G., and Breuer, H. D.: Interstellar Molecules. *44*, 1–81 (1974).

Wittig, G.: Old and New in the Field of Directed Aldol Condensations. *67*, 1–14 (1976).

Woenckhaus, C.: Synthesis and Properties of Some New $NAD^{\oplus}$ Analogues. *52*, 199–223 (1974).

Woltermann, G. M., see Wasson, J. R.: *35*, 65–129 (1973).

Wrighton, M. S.: Mechanistic Aspects of the Photochemical Reactions of Coordination Compounds. *65*, 37–102 (1976).

Yates, R.L., see Epiotis, N.D.: *70*, 1–242 (1977).

Yeager, H. L., see Kratochvil, B.: *27*, 1–58 (1972).

Yen, S. C., see Gelernter, H.: *41*, 113–150 (1973).

Yokozeki, A., see Bauer, S. H.: *53*, 71–119 (1974).

Yoshida, Z.: Heteroatom-Substituted Cyclopropenium Compounds. *40*, 47–72 (1973).

Zahradník, R., see Čársky, P.: *43*, 1–55 (1973).

Zeil, W.: Bestimmung der Kernquadrupolkopplungskonstanten aus Mikrowellen-spektren. *30*, 103–153 (1972).

Zimmermann, G., see Jahnke, H.: *61*, 133–181 (1976).

Zoltewicz, J. A.: New Directions in Aromatic Nucleophilic Substitution. *59*, 33–64 (1975).

Zuclich, J. A., see Maki, A. H.: *54*, 115–163 (1974).

Lecture Notes in Chemistry

Managing Editors: G. Berthier, M.J.S. Dewar, H. Fischer, K. Fukui, H. Hartmann,
H.H. Jaffé, J. Jortner, W. Kutzelnigg, K. Ruedenberg, E. Scrocco, W. Zeil

Volume 1: G.H. Wagnière
**Introduction to Elementary Molecular Orbital Theory and to Semiempirical
Methods**
33 figures. V, 109 pages. 1976

The aim of these notes is to provide a summary and concise introduction to
elementary molecular orbital theory, with an emphasis on semiempirical methods.
Within the last decade the development and refinement of *ab initio* computations
has tended to overshadow the usefulness of semiempirical methods. However, both
approaches have their justification. *Ab initio* methods are designed for accurate
predictions, at the expense of greater computational labor. The aim of semiempirical
methods mainly lies in a semiquantitative classification of electronic properties and
in the search for regularities within given classes of larger molecules. Applications
to optical activity, concerted reactions and to polymers are included.
(34 references)

Volume 2: E. Clementi
**Determination of Liquid Water Structure, Coordination Numbers for Ions and
Solvation for Biological Molecules**
32 figures, 18 tables. VI, 107 pages. 1976

The structure of liquid water and the solvation of ions and molecules represent an
active field of past and current research. The authors have stressed in particular
the new quantum mechanical developments that constitute the conceptual base
for the recent advancements in this field. They have pointed out, with a variety
of examples, the large amount of information embodied in quantum mechanics.
The study of solvation represents the ideal field to pass from small chemical
systems to large ones, to pass from quantum to statistical mechanics with a first
step towards thermodynamics. Particular attention is given to present a unified
picture the coherently retains technique and assumption in passing from one type
of the description of matter to another. This work is dedicated to Prof. Per-Olov
Löwdin, on the occasion of his 60th birthday.
(27 references)

Springer-Verlag
Berlin Heidelberg New York

Reactivity and Structure

Concepts in Organic Chemistry
Editors: K. Hafner, J.-M. Lehn, C.W. Rees, P.v. Ragué Schleyer, B.M. Trost,
R. Zahradník

Volume 1: J. Tsuji
Organic Synthesis
by Means of Transition Metal Complexes
A Systematic Approach
4 tables. IX, 199 pages. 1975

This book is the first in a new series, Reactivity and Structure: Concepts in Organic
Chemistry, designed to treat topical themes in organic chemistry in a critical
manner. A high standard is assured by the compositon of the editorial board, which
consists of scientists of international repute. This volume deals with the currently
fashionable theme of complexes of transition-metal compounds. Not only are these
intermediates becoming increasingly important in the synthesis of substances of
scientific appeal, but they have already acquired great significance in large-scale
chemical manufacturing. The new potentialities for synthesis are discussed with
examples. The 618 references bear witness to the author's extensive coverage of
the literature. This book is intended to stimulate organic chemists to undertake
further research and to make coordination chemists aware of the unforeseen
development of this research field.

Volume 2: K. Fukui
Theory of Orientation and Stereoselection
72 figures, 2 tables. VII, 134 pages. 1975

The 'electronic theory' has long been insufficient interpret various modern organic
chemical facts, in particular those of reactivity. The time has come for a book
making one realize what is within, and what is beyond, the reach of quantum-
chemical methods. Graduate students and young researchers in chemistry both
theoretical and experimental, will find this book useful in getting accustomed to
the quantum-chemical way of thinking. Theory produces new experimental ideas,
and, conversely, a host of experimental data opens new theoretical fields. A book
such as the present one will constantly keep its value, although the quantum-
chemical approach to the theory of reactivity is, of course, still in the developmental
stage.

Springer-Verlag
Berlin Heidelberg New York